Fossil Conchostraca of the Southern Hemisphere and Continental Drift

Paleontology, Biostratigraphy, and Dispersal

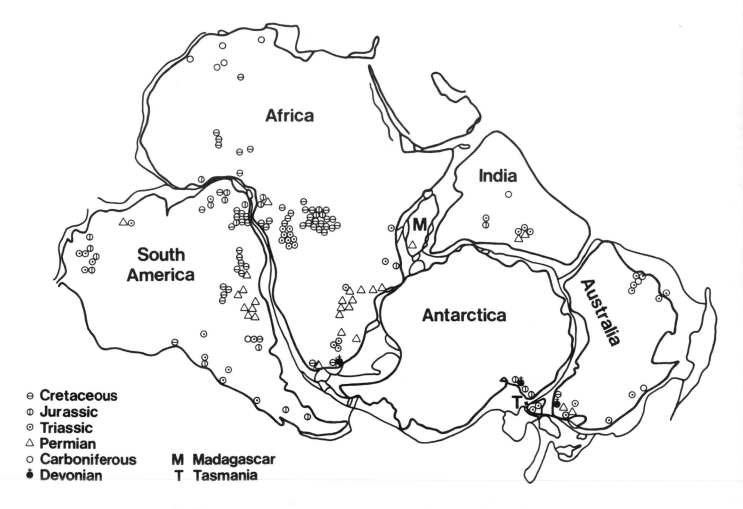

- ⊖ Cretaceous
- ⊕ Jurassic
- ⊙ Triassic
- △ Permian
- ○ Carboniferous
- ● Devonian

M Madagascar
T Tasmania

GONDWANALAND FOSSIL CONCHOSTRACAN DISTRIBUTION

(PLACEMENT OF INDIA AFTER CURRAY AND MOORE, 1974)

The Geological Society of America
Memoir 165

Fossil Conchostraca of the Southern Hemisphere and Continental Drift

Paleontology, Biostratigraphy, and Dispersal

Paul Tasch
Distinguished Professor Emeritus
Natural Sciences/Geology
Wichita State University
Wichita, Kansas 67208

1987

Published by The Geological Society of America, Inc.
3300 Penrose Place, P.O. Box 9140, Boulder, Colorado 80301

Printed in U.S.A.

GSA Books Science Editor Campbell Craddock

Library of Congress Cataloging-in-Publication Data

Tasch, Paul.
  Fossil Conchostraca of the Southern Hemisphere and
continental drift.

  (Memoir / Geological Society of America ; 165)
  Bibliography: p.
  Includes index.
  1. Conchostraca, Fossil.   2. Paleontology—Southern
Hemisphere.   3. Continental drift.   4. Geology—Southern
Hemisphere.   I. Title.   II. Series: Memoir (Geological
Society of America) ; 165.
QE817.C6T37   1987   565'32   87-153
ISBN 0-8137-1165-7

*Memoir (Geological Society Of America).
1987 #165 Tasch, P., Fossil Conchostraca of the Southern
Hemisphere and Continental Drift*

10-21-87

iv

Contents

Figures

Acknowledgments

The present undertaking would not have been possible without the cooperation and support of numerous colleagues and institutions in the United States and abroad. The National Science Foundation, through its Division of Polar Programs, under its program manager, Dr. Mort D. Turner, provided logistical and financial support from 1966 to 1983. In particular, Dr. Turner importantly advanced this Antarctic and other Southern hemisphere research by his broad grasp of the larger picture, which developed from the writer's Antarctic collections. His encouragement and technical experience were also most helpful.

Dr. David Elliot, Institute of Polar Studies, Ohio State University, greatly aided the research reported here through logistical support and as initial field guide to several Antarctic sites, including Storm Peak and Blizzard Heights. He generously provided his own collections and field data from Mauger Nunatak. Other conchostracan collections by Dr. Elliot from the environs of Blizzard Heights and elsewhere, more recently Agate Peak, were contributed for the writer's Antarctic studies.

The late Dr. J. M. Schopf was field guide and companion in the Ohio Range and Sentinel Mountains during the 1966–67 season, and also contributed his collection from Carapace Nunatak. Dietmar Schumacher, a graduate student at the time, was field assistant during the 1969–70 austral season in exploration of Blizzard Heights, Storm Peak, Coalsack Bluff, Carapace Nunatak, and other sites. John Lokey served as maintenance assistant during the same season.

Drs. Borns and Hall of the University of Maine made available their conchostracan insect collection from Carapace Nunatak in the laboratory of Professor Frank M. Carpenter of Harvard University, who was studying the insects. Professor Carpenter and Dr. Jarmila Peck, working in his laboratory at the time, greatly assisted examination of this collection.

George Doumani, Library of Congress, sent his Ohio Range leaiid collection, the study of which alerted the writer to the need and value of further Antarctic and other Gondwana biostratigraphic data and fossil conchostracan collections.

In Australia, several colleagues contributed their time, effort and experience to forward both field and museum research. Doctors C. T. McElroy and Toby Rose of the New South Wales Geological Survey, Sydney, arranged for a field assistant and transportation and provided maps and relevant subsurface data on the Newcastle Coal Measures. Dr. John Pickett, Mining Museum, Sydney, made available for study the conchostracan material on deposit at the Museum, and provided literature as well as guidance to the first Newcastle Coal Measures site studied in 1967. N. R Mauger, Superintendent of Collieries, the Broken Hill Proprietary Company, Ltd., Belmont, New South Wales, helped resolve the puzzle over the exact stratigraphic position of John Mitchell's "old Cardiff Seam."

Dr. J. M. Dickens, Bureau of Mineral Resources, Geology, and Geophysics (BMRGG), Canberra, made available for study at Wichita State University, the conchostracan-bearing cores from western Australia and the Bowen basin that led to two joint papers with Dr. P. J. Jones, (BMRGG). Dr. Edgar F. Riek, Commonwealth Scientific and Industrial Research Organization (CSIRO), collaborated in the identification of a Permian insect collected from the Sentinel Range, Ellsworth Mountains, Antarctica, and contributed valuable specimens from his Australian living branchiopod collection.

At the Australian Museum, Sydney, Dr. Alex Ritchie, Curator of invertebrates and fishes, made available the John Mitchell collection of conchostracan types from the Newcastle Coal Measures and elsewhere, as well as the insect collection, for study on several visits during 1967, 1970, and 1981. He also enlisted the aid of the museum staff for photography, preparation of rubber imprints, etc. Rubber imprints of most types used for taxonomy were prepared by Scientific Technical Officer Robert Jones, assistant to Dr. Ritchie. Ms. Dorothy Jones, assistant, prepared required molds and imprints, and aided my research at the Museum in various ways. Photographs of Mitchell's types and a few other selected nontype specimens were taken by the Museum photographer, John Fields. In addition, Dr. Ritchie provided valuable reference material on the Newcastle Coal Measures and other deposits, and loaned his own Antarctic Devonian conchostracan collection for study back in the United States. Further, he communicated his newest Devonian fish finds in the Antarctic, which helped broaden the writer's understanding of the Antarctic Devonian ecosystem.

Professor John Talent, Macquarie University, was guide and field companion in the exploration of his Buchan Caves Devonian conchostracan collecting site and later, to the site of his Cretaceous conchostracan collections—both of which collectons have added valuable data. The National Museum, Victoria, sent a photograph of Dr. Talent's conchostracan species. Peter Duncan, Melbourne, an amateur collector, contributed conchostracan samples from Talent's Cretaceous site.

Dr. M. R. Banks, University of Tasmania, Hobart, greatly assisted field work in Tasmania; he acted as guide to known localities, provided literature, shared his wide experience of the Tasmanian rock sequence, and arranged loan of fossil conchostracans in the University of Tasmania collection.

Loan of Devonian conchostracan samples (White Collection) from the Antarctic Lashly Mountains was facilitated by Dr. C. A. Fleming of the New Zealand Geological Survey (Lower Hutt).

Many colleagues in South America aided researches there. In Brazil, Professor I. D. Pinto, Universidad Federal Rio Grande do Sul, Porto Alegre, and a graduate student, Ivone Purper, made available their Permian and Triassic fossil conchostracan collections as well as other Mesozoic collections, literature, theses, and photographs of important specimens.

Professor J. J. Bigarella, Universidad Federal Paraná, Curitiba, Brazil, secured a graduate student field assistant, Paul Munoz, for the Rio do Rasto exploration, enlisted the aid of a colleague, Professor F. F. Azevedo and a graduate student to accompany the writer in the field at São Pascoal, where two fossil conchostracan genera not previously reported from the Brazilian rock sequence were found. Professor Bigarella and Professor H. Papp were also field guides to several other Rio do Rasto sites. Professor Papp contributed a copy of his then unpublished topographic profile of the São Pascoal outcrop of the Rio do Rasto Formation.

A former student and now colleague, Professor R. N. Cardoso, Ciudad University (Minas Gerais), Belo Horizonte, Brazil, and Dr. Ioco Katoo made available valuable fossil conchostracan specimens, photographs of important specimens, data on conchostracan sites, and literature. Dr. Katoo contributed a copy of her doctoral dissertation on fossil conchostracans from the Santa Maria and Botucatú Formations.

Professor A. C. Rocha Campos, Universidad Federal São Paulo, São Paulo, made available the Mendes fossil conchostracan collection from the Brazilian Permian and Triassic for study at the university, as well as his own Angola, Africa, collection. Other colleagues at the University, Dr. Mary E. Bernardes Oliviera, contributed a copy of her doctoral dissertation and together with Dr. Oscar Rosler (both paleobotanists) supplied conchostracan fossils found in their floral collections and helped clarify the position of the highest occurrence of the *Glossopteris* flora in the Paraná basin.

Will Maze, graduate student at Princeton University, contributed his collecton of conchostracan fossils from Venezuela and Colombia. Professor Rafael Herbst, Universidad Nacional de Tucaman, Argentina, communicated his find of the first reported Uruguayan fossil conchostracan and sent along a photo-

graph of a fragment of this find, as well as one of his publications.

Dr. Raul R. Leguizamon, Universidad Nacional de Cordoba, Argentina, provided a copy of his dissertation and photographs of his leaiid species and a copy of the paper in which he described it.

In southern Africa, Professor B.J.V. Botha, University of the Orange Free State, Bloemfontein, Republic of South Africa, secured a field associate, Johann Loock, sedimentologist at the University, and made the laboratory facilities of the Department of Geology available to us. Johann Loock's knowledge of Cave Sandstone geology in the Republic of South Africa and Lesotho was a valuable asset during the field exploration.

Dr. James W. Kitching, Bernard Price Institute, University of Witwatersrand, Johannesburg, explained his vertebrate zonation of the Cave Sandstone and also provided other literature.

M. Jean Fabre contributed his and other colleagues' conchostracan-ostracod collection from the central Sahara, Algeria, together with relevant biostratigraphic data. Dr. Phillip M. Oesterlen contributed his Angola conchostracan collection which he obtained during field study for his doctoral dissertation. Fossil conchostracans in the National Collection of Morocco, which includes some Algerian conchostracan fossils, were loaned by the Ministry of Energy and Mines through the Chief of the Geology Division, M. Bensaid, and one of the staff, Ms. Solange Willefert, Curator of Collections, Rabat-Chelah.

Dr. M. R. Cooper, Curator, National Museum of Zimbabwe, Bulawayo, prepared rubber imprints of Professor Bond's leaiid types on deposit there, as well as supplying detailed locality data and his own publications on fossil vertebrates. Dr. Reinhard Förster, Munich, arranged for the writer to examine his Jurassic, Lebombo conchostracans at the Universität-Institut Paläontologie und Historische Geologie. The curator of the Albany Museum, Grahamtown, Republic of South Africa, sent a photograph of Rennie's Devonian species deposited there. Dr. H. W. Ball and S. F. Morris, British Museum (Natural History), London, from time to time loaned fossil conchostracan types from Africa, South America, and other places from their collections.

The late Sri P. K. Ghosh importantly assisted the writer's India researches during 1970–71, through the Geological Survey of India (G.S.I.). With the approval of the Director General, field associates in the various G.S.I. districts that were to be explored were assigned, and transportation, as well as literature, provided. Dr. M.V.A. Sastry, formerly Director of Paleontology, G.S.I., Calcutta, and Dr. S. C. Shah, G.S.I. shared their knowledge of the areas to be studied.

There were three G.S.I. field associates without whose knowledge of the terrain in the Raniganj Basin and environs, and the Pranhita-Godavari Valley sites, and ability to negotiate for sundry supplies with the local people, field work would not have been successful. Sri Shekkar Chandra Ghosh (Raniganj Basin and environs) was an invaluable guide and field colleague in exploration. On another visit in 1979, Shekkar Ghosh, in charge of the S.E.M. Laboratory (G.S.I. Calcutta), made available for study subsurface and surface collections from different coal fields and sites in India. Joint examination of and photography of selected fossil conchostracans led to new findings. Sri C. Nageswara Rao (G.S.I. Hyderabad) was very helpful for part of the Kotá Formation exploration. Sri. B. R. Jagannatha Rao, currently Director of the Regional Paleontological Laboratories, Hyderabad, was a valuable field companion and associate in the common enterprise during the 1970–71 exploration in the states of Maharashtra and Andhra Pradesh.

As a guest of the Nanking Institute for Paleontology and Stratigraphy in 1979, through the invitation of the Academia Sinica and colleagues at the Institute, I was able to study the important and extensive Mesozoic, and some Paleozoic, conchostracan collections. Professor Chang Wen-tang, Chen Pei-chi, and Shen Yan-bin accompanied the writer in the field (Le Chang District) to examine and collect from the first, and at the time, the only known Devonian leaiid-bearing bed. This last find had relevance for the Australian Carboniferous leaiids and other ribbed forms. Subsequently, Professor Chang informed the writer of new collections by Shen Yan-bin of the first found Permian leaiids from Hunan, Kwangsi, and Kansu provinces (1984, p. 511, pt. 1) and of more Devonian sites with leaiids.

Many laboratory assistants at the Geology Department of Wichita State University participated in this project, measuring specimens, curating, conducting literature searches, and doing other chores. Included were: Karen Willits, Don Deaton, Dennis Carlton, Rod Anderson, Shari Simon, Tom Matzen, and Robert Kennedy. Drafting and sometimes researching obscure geographic place names in the literature was competently performed by Don Deaton. Don Rosowitz efficiently, and with dispatch, drafted some seventy-five percent of the line drawings. Dr. Edward L. Gafford, a former student, collaborated in geochemical study of Antarctic samples. Dr. James M. Lammons, a former student, participated in joint palynological studies of Antarctic samples from several Paleozoic/Mesozoic sites. The following students were photographers: T. J. Petta, Scott Finch, Mark Rude, Terry Hutter, and Shari Simon.

Wichita State University has been most supportive of researches on this project over the years, from the offices of President Clark Ahlberg and John Breazeale, Vice-President for Academic Affairs, as well as the Office of Research and Sponsored Programs under Director Frederick Sudermann, and his staff. Excellent cooperation and assistance on interlibrary loans and searching out needed reference material was afforded by the staff of the Interlibrary Loan and Reference Section of the Wichita State University Ablah Library.

Three colleagues, Campbell Craddock, Kenneth E. Caster, and Peter Webb, generously proferred constructive criticism of the original manuscript.

Geological Society of America
Memoir 165
1987

Fossil Conchostraca of the Southern Hemisphere and Continental Drift

ABSTRACT

Biostratigraphic and taxonomic studies were carried out to determine whether the fossil conchostracan record of the southern continents contained credible evidence of *nonmarine* dispersal between them during portions of Paleozoic and Mesozoic time. Fossil collections were made toward that end in the following: *Africa,* Cave Sandstone, Triassic of the Republic of South Africa and Lesotho; *Australia,* Newcastle Coal Measures, Permian (Tartarian); *Antarctica,* Ohio Range, Permian, and Queen Alexandra Range and southern Victoria Land, Lower Jurassic; *India,* Raniganj Formation, Upper Permian, Panchet Formation, Triassic, and Kotá Formation, Lower Jurassic; *South America,* Brazil, Rio do Rasto Formation, Upper Permian.

Other conchostracan taxa were contributed and/or loaned by colleagues or museums, or came from the writer's collections from Africa, Antarctica, Australia, India, and South America. These included conchostracan fossils from Morocco and Algeria, Carboniferous and Cretaceous; Angola, Triassic; Devonian and Jurassic of southern and northern Victoria Land, Antarctica, respectively; upper Paleozoic and Triassic of western and eastern Australian basins; the Cretaceous of Victoria, Australia, and the Tasmanian Triassic; the Mesozoic of Brazil, Chile, Argentina, Bolivia, Colombia, and Venezuela; and conchostracan-bearing cores from the Triassic of India. For all conchostracan taxa studied (where preservation permitted) a standardized series of measurements and ratios were used to facilitate inter- and intra-continental comparisons between species. Mitchell's Australian conchostracan types are here revised and refigured, as are T. R. Jones conchostracan types from the southern continents (exclusive of Antarctica), as well as other taxa described in the older and modern literature.

Findings.

New Taxa

Africa. Cyzicus (Euestheria) lefranci (Algeria); *(Palaeolimnadia (Grandilimnadia) oesterleni* (Angola); *Cornia angolata* (Angola); *Palaeolimnadia (Grandilimnadia) africania* (Angola); *Cyzicus (Eu.) thabaningensis* (Lesotho); *Leaia (Leaia) bertrandi* (Morocco); *Afrolimnadia sibiriensis* (Republic of South Africa–RSA); *Leaia (Hemicycloleaia) cradockensis* (RSA).

Antarctica. Cyzicus (Eu.) lashlyensis (Lashly Mountains); *Cyzicus (Eu.) ritchiei* (Portal Mountains); *Cyzicus (Lioestheria) rickeri* (Brimstone Peak); *Cyzicus (Eu.) juravariabalis* (Mauger Nunatak [MN]); *Cyzicus (Lio.) maugerensis* (MN); *Cyzicus (Lio.) longacardinis* (MN); *Cyzicus (Lio.) longulus* (Blizzard Heights [BH]); *Cyzicus (Eu.) beardmorensis* (BH); *Cyzicus (Eu.) ellioti* (BH); *Cyzicus (Eu.) juracircularis* (BH); *Cyzicus (Eu.) formavariabalis* (BH); *Glyptoasmussia prominarma* (BH); *Glyptoasmussia meridionalis* (BH); *Estheriina (Nudusia) brevimargina* (BH); *Glyptoasmussia australospecialis* (Storm Peak [SP]); *Estheriina (Nudusia) stormpeakensis* (SP); *Cyzicus (Lio.) disgregaris* (SP); *Cyzicus (Eu.) ichthystromatos* (SP); *Cyzicus (Eu.) crustapatulus* (SP); *Cyzicus (Eu.) rhadinis* (SP); *Cyzicus (Eu.) transantarctensis* (SP); *Cyzicus (Eu.) casteneus* (SP); *Palaeolimnadia (Grandilimnadia) minutula* (SP); *Pseudoasmussiata defretinae* (SP).

Australia. Cyzicus (Eu.) talenti (Victoria); *Leaia (Hemicycloleaia) kahibahensis* (New South Wales [NSW]); *Leaia (Hemicycloleaia) magnumelliptica* (NSW); *Palaeolimnadia (Grandilimnadia) arcadiensis* (Queensland); *Cyzicus (Eu.) triassibrevis* (Garie Beach).

1

India. Cyzicus (Eu.) basbatiliensis (Raniganj Coal Field [RCF]); *Pseudoasmussiata bengaliensis* (RCF); *Cyzicus (Lio.) miculis* (RCF); *Cyzicus (Eu.) raniganjis* (RCF); *Cornia panchetella* (RCF); *Cyclestherioides (Cyclestherioides) machkandensis* (RCF); genus, *Cornutestheriella* (East Bokaro Coal Field [EBCF]); *Cyzicus (Lio.) bokaroensis* (EBCF); *Pseudoasmussiata indicyclestheria* (Mangli); *Estheriina (Nudusia) adilabadensis* (Sironcha Forest Preserve [SFP]); *Estheriina* (Nudusia) *indijurassica* (SFP); *Cyzicus (Eu.) crustabundis* (southeast of Metapalli); *Estheriina (Estheriina) pranhitaensis* (southwest of Kotá Village); *Estheriina (Nudusia) bullata* (SFP).

South America. Cyzicus (Lio.) malacarensis (Santa Cruz, Argentina); *Pseudoasmussiata katooae* (Brazil, "Belvedere"); *Gabonestheria brasiliensis* (São Pascoal, Brazil); *Cyzicus (Lio.) bigarellai* (Brazil); *Cyclestherioides (Cyclestherioides) pintoi* (Brazil); *Cyzicus (Lio.) mazei* (Venezuela).

Dispersal

Paleozoic/Mesozoic. Each of the five Gondwanaland continents had fossil species of the genera *Palaeolimnadia, Estheriina, Cornia,* and *Cyclestherioides.* Four of the southern continents had species of *Leaia* (all except India) and *Pseudoasmussiata* (all except Australia).

Eggs of species of *Palaeolimnadiopsis* and related genera were dispersed to three southern continents (South America, Africa, and Australia), while those of *Estheriella* and related genera, as well as *Gabonestheria,* were dispersed to India, Africa, and South America.

Some species dispersed to several southern continents are represented by identical or very close morphological equivalents that share many valve parameters and features. Among others, examples include: African species *Cyzicus (Lioestheria) malangensis* (Marlière) from the Upper Triassic of Angola and Zaire, found in the Lower Jurassic of Blizzard Heights, Antarctica; *Palaeolimnadia (Palaeolimnadia) wianamattensis* (M.) from the Middle to Upper Triassic of Australia has a very close equivalent in the Upper Triassic Phyllopod Beds of northern Angola, as does another Australian species from the Bowen Basin, *Estheriina (Nudusia) rewanensis* Tasch; *Palaeolimnadia (Grandilimnadia) glenleensis* Mitchell from the Australian Triassic Wianamatta Group has a close equivalent in the Lower Jurassic Storm Peak, Upper Flow interbed, Antarctica; the *Estheriina (Nudusia) indijurassica* n. sp., from the Lower Jurassic of India and *Estheriina (Nudusia) stormpeakensis* n. sp., from Storm Peak, Upper Flow interbed, Antarctica, share several important valve characters and parameters.

Correlations

Antarctica. There are correlations between Blizzard Heights and Storm Peak, Queen Alexandra Range, on the basis of shared conchostracan taxa; between Storm Peak, from which a dimorphic species *Cyzicus (Lioestheria) disgregaris* n. sp. was first described, Carapace Nunatak, southern Victoria Land, where this species was dominant in all measured sections, and Agate Peak, northern Victoria Land, where it also occurs.

India. There is a delineation of a traceable conchostracan horizon (Triassic, Panchet Formation) from the East Bokaro Coal Field subsurface; there is also correlation of northern and southern outcrops of the Kotá Formation (Lower Jurassic), on the basis of shared conchostracan and insect fossils.

Australia. Mitchell's types were placed in a biostratigraphic context when equivalent species were uncovered during field study in the Lake Macquarie district, New South Wales, Australia; in addition, a traceable conchostracan horizon was mapped.

Southern Africa. There is correlation between the Cave Sandstone of Lesotho and the Republic of South Africa on the basis of a shared conchostracan species.

The original query on *nonmarine* conchostracan dispersal between the southern continents can be answered affirmatively. Based on documentation of this study, such dispersal did occur. That, in turn, implies continental proximity during late Paleozoic and parts of Mesozoic time (Triassic to Cretaceous).

Introduction

Two outstanding investigators of living conchostracans made important contributions to our knowledge of the taxonomy and basic anatomy of the genera and species known at the time (Daday de Deés, 1910, 1915, 1923, 1925–1927) and to the developmental history, from egg through the several growth stages, of species of *Leptestheria, Cyclestheria,* and *Eulimnadia* (Sars, 1867, 1888, 1899, among others). These studies proved to be of great utility for work on conchostracans in the fossil record, and yielded many insights.

Study of the scattered world literature on fossil conchostracans, recorded in perhaps a dozen languages in addition to English, has placed in focus the many gaps in the data base. No one investigator had previously conducted biostratigraphic field researches in *all* southern continents. Several prominent investigators have extended knowledge of fossil conchostracans: Jones (1862–1901), described species from England and Europe, South America, Africa, and India, most of which had been collected by others; Novojilov (1958a–e, 1976), described abundant fossil conchostracan species and higher categories from the USSR, Asia, Africa, and elsewhere; Kobayashi (1954, 1973), published extensively on conchostracan classification and Asian conchostracan-bearing beds; and Defretin-Le Franc (1967), who contributed extensively to the taxonomy of fossil conchostracans from several African countries, Defretin and others, 1953, 1956 (Zaire, Niger, Cameroon, Morocco, and Algeria), as well as Europe and Greenland. As a consequence of Operation Deep Freeze during the International Geophysical Year, there was a need to reenforce this perspective by field work in other southern continents and countries. The present study fills that gap.

Field exploration in the Ohio Range fossiliferous deposits (*Leaia* zone) and subsequently in the Transantarctic Mountains prompted thinking about hemispheric fossil conchostracan distribution. For example, *Leaia gondwanella,* which this writer described (1965) from the Ohio Range, led to the realization that here was a "missing link," in as much as the genus had been previously known from South America, and Africa (North and South), as well as Australia. The Antarctic find provided a new link in the Gondwanaland reassembly of continents. The question arose as to how many other fossil conchostracan genera and/or species could be linked to the several southern continents. Being chiefly freshwater forms (occasionally brackish), Paleozoic/Mesozoic conchostracans seemed to be an ideal group for paleogeographic investigation, because the appearance of a given genus or species in deposits of several continents would require *nonmarine dispersal.* (This theme will be explored further in a subsequent section.)

This memoir was undertaken initially to record this writer's Southern Hemisphere findings on conchostracan biogeology, paleoecology, and systematic taxonomy, and has been extended to include those of many other investigators around the world. (For many of the last noted, see Acknowledgments and References Cited).

CONCHOSTRACAN TAXONOMY

Misconceptions

Hitherto there has been a parochial approach to the study of conchostracan distribution, taxonomy, and interpretation of new finds in the Southern Hemisphere. This present study includes new measurements, new photographs, and emended descriptions, as well as more comprehensive comparisons. Some of the misconceptions of major import are:

1. Horizontal Distribution. Conchostracan taxa in any country or district of a given country cannot be identical with, or closely related to, a given genus or species in another country and/or district of said country; the greater the geographic separation involved, the more likely this is.

2. Vertical Distribution. No conchostracan genera or species could persist across geologic systematic boundaries. This concept is, and was, probably based on the ephemeral small water situations which conchostracans inhabit(ed).

A confused taxonomy has resulted from these two misconceptions because many workers, from both the southern and northern continents, have created their own indigenous suites of new fossil conchostracan taxa. In brief, creation of a whole new national or regional taxonomy, unlike any prevailing elsewhere, often has been the practice. This may be attributed, in part, to lack of acquaintance with, or unavailability of, the world literature.

The first of the two concepts can be shown to be erroneous by reference to the writer's collections in Africa and India. The two genera *Cornia* and *Gabonestheria* occur together or separately in beds of Triassic age in Africa (Phyllopod Beds, Angola) and the subsurface Panchet Formation of India. They also occur in the Upper Permian of Brazil, which bears on the second mis-

3

conception. *Cornia* alone occurs in each of the five southern continents.

The second conception can be shown to be fallacious, because many fossil conchostracan genera and some species persisted through great geologic time spans; for example, besides *Cornia* which persisted from Permian through Jurassic time, there are, among many others, *Palaeolimnadiopsis* (Lower Carboniferous through Cretaceous), or *Palaeolimnadia* (Paleozoic/Mesozoic), and ubiquitous *Cyzicus* (Devonian through Recent). An example of a long-ranging Australian conchostracan species is *Palaeolimnadia (Grandilimnadia) glenleensis* Mitchell (late Paleozoic and Upper Triassic into the Jurassic of the Antarctic).

Dispersal of living conchostracan species is pertinent to the above observations.

". . . Most species have a wider distribution than previously judged, extending in most cases from Eastern Asia to Middle Europe" (referring to the family Leptestheriidae); and "most species examined seem to inhabit large areas" (Straśkraba, 1966, p. 584–585).

The living species *Limnadia lenticularis* is reported from most European countries as well as the USSR, Japan, and Alaska (Straśkraba, 1965a, p. 265–266, see also 1965b). If this modern species were to become extinct today over its entire range, it would obviously appear fossilized on the same time horizon in the respective rock column of the various places; across a vast, continuous land mass, in an island country, and on another continent.

It is equally true that "due to endemism, many living estherians are local species." (Kobayashi, 1954, p. 39). Study of the data of Daday de Deés (1915, 1923, 1925, 1926, 1927) living forms led Kobayashi and Kido (1943, p. 37) to conclude that "The distribution of living estherians is intimately related to 'the hot season' . . . in July in the northern hemisphere and in January in the southern hemisphere."

Conchostracan distribution is also dependent on suitable passive dispersal opportunities such as the wind currents that will transport conchostracan eggs from the dried-out small-water situations they inhabited. Where these eggs fall into a viable environment they will hatch. This also explains the way in which local species originate.

If, to the misconceptions discussed above, we add a poor grasp of the biology and ecology of living conchostracans (characterstic of many workers on fossil forms), inadequate generation of numerical data, and faulty taxonomic procedures, the result is the proliferation of synonymous generic names. That has deprived us of the very important bearing of the record of fossil conchostracans to global geology. This potentially ranges from tracing of numerous inter- and intracontinental dispersals during Paleozoic/Mesozoic time to casting new light on such questions as sea-floor spreading and placement of southern continents through geologic time; the record of the evolution of the Conchostraca is also at stake.

In this last connection, morphological changes in the conchostracan carapace through time yields evolutionary insight.

Some of these modifications are carapaces with ribs or spines, the development of recurved posterior margins, variation in size and placement of umbos, diverse configuration without singular structures, and secondary modifications of ornamentation.

SHELL MEASUREMENTS

One of the major weaknesses in fossil conchostracan taxonomy has been the absence of an extensive and standard series of measurements. Here, the system of Mme. Defretin-Le Franc (1967) has been used with some modification (see Fig. 18). Failure to have uniform and comprehensive numerical data on fossils makes comparison of taxa extremely difficult. Herein will be found as complete a system of measurements on many type specimens as possible. All workers, with rare exceptions, have given only the following dimensions: length and height of the valve (unribbed forms), occasionally a postero- and/or an anterodorsal angle and umbonal dimensions for large umbos, and for rib types, alpha and beta angles (see Fig. 18 for angles).

A common number used and usually cited in species diagnosis is the number of growth bands. These have limited taxonomic value even though they can denote relative age of an individual conchostracan in a population. Growth bands have an upper and lower margin; each of these is a potential growth line on the carapace. However because successive growth bands commonly overlap, the posterior margin of any younger growth band will not be visible. Less common, but known in some species, are *double growth lines* which appear to denote a mutant condition, and hence could have taxonomic value. Moreover, in some living species (Kusumi, 1961) the growth-line number may vary between males and females. In the fossil record of conchostracans such evidence of dimorphism is often too difficult to determine due to erosion or small sample size; rather, dimorphism is most frequently determined by relative size and shape and other comparative valve measurements.

Field Measurements

In order to trace seasonal events in a given small water situation in the geologic past, it is preferable to sample the conchostracan-bearing beds in centimetres; however, when back in the laboratory one usually needs to slice or split slabs so that the fossiliferous layers can be measured in millimetres. Precise data of this kind can permit tracing the history of the ecosystem through the time represented (Tasch, 1977a). Paleosalinity and other geochemical changes (trace elements) can thus be profiled (Tasch and Gafford, 1968), and evolutionary changes documented, among a host of other possibilities.

ENDEMISM

Predrift assembly of southern continents was one of contiguousness. That, in turn, facilitated dispersal of given conchostracan bioprograms between continents. While there was a great deal of endemism on the specific level, considerable inter-continental spread on the generic level took place, but has not always been recognized due to parochial taxonomic practices. As of this writing, endemism on the generic level appears to have existed in five indigenous genera in the South American Mesozoic: *Macrolimnadiopsis, Graptestheriella, Unicarinatus, Aculestheria,* and *Acantholeaia;* in the Carboniferous of Australia, *Limnadiopsileaia;* and in the Triassic of India, *Cornutestheriella.* Obviously, the origin of endemic genera is traceable back to endemic species.

Continental Overview of Gondwana Estheriids

The nonmarine fossil conchostracan fauna, and related biotic elements, newly found (and/or previously reported) from the five southern continents, are briefly reviewed below. Highlights of other aspects, including taxonomy, stratigraphy, correlation, paleoecology, and dispersal, are succinctly indicated, but more completely explored and documented in the text and appendix.

AFRICA

Nineteen African countries are now known to possess Paleozoic–Mesozoic nonmarine, conchostracan-bearing deposits: Algeria, Angola, Cameroon, Chad, Gabon, Ghana, Kenya, Lesotho, Malagasy, Malawi, Morocco, Mozambique, Niger, Republic of South Africa, Rio Muni, Zaire, Tanzania, Zambia, and Zimbabwe.

The greatest concentration of such deposits occurs below the equator, as is the case with South America. Zaire has the greatest concentration of such fossiliferous sites (see Fig. 1; and Fig. 20 for location of sites). Angola has the next highest concentration, followed by the Republic of South Africa (RSA), Zimbabwe, and Lesotho P. 48, Figs. 1, 2). North of the equator, Gabon, Niger, Algeria, and Morocco each have several sites. From most of the remaining countries, reports of *"Estheria"* are mere mentions.

In this section, some new fossil conchostracan data are presented. This includes description of a new genus, *Afrolimnadia,* from the R.S.A. (Pl. 3, Fig. 1, 2), and a new species of *Pseudoasmussiata;* this new genus replaces an occupied name, and is described in the South American section. There are a few new species from Lesotho and Angola, new species of *Leaia* from Morocco, and one from the Republic of South Africa; there are also new descriptions and measurements of British Museum types from Africa (including new photographs), as well as Professor Bond's types from the National Museum of Zimbabwe. All previous relevant descriptions are reviewed and revised on the basis of restudy of the available evidence.

Thirteen conchostracan genera in the African rock column are recognized in this memoir. There are more conchostracan genera in the African Triassic (ten in all) than in the Jurassic and Cretaceous combined, or in the Pennsylvanian and Permian combined. Where age is bracketed as Upper Triassic–Jurassic, conchostracan-bearing beds are found in Zaire's Haute Lueki Series, as well as in Gabon and Rio Muni.

Viewed in terms of persistence in the rock column of Africa,

Cyzicus is ubiquitous, ranging from Pennsylvanian to Cretaceous time; *Leaia* ranges from Pennsylvanian to Permian, while *Gabonestheria* extends from Permian to Triassic time. At the beginning of the Mesozoic (Triassic), several genera appeared that were not known in the African Paleozoic: *Afrolimnadia* n. gen., *Estheriella, Cornia, Palaeolimnadia, Estheriina,* and *Asmussia. Asmussia,* along with *Palaeolimnadiopsis, Echinestheria,* and *Estheriella* persisted into the African Jurassic, and *Estheriella* persisted into the Cretaceous. Among the forms that occur in Jurassic and older beds are *Cyclotherioides, Pseudoasmussiata,* and *Glyptoasmussia.*

Of interest are the characteristics of several genera that persisted in Africa from one geologic period to another: *Leaia* (Pennsylvanian–Permian) a ribbed type; *Gabonestheria* (Permian–Triassic) which has a prominent umbo that terminates in a spinous projection; *Echinestheria* (Triassic–Jurassic) with a comparatively thick, umbonal spine rising above the dorsal margin, and spinous projections from the dorsal-posterior as well as the dorsal terminations of successive growth bands; and *Estheriella* (Triassic–Cretaceous) a multicostate form.

Spines, corrugated nodes (seen by S.E.M.), ribs, or ridges on the valve face, and recurved postero-dorsal margins forming cutting edges *(Palaeolimnadiopsis)* appear to have had adaptive value during Paleozoic and Mesozoic times in the restricted, ephemeral, shallow, nonmarine habitats that conchostracans inhabited. Such puddles, ponds, lake margins, and river floodplains dried up or were exposed seasonally by lowering of the water level. Conchostracans buried in bottom mud, their ventral margin flush with the substrate and slightly ajar for feeding, were vulnerable when the water level was lowered in their lacustrine or fluvial environment. The specialized valve characters could have served effectively to break the mud seal of the almost dried bottom of their respective habitats. These traits may also have facilitated movement through vegetal and other bottom debris (Tasch, 1965).

Living *Limnadiopsis* species (Spencer and Hall, 1896) and *Caenestheriella belfragei* valves (Packard, 1883; Tasch, 1965, for photograph) have a prominent or subdued, serrated dorsal margin, respectively. These cutting-edge, toollike structures of contemporaneous conchostracan species suggests their current adaptive value. *Limnadiopsis* species, for example, show marked differences in the expression of serration (Spencer and Hall, 1896) Tasch, 1965.) The long persistence of the several valve

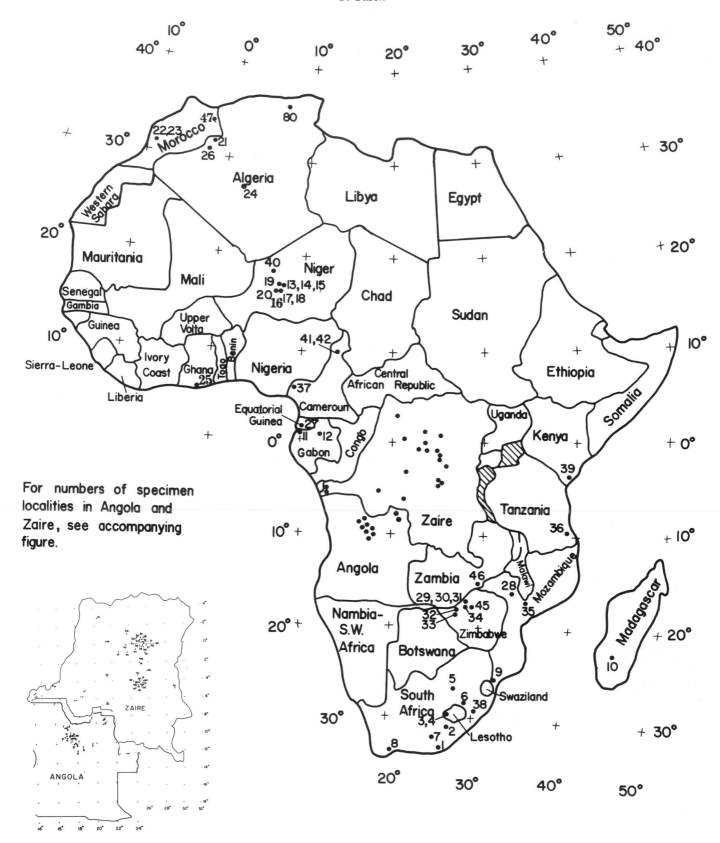

Figure 1. Location Map of Africa. Conchostracan-bearing beds, with detail of sites in Zaire and Angola. Enlarged figures with site locations for Zaire and Angola in Figures 20 and 20 A.

SCHEDULE OF FOSSIL CONCHOSTRACAN SITES IN AFRICA

[Numbers correspond to numbers on map of Africa, Figure 1. Sequence of numbers at a given site denotes additional species at that site. (See also Fig. 20 and 20A schedules.)

1. Port Elizabeth, Republic of South Africa (R.S.A.)
2. Coetzershoek, R.S.A.
3. Thabaning, Lesotho
4. Mofoka's Store, Lesotho
5. Vereening, Transvaal, R.S.A.
6. Plattsburg near Harrismith, east end of Drakensberg, R.S.A.
7. Near Cradock, Cape Province, R.S.A.
8. Approximately 48 km east of Swellendam District, R.S.A.
9. Namahacha, west of Lorenço Marques, Mozambique
10. Malagache (formerly Madagascar)
11. Coco Beach, Gabon
12. River M'Youm, Gabon
13. 95 km southwest of Agades, Niger
14. 93 km southwest of Agades, Niger
15. 4 km east of In Gall, Niger
16. 95 km southwest of Agades, Niger
17. 93 km southwest of Agades, Niger
18. 4 km east of In Gall, Niger
19. Mount Arli, 50 km southwest of Agades, Niger
20. 18 km southeast of Assaouas, Niger
21. Columb-Becher, Algeria
22. Region of d'Argona-Bigoudine, Haut Atlas Occidental, Morocco
23. Ida ou Zal, Morocco
24. 100 km north of In Salah, Algerian Sahara
25. Along coast, near mouth of the Amissa River, between Salt Pond and Winneba, Ghana
26. Guir Basin, Algeria
27. Coco Beach Series, Rio Muni (formerly Spanish or Equatorial Guinea)
28. Tête, Mozambique
29. Sengue Valley, Cliff Section, K^{5d}, lat 17°16′S, 28°15′E, Zimbabwe
30. Sengue Valley, Cliff Section, K^4, lat 17°27′S, long 28°18′E, Zimbabwe
31. Sengue Valley, roadside exposure, K^1, lat 17°27′S, long 28°18′E, Zimbabwe
32. Sebingwe Valley, Cliff Sections, K^{5c}, lat 17°57′S, long 27°15′E, Zimbabwe
33. Gwaai Valley, Chimwava Ranch, K^{5c}, lat 18°42′S, long 27°15′E, Zimbabwe
34. Sessami Valley near Madziwadzide Kraal, Cliff Section, K^{5c}, lat 17°38′S, long 28°30′E, Zimbabwe
35. Confluence of Shire and Zambesi rivers, Mozambique
36. Northwest of Lindi, Tanzania
37. Mamfe, Cameroon
38. 3.2 km south of Moos River, Natal, R.S.A.
39. Majiya Chumvi, Kenya
40. West of Air, Niger
41. Pont du Mayo, Louti, Cameroon
42. Pont du Mayo, Tafal, Cameroon
43. Not on map
44. Not on map
45. Mafungatusi Plateau, Zimbabwe
46. Luane Valley, Zambia
47. Djerada, Morocco
80. North Constantine, Algeria

features discussed above, some known only in extinct species, others such as recurved posterior margin in *Palaeolimnadiopsis* and living *Limnadiopsis,* clearly point to some adaptive value over geologic time spans to the present time.

As far as the nonpersistence of any African Cretaceous form into post-Cretaceous time (*Cyzicus* excepted), one may attribute it, in part at least, to continental drift and changing temperature zones. The same nonpersistence is also duplicated in the South American conchostracan fauna.

Several conchostracan genera and species have general southern distribution (Gondwanaland): i.e., Africa, Antarctica, and Australia. *Palaeolimnadia (Palaeolimnadia) wianamattensis* (Mitchell) is known from the Australian Permian and Triassic, and a very closely related form *Palaeolimnadia (Palaeolimnadia)* cf. *wianamattensis* (Mitchell) is known from the Triassic of Angola and is described below. The Antarctic Jurassic has *Cyzicus (Lioestheria) malangensis* (Marlière) which was originally described from the Zaire Upper Triassic or Jurassic (Defretin-Le Franc, 1967). *Gabonestheria* occurs in the Upper Permian of Brazil (described below) and the Lower Triassic of Angola (an Angola specimen of Marlière's species is described below). *Cornia* species occur in the Lower Permian of Gabon (Marlière, 1950a, Novojilov, 1958a), the Upper Permian of Australia (described below), and the Upper Triassic Cave Sandstone of Lesotho (Tasch, 1984). Dispersal of these and other conchostracans through the southern continents during Paleozoic and/or Mesozoic time will be more fully discussed in a subsequent section.

ANTARCTICA

Fossil conchostracans are now known from several Paleozoic/Mesozoic sites of Antarctica: Lashly Mountains and Portal Mountain (Devonian) of southern Victoria Land; Ohio Range (Permian); Transantarctic Mountains, including Blizzard Heights, Blizzard Peak and environs, Storm Peak, and Mauger Nunatak; Brimstone Peak (Lower Jurassic), and Carapace Nunatak, of Southern Victoria Land (Lower Jurassic); Agate Peak, Mount Frustum, and Gair Mesa, in northern Victoria Land (Lower Jurassic) (Figs. 2 and 3).

Lashly Mountains and Portal Mountain

Lashly Mountains. The Beacon Sandstone (Middle–Upper Devonian) of the McMurdo Sound area contained a single species, *Cyzicus (Euestheria)* n. sp., (described herein) from a laminated, dark-grey siltstone. A carbonized syntype occurred on a slab bearing a fair-sized arctolepid fish plate. Other conchostracan-bearing slabs also contained fish fragments. These fossils were collected by the New Zealand Geological Survey.

Portal Mountain. The Aztec Siltstone (Upper Devonian) crops out on the east of Portal Mountain; it was collected by Alex Ritchie while in search of Devonian fish. Besides new data on fish fossils, he found several finely laminated, hard, light-grey siltstone slabs which were covered with a coquina of a new species of *Cyzicus (Euestheria)* (herein described) and which differ from the Lashly Mountains species in almost all parameters. These two

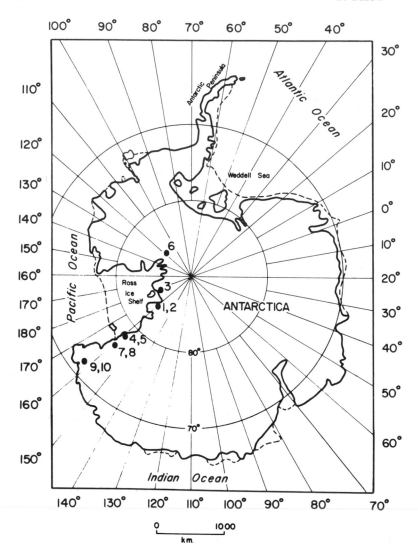

ANTARCTIC LOCALITIES:

1. Blizzard Heights (84° 37'S, 163° 53'E)
2. Storm Peak (84° 35'S, 163° 55'E)
3. Maugar Nunatak (85° 44'S, 176° 40'E)
4. Lashly Mts. (77° 54'S, 159° 73'E)
5. Portal Mtn. (78° 07'S, 159° 15'E)
6. Horlick Mts. (84° 45'S, 114° 00'E)
7. Carapace Nunatak (76° 54'S, 159° 27'E)
8. Brimstone Peak (75° 38'S, 158° 33'E)
9. Agate Peak (72° 57'S, 163° 48'E)
10. Mt. Frustrum (73° 22'S, 162° 57'E)

Figure 2. Antarctica. Map of conchostracan-bearing sites. 4, 5, Devonian; 6, Permian; 1–3, 7–10, Jurassic.

Devonian collecting sites are some 24 km apart. (See systematic descriptions for further discussion.)

Ohio Range

Only the *Leaia* zone of the Mt. Glossopteris Formation, Mercer Ridge, Ohio Range has thus far yielded Permian conchostracan fossils. The *Leaia* zone is about 1 m thick and traceable laterally for approximately 33 m, being faulted out at both ends. Fifteen successive generations of ribbed *Leaia gondwanella* Tasch and two generations of *Cyzicus (Lioestheria) doumanii* Tasch were found in the Mercer Ridge collections.

(George Doumani originally found this zone and contributed his samples.) Microstratigraphic study of these fossiliferous slabs indicated that conchostracan occupancy was intermittent and extended over some 136 y. (For further data on microstratigraphy, measured sections, Gondwana correlations, and Permian nonmarine ecosystems, supplemented by other data, see Tasch,

1977a (Table 2 and text, and Tasch, 1970a, p. 185–194). The conchostracans occur with a *Glossopteris* flora and several insects pertaining to *Plecoptera,* or *Protorthoptera* and *Uralonympha;* there is also a single, probable mollusk in a peel section. The Mt. Glossopteris Formation at the same site yielded at more than 480 m below the *Leaia* zone probable molluscan and ostracodal shells; and at approximately 500 m, a vermiform trace fossil (*Cochlichnus* sp.) (Tasch 1968).

Jurassic Conchostraca of the Antarctic

Most conchostracan-bearing beds are of Jurassic age and include collections from Blizzard Heights and environs in the Transantarctic Mountains, both in the Queen Alexandra Range, and from southern and northern Victoria Land (see Fig. 3). The fossils generally occur in sedimentary interbeds of basalt flows. The genera known from Antarctic Jurassic beds are for the most part known from rocks of the same age on other Gondwana

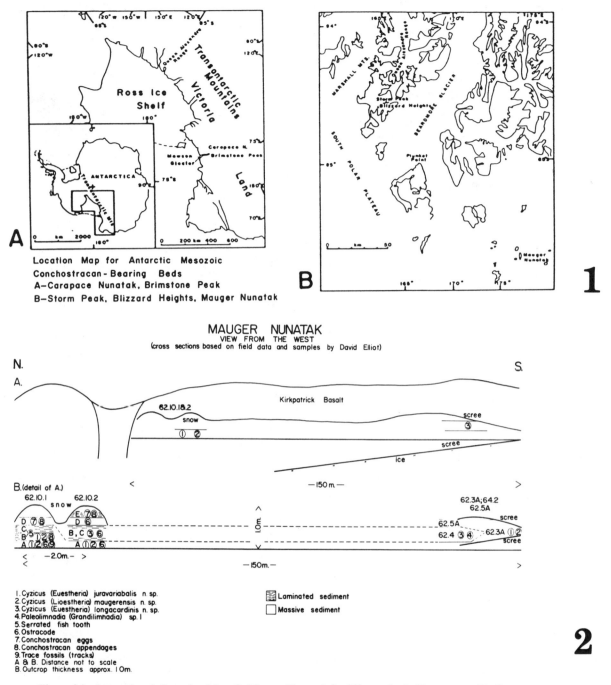

Figure 3A. Antarctica. 1: Location Map. 2: Mauger Nunatak fossiliferous beds. Data on section from David Elliot; conchostracan species and other biota (Tasch), Figure A after Elliot, 1972.

continents. Their now-isolated distribution is best explained by the concept of continental drift (i.e., plate tectonics), which is further discussed below. (See Fig. 2.)

Blizzard Heights and Environs

Fossils in the Tasch Jurassic collection from Blizzard Heights and David Elliot's collections from Blizzard Peak and environs include the following genera: *Cyzicus* (12 species, 9 of which are new), *Estheriina* (4 species), *Glyptoasmussia* (3 species), *Cyclestherioides* (5 species), and *Cornia* (2 species).

Worthy of special note in the collections above is the occurrence of *Cyzicus (Lioestheria) malangensis* (Marlière) described first from the Upper Triassic or Jurassic of Zaire, Africa. This species was found in three beds of Blizzard Heights (Station 1, beds 2, 3, and 7). A Blizzard Heights (Station 1 and 5) *Glyptoas-*

BLIZZARD HEIGHTS

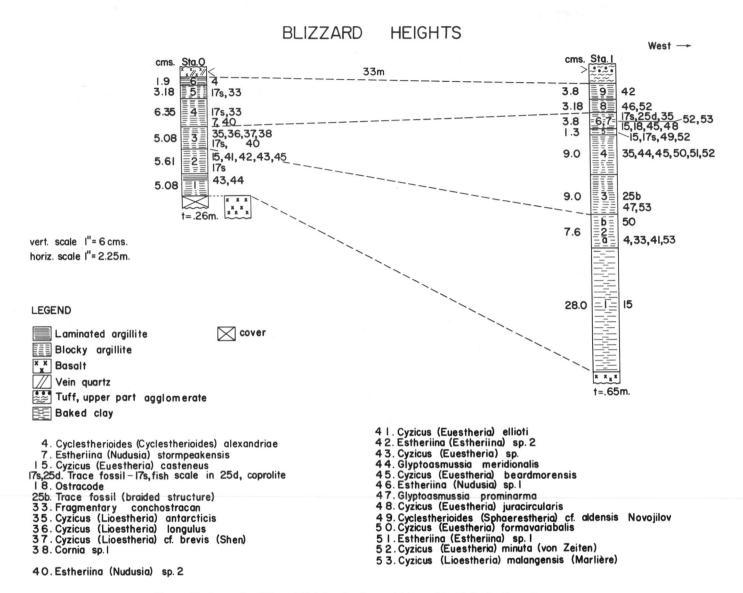

Figure 3B. Antarctica. Blizzard Heights. Section and biota with chiefly fossil conchostracans.
[35. *Cyzicus (Lio.) antarctis non antarticis*]

mussia prominarma n. sp. is closest in several measured parameters and ratios to *G. corneti* Defretin-Le Franc from the Upper Jurassic of the Congo Basin. They overlap in several ratios, but differ in configuration.

Other occurrences from Blizzard Heights and environs that are known elsewhere include *Cyzicus (Euestheria) minuta* (von Zieten), recorded from the Triassic of Greenland, Europe, and a few other scattered areas. Defretin-LeFranc's measurements of the Greenland "minuta" compare most closely with those from Blizzard Heights. *Diaplexa? brevis* Shen 1976 from the Upper Mesozoic of mainland China is comparable to a form from Blizzard Heights (Station 0, bed 2), *Cyzicus (Lioestheria) cf brevis* Shen 1976. These relationships of Antarctic and exotic species will be further explored subsequently.

Differences and Similarities between the Two Blizzard Heights Stations

The conchostracan faunas at the two Tasch Blizzard Heights stations (Station 0 and 1) differed in some respects: (1) Station 1 has several species recurrent through the section: *Cyzicus (Euestheria) castaneus* n. sp., *Cyzicus (Euestheria) formavariabalis* n. sp., *Cyzicus (Lioestheria) malangensis* (Marlière), *Cyzicus (Lioestheria) antarctis* n. sp., *Cyzicus (Euestheria) minuta* (von Zieten), and *Cyzicus (Euestheria) beardmorensis* n. sp. In contrast, approximately 33 m to the east at Station 0 there were no recurrent species. (2) *Cornia* sp. 1 was found at Station 0, bed 3, Blizzard Heights and its correlate, D.E. Station 23.85 (See Fig. 4A) yielded *Cornia* sp. 2. Station 1 had exotic species not found in Station 0.

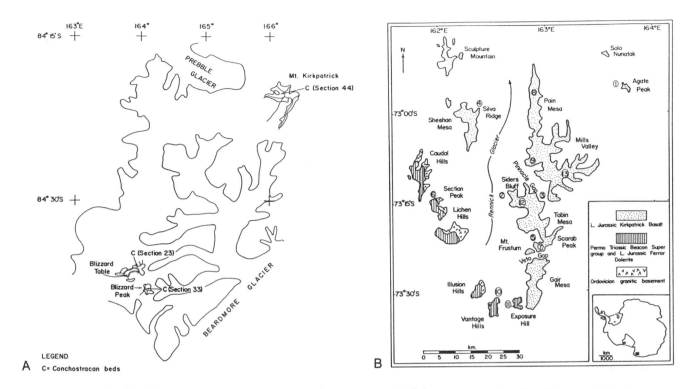

Figure 4. Antarctica. A: Blizzard Heights environs with conchostracan-bearing beds. B: Agate Peak, Mt. Frustrum, and Gair Mesa. C: Blizzard Heights conchostracan-bearing bed. Tasch Station O around the bend. (Photograph by David Elliot.)

The two Blizzard Heights stations shared six species: *Cyzicus (Euestheria) ellioti* n. sp., *C. (Eu.) beardmorensis* n. sp., *C. (Lioestheria) antarctis* n. sp., *Cyclestherioides (Cyclestherioides) alexandriae* n. sp., *Glyptoasmussia meridionalis* n. sp., and *C. (Eu.) casteneus* n. sp.

There are two possible explanations for the difference at the two Blizzard Heights stations: geographic isolation and different levels.

Geographic Isolation. In life, the pond represented by deposits at the two stations deepened to the west (i.e., in the direction of Station 1) (See Fig. 4C for photograph and Fig. 3B). The conchostracan fauna in the area of Station 0 inhabited the shallower part of the pond even though the deeper portion (Station 1) was also shallow.

During seasonal drying of the pond, isolated pools formed (Station 0). This is a common occurrence in many modern ponds with living conchostracans, as the shoreline lowers and a mud margin with restricted, relict water bears conchostracans among other biota (Tasch and Zimmerman, 1961). However, shared species between Stations 0 and 1 (Blizzard Heights) denote repeated, though noncontinuous, dispersal when the pond was seasonably refilled.

Different Levels. Within a given bed, even if only a few centimetres thick, sedimentary layers of a few millimetres thickness can contain multiple levels representing successive pond deposits and the given season's biota. A barren interval may follow, and subsequently a new series of levels follows bearing pond biota, including conchostracans.

There are two main sources to account for known and new conchostracan species in any interbed deposit such as encountered in the present study and described in the preceding paragraph: (1) direct migration of individuals from the farthest, undetermined reaches of the given three-dimensional water body of the time now represented by a particular sedimentary interbed; and (2) wind-dispersal of conchostracan eggs from dried substrates of larger or smaller water situations of a given time. Such egg dispersal need not be viewed as an immediate transit from a to b or c, however close or distant. Rather, it could be a kind of "stepping stone" dispersal (Tasch, 1979c, p. 38, Fig. 4 for an example between eastern and western Australia). Such dispersal could occur in a single season (as on modern floodplains dotted with pools and puddles) or require multiple seasons to reach a given water body in the Antarctica of the time. Whatever the transport time, the same level in an interbed can bear species of different genera, or different levels can bear several species of the same genus.

Presence of exotic species in Antarctic beds suggests that they could have reached the region along the route outlined above.

Conchostracan-bearing beds contained fish scales at Station 0, beds 2 through 5 (Blizzard Heights). At Station 1 (B.H.) only beds 5 and 7 had scales. The bed 7 occurrence is in a probable fish coprolite bearing other fish debris (Tasch, 1976); this indicates cannibalism by larger fish on smaller ones.

Storm Peak

Storm Peak has a Lower Flow interbed with three stations and an Upper Flow interbed station that is represented by a shattered fossiliferous block surrounded on three sides by diabase basalt talus. Fossiliferous conchostracan-bearing fragments were broadly scattered in front of the block and over the underlying basalt talus as the block slumped downslope (Plate 45, left).

The three Lower Flow interbed stations (Stations 2, 0, 1) indicate that the shallow basin deepened toward the east. One of the prominent features of the Storm Peak section is a Fish Bed with the actinopterygian *Oreochima ellioti* Shaeffer and in which the conchostracan *Cyzicus (Euestheria) ichthystromatos* n. sp. is also present. This fish horizon is seen at Station 2, bed 4, and Station 0, bed 3, and is traceable to Station 1, bed 4, some 50 m distant from Station 0 (Fig. 5.) Another prominent feature of the Storm Peak strata is a fossil wood horizon (large segments of logs) traceable between Stations 0 and 1, and apparently present at Station 2, but without logs; the basal bed at all three stations contains palynomorphs. It should be noted, however, that Stations 0 and 1 share the same palynomorphs while others were found at Station 2.

In Storm Peak Lower Flow interbed, all stations had one or more beds that were barren of fossils (Station 2, bed 1; Station 0, beds 1 and 4; Station 1, beds 2, 3, and 7). The Upper Flow had only one bed that lacked fossils (Upper Flow, bed 3). Only a single bed of one station at Blizzard Heights was barren of fossils (B.H. Station 1, bed 9).

Species shared between Storm Peak Stations 2 and 0 besides *Cyzicus (Euestheria) ichthystromatos* n. sp. included *C. (Euestheria) castaneus* n. sp. This last species also recurs at Station 1 (beds 1, 5, and 6). Only *Cyzicus (Euestheria) crustapatulus* n. sp. (Lower Flow, Station 0, bed 2) persisted into the Upper Flow interbed (bed 11, lower).

The single station of the Upper Flow interbed had a much more varied conchostracan assemblage than the Lower Flow stations, as indicated by the genera: *Cyzicus, Cyclestherioides, Palaeolimnadia, Asmussia, Glyptoasmussia, Pseudoasmussiata,* and *Estheriina.* Four of the genera of the Upper Flow interbed are also found in the Blizzard Heights sections, as well as four species: *(Estheriina (Nudusia) stormpeakensis* n. sp., *Cyclestherioides alexandriae* n. sp., *Cyzicus (Euestheria) beardmorensis* n. sp., and *Cyzicus (Euestheria) castaneus* n. sp.

Several Storm Peak Upper Flow species recurred: *Palaeolimnadia (Grandilimnadia)* cf. *glenleensis* (Mitchell) (Upper Flow, beds 2, 7, and 8), *Cyzicus (Lioestheria) disgregaris* n. sp. (beds 6 and 11 lower), and *Cyzicus (Euestheria) crustapatulus* n. sp. (bed 5 lower, bed 11 lower). Noteworthy is the absence of *Cornia* species at Storm Peak stations and *Palaeolimnadia* at Blizzard Heights stations and likewise *Pseudoasmussiata, Asmussia,* and *Glyptoasmussia,* except for one species (at B.H.). Although *Cyzicus (Euestheria) castaneus* n. sp. recurs in the Lower Flow, it does not appear in the Upper Flow.

The data reviewed above are of interest when related to the

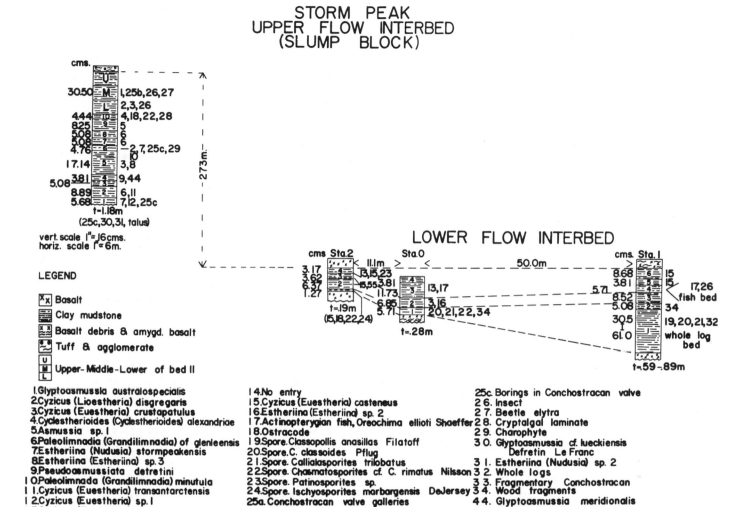

Figure 5. Antarctica. Storm Peak. Upper and Lower Flow interbeds; fossil conchostracan species and associated biota of insects, ostracods, palynomorphs, trace fossils, and fossil wood. [9. *defretinae non defretini*; 30. *luekiensis non lueckiensis*].

273 m of diabasic basalt that separates the Lower from the Upper Flow interbed at Storm Peak. Only the Lower Flow interbed has recurrent *Cyzicus (Euestheria) castaneus* n. sp., as does Station 1 of Blizzard Heights. That indicates that the Lower Flow interbed shared a common bioprogram for this species with Blizzard Heights Station 1. But the evidence is more complex, as noted above, for while the Upper Flow shares four genera with the Blizzard Heights sections, only *Cyzicus* of those four is in the Lower Flow interbeds. The persistence of *Cyzicus (Euestheria) crustapatulus* n. sp. from Lower to Upper Flow interbed also needs to be taken into account.

David Elliot, who mapped the Kirkpatrick Basalt of the Marshall Mountains, correlated the Upper Flow interbed at Storm Peak with the Blizzard Peak interbed. His diagram showed both of these interbeds directly underlying diabasic basalt flows (1970, p. 311, Fig. 7, columns 23, 27). A Lower Flow of hyalo-ophitic basalt at Storm Peak was also shown, but no Lower Flow interbed. This correlation can be attributed to the fact that the paleontology of the Lower Flow interbed had not yet been studied at that time. With the 1970–71 faunal data described and discussed in this section of the memoir, Elliot's interpretation needs to be revised. If the 273 m of flows intervening between the Lower and Upper Storm Peak interbeds are taken into account, then, from a fossil faunal perspective, *both* interbeds appear to be correlates of the Blizzard Heights interbed.

Recurrent genera and species are known for the interbeds at both sites and that is the probable explanation for the shared taxa of the Storm Peak Upper Flow and the Blizzard Heights sections. However, to account for the gap separating both interbeds at Storm Peak, in light of the recurrent taxa, one must take into consideration the persistence elsewhere of the biograms of these taxa. They appear first in the pool represented by the Blizzard

Heights interbed. Subsequently, they dispersed via wind-borne eggs to the pond represented by the Storm Peak Lower and Upper Flow interbeds. The detailed mechanism is unclear; however, there is evidence of widespread dispersal of the bioprograms of conchostracan taxa in the Transantarctic Mountains during Jurassic time.

Some aspects of the Storm Peak conchostracan occurrences merit further comment. (1) Although no Australian Jurassic conchostracans have been reported as of this time, an Australian palaeolimnadid of Permian and Triassic age has a close relative in the Storm Peak Upper Flow interbed, beds 2, 7, and 8 *Palaeolimnadia (Grandilimnadia)* cf *glenleensis* Mitchell, which is of Jurassic age. (2) *Glyptoasmussia* cf *luekiensis* Defretin-LeFranc (TC 9097) is here identified from a loose piece from the Storm Peak Upper Flow; that species was originally described from the Congo Basin (Haute Lueki Series, which is either Upper Triassic or Jurassic (pre-Oxfordian), according to Defretin-LeFranc). (3) *Pseudoasmussiata defretinae* n. sp., occurs at Storm Peak, Upper Flow, bed 4 (see Fig. 5) and also has comparable, geologically younger equivalents (not identical) in the Congo Basin (Loia Series, Upper Wealden) particularly *Ps. ndekeensis* D.-LeFranc and *Ps. banduensis* D.-LeFranc (Defretin-LeFranc, 1967) (see taxonomic descriptions below). (Items 2 and 3 should be viewed in the context of the Blizzard Heights occurrences of *Cyzicus Lioestheria malangensis* Defretin-LeFranc; i.e., the same species originally described from the Upper Triassic or Jurassic of Zaire, Africa.)

Biotic Associates. The biotic associates of conchostracans in the Storm Peak beds include a rare charophyte (Chlorophyta) and a cryptalgal laminate; palynomorphs occur in all but the basal bed of Storm Peak Station 2. In the Storm Peak record there are, besides a fossil wood horizon and fishes (Actinoptergii), insects (beetle elytra and other wing fragments), trace fossils, and ostracods.

Mauger Nunatak

At one Mauger Nunatak (MN) site (62-10-1), David Elliot found what turned out to be a traceable conchostracan-bearing interbed 1 m thick. (See Fig. 3A for sketch of section and biota, and Plate 45 for labelled photograph). This zeolitic silty mudstone is overlain and surrounded by a scree slope about 15 m thick which lies on a massive basalt flow. The conchostracan-bearing beds were traced through the scree cover, either by digging under it or by a short lateral exposure through the scree at two other stations (62-10-2 and 62-3, 4, and 5A). Station 62-10-2 is some 2 m south of 62-10-1 and is divided into A, B, C, D sampling units. Station 62-3 and the others occur towards the north end of the more southerly facing cliff, which is about 150 m from Station 62-10-1 (D. Elliot, 1971, personal communication).

There are three species of *Cyzicus* present and one species of *Palaeolimnadia* (see Fig. 3A for species names and biostratigraphy). Two of three cyziciid species occur in the same bed

(62-10-BC) at the three stations, while one occurs only at 62-10-2BC. The palaeolimnadid species is restricted to Station 62.4, which can be correlated with 62-10-2BC because they both contain a genus or species not found in an underlying bed.

Biotic Associates and Spoor. Also present are a single, broken, marginally serrated, rather flat (probably fish) tooth, ostracods, and what are likely conchostracan eggs attached to valves, and unidentified possible food-search tracks (bed 62-10-1A). The sparse occurrence of fragmentary floral remains is striking and suggests that although the Nunatak beds are contemporaneous with those at Storm Peak and Blizzard Heights, the wooded area did not extend to the Mauger Nunatak area.

Appendages. A slab from bed 62-10-1B of the Nunatak section preserved dispersed, setose and lobate conchostracan body appendages, and the segmented rami of more than one second antenna, all concentrated in a small area (see Plate 18, Fig. 4). J. J. Schlueter of Wichita, Kansas, made X-ray examination of specimens 8305 and 8400 in the writer's collection, but this failed to reveal further soft-part anatomy. This may be attributed to the high igneous heat to which the bed had been exposed during the overlying lava flow (Tasch, 1982b).

Brimstone Peak

John Ricker sampled and described a single fossiliferous interbed more than 151 m above the base and above the eighth basalt flow at Brimstone Peak. The site is situated between the Mawson and Priestley glaciers, southern Victoria Land. This is the only fossiliferous horizon reported to occur from base to summit. The conchostracan fossils described below have been assigned to *Cyzicus (Lioestheria) rickeri* n. sp. (Plate 6, Fig. 1) and are all carbonized and embedded in lenses of siltstone and fireclay; the interbed is overlain by a basalt flow (the ninth). (For data on the temperature indicated by the baked and carbonized fossils, see Tasch, 1982b).

Agate Peak

At Agate Peak in north Victoria Land (see Fig. 2), an interbed below the Lower Jurassic Kirkpatrick Basalt contained specimens assignable to a single conchostracan species *Cyzicus (Lioestheria) disgregaris* n. sp. (described under Storm Peak entries). A few partial valves of this species also were found on the same bedding plane along the outcrop and on other slices in the area. This is the first reported fossil conchostracan from north Victoria Land and extends the Lower Jurassic conchostracan spread in Antarctica. The only other biota (other than wood fragments) accompanying the conchostracans are ostracods and fish fragments.

Of special interest is the fact that the fossils described for Agate Peak were in sediments in contact with pillow lavas. (Pillow lavas are also known from Carapace Nunatak.) The deep red-brown color of the conchostracan fossils, in some instances with a carbonized underlying cuticle, corresponds to the altered condition of the shells at the occurrence at Carapace Nunatak.

CARAPACE NUNATAK – SOUTHERN VICTORIA LAND

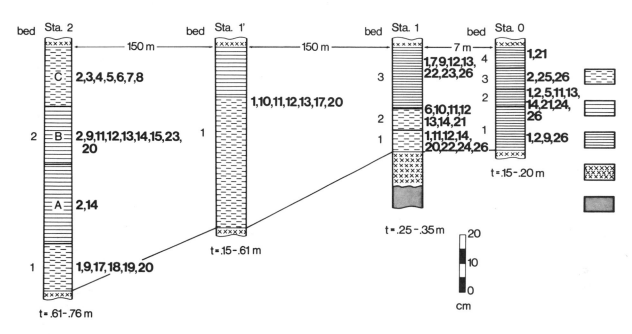

1. Cyzicus (Lioestheria) disgregaris n. sp.
2. Araucaracites australis Cookson
3. Undulatosporites sp.
4. Cibotiumspora juriensis
5. Classopollis simplex
6. Laricondites triquetrus
7. Carbonized wood
8. Notostracan caudal filaments
9. Carbonized needles
10. Carbonized wood fragments
11. Syncarids
12. Mayfly Nymph
13. Ostracodes
14. Notostracan
15. Isopods
16. Plant debris
17. Fish scales/teeth
18. Classopollis chateaurovi Reyne
11. Exesipollenites cf. E. tumulus Balme
20. Beetle elytra
21. Scalariform Tracheids
22. Abundant triletes
23. Unidentified insects
24. Carbonized plant fragments
25. Inaperturopollenites Balme
26. Classopollis sp.

Figure 6. Antarctica. Carapace Nunatak. Measured sections (Tasch and Schumacher, unpub. data). The contained biota included a single conchostracan species in addition to notostracans, syncarids, insects, fish fragments, several palynomorphs, fossil plants, and wood.

Both are indicative of an effective temperature induced by igneous heat of 450°–600°C at the two sites (Tasch, 1982b). The carbonized condition of wood fragments occurring on conchostracan-bearing slabs also denotes a low effective temperature. The pillow lavas denote under-water flows at both sites, showing the mode of termination of the respective water basins bearing conchostracans and other biotic elements.

Carapace Nunatak

There are several prior reports of conchostracans and other biota at Carapace Nunatak (Gunn and Warren, 1962; Tasch, 1971, 1973; Borns and others, 1979; see Fig. 2). Ballance and Watters (1971) provided a very useful, generalized stratigraphic section of the Mawson diamictite and the Carapace Sandstone; according to them, the latter includes "2.5 m of chert lenses with

fossils, and the capping Kirkpatrick Basalt." (See Appendix for Tasch measured sections and other data.) Ball and others (1979) described fossiliferous sections from the south and north spurs. The south spur fossiliferous bed included a discontinuous chert lens (F_1), a few centimetres thick, and higher up in the same section, another chert lens (F_3) some 60 cm thick and extending laterally some 30 m. The north spur had a nonsiliceous, fossiliferous bed (F_2) some 2 m thick, with graded bedding (turbidite) and an average bed thickness of 0.6 cm (range 0.2–1.5 cm).

Our party studied the Carapace Nunatak sedimentary interbed, which is discontinuous but traceable over a distance of 307 m and directly overlain by the Kirkpatrick Basalt (Pls. 46, 49). It is located along the inner southern wall of an amphitheatre-shape exposure. Sections at four stations were measured and fossils precisely located stratigraphically (Fig. 6). All sections were less than 1.0 m thick (range 15–76 cm). The conchostracan-bearing bed reported by Gunn and Warren (1962) and Ballance and Watters (1971) was to the southeast of the Tasch stations. The fossiliferous beds reported by Ball and others (1979) were farther south on the Southern spur than Tasch Station 0. The north spur was not sampled by Tasch, but samples collected by the late J. S. Schopf are available. Other Schopf collections from the southeast were taken from above the base of the interbed. Schopf's southeast spur collection is very likely from the same site as that of Gunn and Warren (1962) and Ballance and Watters (1971).

Three types of fossiliferous beds have been reported in the cited studies: chert lenses (south spur), beds only locally silicified and cherty (north spur), and lenticular, fossiliferous siltstone, shale, and chert horizons with graded beds (turbidite) (north spur, F_2 station). At the Tasch stations, baked, upwarped, fragmented and eroded, hard carbonaceous shales that were thin to massive bedded constitute the dominant lithology. Since none of the slabs reacted positively to dilute HCl, secondary silicification must have been widespread along the entire interbed that extended for a distance of some 307 m.

The effective temperature that baked these beds and their contained fossils was between 573° and 600°C (Tasch, 1982b, p. 667).

At the four Tasch stations, conchostracans, all dimorphs of a single species, *Cyzicus (Lioestheria) disgregaris* n. sp. occurred as follows: Station 2, bed 1, in massive black shale, but not in the thicker overlying laminated shale; at Station 2, bed 2, unit C, in a massive black shale; Station 1', 150 m west of Station 2, had conchostracans in laminated hard shale that graded down to massive shale; at Station 1, 150 m south of Station 1', conchostracans occur in three distinct beds; bed 1, laminated shale, bed 2, massive black shale, and bed 3, fissile shale in contact with the capping basalt sill; at Station 0, 7 m south of Station 1, only fissile shale occurs, and of the four sampling subdivisions, conchostracan fossils were found in three (see Fig. 6).

The arthropod fauna at Carapace Nunatak contains elements absent from the other Jurassic conchostracan sites in Antarctica; for example, syncarids, isopods, and notostracans. However, beetles were found at both Storm Peak and Carapace Nunatak, and the flora at Carapace Nunatak has been shown to have some analogues to the flora of Hope Bay, Antarctic Peninsula (Halle, 1913; Plumstead, 1964; Volkheimer, *in* Tasch and Volkheimer, 1970; Tasch, 1970b). Zeuner (1959) described some beetle elytra from Jurassic beds of the peninsula that contain plant material of the sort noted below.

The associated biota is more varied than at the other Jurassic conchostracan-bearing interbeds in Antarctica (Storm Peak, Blizzard Heights and environs, Mauger Nunatak, Brimstone Peak, and Agate Peak). Arthropoda constitute the largest component of the fauna: present are conchostracans, notostracans, and ostracodan Crustacea; syncarids and isopods, Malacostraca; and Insecta represented by beetle elytra (single and paired), mayfly nymphs, and in the collection of Borns and others, in addition to mayfly nymphs, there are dragonflies (Carpenter, 1969).

This writer noted (Tasch, 1973) that in sections described from Carapace Nunatak the faunal element (mayfly nymphs and conchostracans) contained species identical to those in the Borns and Hall collection from their locality F_2. That was based on examination of their original collection in Professor Carpenter's laboratory (Harvard) and is now extended to include the same dimorphic conchostracan species. This new information suggests that the north and south sections were probably once connected by a continuous water body having shallower southern sections (Ball and others, 1979).

Floral elements include wood fragments, carbonized needles, several species of palynomorphs (Tasch and Lammons, 1978), and other fragments. Fish are also represented, chiefly by scales, but there are also other fragmentary pieces. The floras in other collections from this site have been described by Plumstead (1962), Townrow (1967), and Delevoryas (*in* Ball and others, 1979), and included ferns, conifers, and cycadophytes. Ball and others (1969) suggested that north and south spurs were offset by a fault. (The structural aspects are beyond the scope of the present study.) If so, that could support the faunal evidence for a shared fauna at their sites and this writer's and hence for continuity of the lake in Jurassic time.

Radioactive dating indicates a Lower Jurassic age (approximately 180 Ma) for the fossil-bearing beds at Carapace Nunatak. The faunal data sustain that age independently (Jain, 1980).

The question might arise as to how a single conchostracan species dispersal to Carapace Nunatak (lat 76°54'S, long 159°27'E) might be explained. Storm Peak (lat 84°35'S, long 163°55'E) not only has the common conchostracan species, *Cyzicus (Lioestheria) disgregaris* n. sp., but several other rarer *Cyzicus* species; clearly, conditions for speciation recurred there. Accordingly, it is more likely that the Storm Peak *"disgregaris"* species was dispersed to Carapace Nunatak than the reverse.

This inference has support of the available age determinations. Thus, Ball and others (1979), on the basis of $^{40}Ar/^{39}Ar$ ages, noted an age for the fossiliferous lacustrine beds at Carapace Nunatak and the overlying hyaloclastite, of about 180 Ma; Lower Jurassic. The actual readings were 177 ±4.4 Ma and about 170.3 Ma. At Storm Peak, Elliot (1970) recorded three readings on the basis of whole rock K/Ar analysis of the basalt. Two of these are quite close to those at Carapace Nunatak: 179 ±7 Ma and 170 ±7 Ma. The conchostracan bed this writer sampled is directly overlain by the basalt. Both of these sets of figures indicate Toarcian to Aalenian (Lower Jurassic) age (Cohee and others, 1978, p. 291, Fig. 2); the Storm Peak fossiliferous deposit is apparently older. However, we have compared two sets of figures derived by different methods; the two deposits (Storm Peak and Carapace Nunatak) may prove to be nearly contemporary. But, how to account for only one of several Storm Peak species occurring at Carapace Nunatak?

C. (Lioestheria) disgregaris occur in only two interbed horizons at Storm Peak Upper Flow. In the older of the these, bed 6, its near-contemporary was *Estheriina (Nudusia) stormpeakensis* n. sp., whereas in the younger bed 2, it was *Cyzicus (Euestheria)* crustapatulus n. sp. The oldest occurrence of *"disgregaris"* at Carapace Nunatak was in the basal bed at all stations except Station 1. Thus, if this stratigraphic position is viewed as somewhat younger or possibly contemporaneous with that of this species in Storm Peak, bed 2, then only the estheriinid species mentioned above needs to be considered. Perhaps both species were dispersed via eggs in dried-mud substrates at the same or nearly the same time, but only *"disgregaris"* survived the discontinuous dispersal and reached the Carapace Nunatak area.

The Agate Peak site (lat 72°58′S, long 163°48′E) also has the common conchostracan species *"disgregaris"*. Although present data are sparse from this site, the conchostracans that occur there may originally have experienced egg dispersal from the Carapace Nunatak area. Dispersal should not be viewed as a one-step event. Rather, it could have been a very slow process, a stepping-stone type of dispersal through wind-blown eggs (cf. Tasch, *in* Tasch and Jones 1979c, p. 38, for what appears to have been precisely this type of dispersal between east and west Australia). There is also on record (Tasch and Zimmerman, 1961) the widespread occurrence of living *Cyzicus mexicanus* (Claus), apparently due to wind-blown eggs from dried mud substrates.

AUSTRALIA: OBSERVATIONS ON THE PALEOZOIC/MESOZOIC CONCHOSTRACAN FAUNAS

Devonian

The only Australian Devonian (Fig. 7) conchostracans presently known came from the Fairy Beds of Buchan Caves Park, Victoria. These are of lower Middle Devonian age. In company with Professor John Talent (Macquarie University, New South Wales), who discovered these fossiliferous beds, this writer collected conchostracans here described as *Cyzicus (Euestheria) talenti* n. sp. (Plate 4, Fig. 1; also see Systematic Descriptions, "Australia").

Carboniferous

Two Lower Carboniferous conchostracan faunas are known from Australia, from the Raymond Formation of the Drummond basin (Upper Tournasian-Visean) and from the Anderson Formation of the Canning basin (Lower Visean to Early Namurian[?]).

Only a single genus, *Leaia,* is known from the Raymond Formation. This ubiquitous genus also occurs higher in the Lower Carboniferous and continues through the Permian. In the Anderson Formation, several species of *Leaia* and its two sub-genera, *Leaia (Leaia)* and *(Leaia (Hemicycloleaia)* appear. In addition to *Leaia,* four other genera were found in the Anderson Formation borings: *Rostroleaia, Limnadiopsileaia, Monoleaia,* and *Cyzicus.* The first three of these supplemental genera do not appear above the Lower Carboniferous; the fourth, *Cyzicus,* does. In fact, *Cyzicus* and its subgenus *Lioestheria* persisted in Australian basins through the Lower Cretaceous, although there is no Juras-

sic record to date. (See Fig. 26, and see below for further details on Mesozoic conchostracans.)

Of special interest is the appearance of *Rostroleaia* in the Lower Carboniferous of Australia. This genus was previously unknown outside the Kazanian (Upper Permian), Nikolaesk Gorge, USSR. An older occurrence of *Rostroleaia* in the Middle Devonian Gitou Formation (Le Chang, People's Republic of China) suggests China as a probable source for the Canning basin *Rostroleaia* of Lower Carboniferous age in west Australia (Fig. 26; see Tasch, 1980; cf. Shen, 1978).

Permian and Triassic

Biostratigraphy. There is a considerable record of conchostracans in the Australian Permian and Triassic. Plotting all Tasch present stations and one of Mitchell's (M15) on the Pryor and Mursa map (1968) of the surface geology of the Newcastle Coal Field (Fig. 8) indicated the most proximate underlying coal seam. Figure 26 records the conchostracan genera and species occurring at each station and the closest underlying coal seam. In addition, it indicates the formation in which the fossiliferous beds were sampled. The vertical distribution of the conchostracan fauna was thus found to be related to three coal seams, (from older to younger) Dudley, Upper Pilot, and Fassifern.

The bulk of the Tasch stations were found to be underlain by the Upper Pilot seam (Stations 1, 2, 3, 4, 5, 7, 7A, 8, 9, 9A, 11, 12A, 12B, and 13) and the measured sections were part of the Croudace Bay Formation. Stations 6 and 10 were underlain by the Fassifern seam, and the measured sections were part of the Eleebana Formation (Fig. 9).

One of Mitchell's localities, the Old Cardiff seam is the same as the Fassifern seam (the writer's locality number M15; N. R. Mauger, Superintendent of Collieries, The Broken Hill Proprietary Company Ltd., Belmont, New South Wales 1980, personal commun.) Mitchell's fossil conchostracans came from the Awaba Tuff member of the Eleebana Formation above the exposure of the Fassifern seam on the northwest shore of Belmont Bay of Lake Macquarie, New South Wales (Fig. 8).

Knight (1950) and others collected from the insect-bearing bed of the Croudace Bay Formation and noted its geographic spread. The present study expands this information by providing a series of measured sections in the same or adjacent areas. It also established an areal correlation of the conchostracan-bearing beds (Fig. 9). Conchostracan fossils were found together in the same bed with fossil insects at several Tasch stations.

Most Tasch stations in the Croudace Bay Formation are clustered in an area approximately 18.3 km² (Pl. 47). Accordingly, wind dispersal of conchostracan eggs from dried out pools and/or lake margins was possible. Distances between most proximate stations ranged from 0.32 to 1.52 km (Fig. 10, top). As can be seen in both parts of Figure 10, despite relative proximity, the bulk of the conchostracan fauna that inhabited such water bodies, namely the ribbed types and palaeolimnadids, were not dispersed to each such neighboring pond or lake. Rather, selective wind

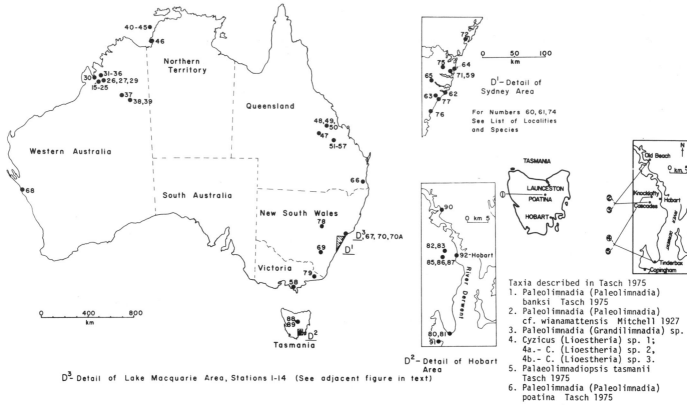

Figure 7. Australia. Location map of conchostracan-bearing beds with detail of Sydney area (D^1) and Tasmania (D^2).

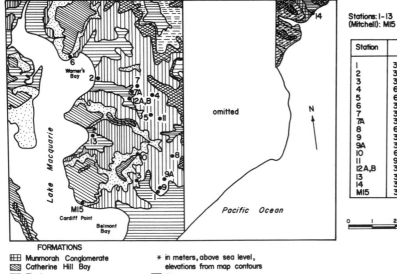

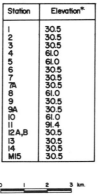

Figure 8. Australia. Lake Macquarie area, New South Wales; Tasch collecting sites (D^3) and conchostracan-bearing beds by formation.

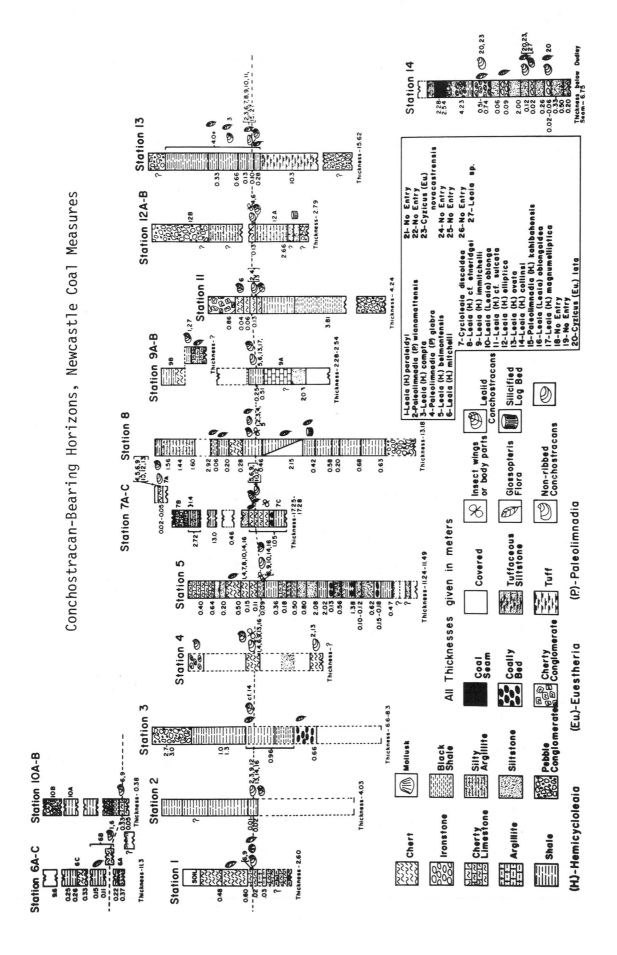

Figure 9. Australia. Lake Macquarie area (New South Wales). Measured sections of conchostracan-bearing horizons in the Newcastle Coal Measures.

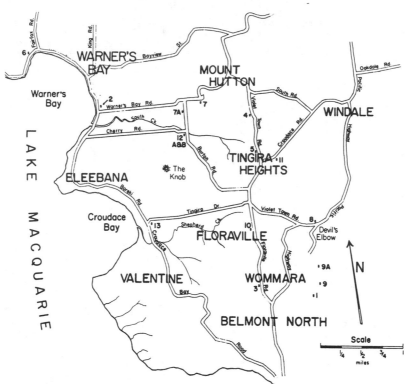

Distribution of Conchostracan Species in Newcastle Coal Measures

Stations: *

Conchostracan Species	14	MI4	1	2	3	4	5	7	8	9	11	12	13	6	10	MI5
Cycloleaia discoidea							●						●			
Leaia (H.) mitchelli			●	●		●	●	●	●?	●	●	●	●	●	●	
Leaia (H.) collinsi				●	●		●									
Leaia (L.) oblongoidea				●		●	●									
Leaia (H.) immitchelli			●	●		●	●	●					●		●	
Paleolimnadia (P.) wianamattensis				●		●			●		●					
Leaia ovata				●		●	●	●		●	●					●
Paleolimnadia (P.) glabra						●	●	●	●		●	●				
Leaia (H.) compta		●				●		●?								
Leaia (H.) oblonga						●	●									
Leaia (H.) elliptica						●	●									●
Leaia (H.) paraleidyi						●		●?	●				●			●
Leaia (H.) belmontensis								●		●						
Leaia (H.) magnumelliptica										●						
Leaia (H.) kahibahensis													●			
Leaia (H.) cf. etheridgei													●			
Leaia (H.) latissima																●
Cyzicus (Lio.) lenticularis		●														
Leaia (H.) lata	●	●														
Cyzicus (Lio.) trigonellaris		●														
Cyzicus (Lio.) obliqua		●														
Cyzicus (Eu.) novacastrensis	●	●														
Cyzicus (Eu.) sp. indet.													●			
Leaia (H.) cf. sulcata													●			
Leaia (H.) pincombei																●

* Stratigraphically younger beds ⟶

Mitchell Localities
MI4 = Tasch
         Station 14
MI5 = NW Shore
         Belmont Bay
(Lio.) = Lioestheria
(Eu.) = Euestheria
(H.) = Hemicyclo-
         leaia
(P.) = Paleolimnadia
(G.) = Grandilimnadia
(L.) = Leaia

MI5
6,10
1 to 13
14, MI4

Vertical Sequence of Beds

Figure 10. Australia. Top: Tasch Collection sites Lake Macquarie area, New South Wales. For street and district maps, see Robinson's Newcastle Street Directory, 4th edition. Bottom: Stratigraphic placement of Mitchell's species and a few new species by station and distribution from older to younger bed. [*Cyzicus (Eu.) lata non Leaia (H.). Cyclestherioides (Cyclestherioides) lenticularis, non Cyzicus (Lioestheria)*].

dispersal of conchostracan eggs is indicated between the clustered stations where outcrops of the Croudace Bay Formation occurred. Also, where it was possible to uncover successive conchostracan-bearing beds (Fig. 11), at any given time only certain species out of the species spectrum were available for dispersal (e.g., Tasch Stations 4, 5, 8a, and 13; Fig. 10, bottom).

The geologically oldest conchostracan-bearing outcrop in the Coal Measures (Station 14, Merewether Beach) had a unique suite of cyziciids, including *Cyzicus (Euestheria) lata* M., among other conchostracans endemic to the area. Among ribbed forms, *Leaia (Hemicycloleaia) compta* Mitchell not only occurred at Station 14, but also in geologically younger beds (Tasch Stations 4, 8 and probably 7). If we assume that the outcrops at these last-named stations are contemporary or nearly contemporary, as seems to be the case (see correlation chart, Fig. 9), then that particular distribution can be explained by the proximity discussed above.

Since *Leaia (H.) compta* M. occurred in the Bar Beach Formation above the Dudley seam, which is an estimated 250 m below the Croudace Formation (Fig. 9), a considerable time apparently elapsed between the older and younger appearances of "*compta*," based on available data. The remaining question concerns this species distribution through the indicated time gap as well as any barren intervals for which presently no data are available.

Mitchell's Localities. Mitchell (1925, 1927) reported 13 species of chiefly ribbed leaiids from the chert (tuff) quarry near Belmont. Another 6 species, two of which were ribbed forms, came from Merewether Beach, below the Dudley Coal Seam (Tasch Station 14). A single species, *Leaia (H.) compta* Mitchell, which also occurs elsewhere in the Coal Measures, was reported from Croudace's Hill, a locality sampled in the present study, but no fossils were found.

The location of the chert (tuff) quarry noted above, was on the northwestern shore of Belmont Bay. It was from the Awaba Tuff Formation at that site that Mitchell (1925, 1927) reported the conchostracan species, studied in a separate section of this memoir. These fossiliferous beds are above the Fassifern Coal Seam (formerly Old Cardiff Coal seam).

A tuffaceous shale and plastic clay, capped by a pebble conglomerate (chert) quarry was explored at Tasch Station 3, beds 3 and 4 (see Figs. 8 and 10). Careful examination of these deposits yielded a single, incomplete specimen of *Leaia (H.) collinsi* Mitchell, a species that also occurred in the chert quarries near Belmont; at both localities the occurrence was in association with *Glossopteris* flora (Mitchell, 1925, p. 446).

In addition, Mitchell described several Triassic species, which are treated separately below.

Thus, the bulk of Mitchell's 20 species from the Upper Permian Newcastle Coal Measures came from two localities. Fifteen were collected in this biostratigraphic study, plus four new species. Three of Mitchell's species have been reassigned to synonomy under *Leaia (H.) mitchelli* Etheridge Jr., reducing the actual number to seventeen.

Palaeolimnadia (Palaeolimnadia) glabra (Mitchell) first occurs above the Upper Pilot seam in the Croudace Bay Formation at Tasch Stations 4, 5, 7, 8, 11, and 12, along with *Palaeolimnadia (Palaeolimnadia) wianamattensis* (Mitchell). The last-named species was collected at Tasch Stations 2, 4, 8, and 11 (see Fig. 10 for location map of stations), which all occur above the Upper Pilot seam.

P. (P.) wianamattensis (M.) persisted throughout the Triassic in the Blina Shale and younger Rewan Group (Lower Triassic), in the Hawkesbury Sandstone (Middle Triassic), and in the Wianamatta Group (Middle through Upper Triassic). This last was the source of Mitchell's type. In contrast, *P. (P.) glabra* (M.) did not continue into the Triassic, according to presently available collections. However, other palaeolimnadiid species did occur in the Triassic (see Fig. 26).

Leaia (H.) mitchelli Etheridge Jr. is the most widely distributed ribbed conchostracan in the Tasch collection; it is absent in only three or four stations (3, 8(?), 13, 14) out of fourteen. In contrast, several conchostracan species are known only from a single Tasch station or Mitchell locality: *Cyzicus (Eu.) lata* M. (Station 14), *Leaia (H.) latissima* M. (M15), *Leaia (H.) kahibahensis* n. sp. (Staton 13), *Leaia (H.)* cf *etheridgei* (M15), *Cyzicus (Lioestheria) trigonellaris* (M.) (M15), *Cyzicus (Euestheria) novocastrensis* (M.) (Station 14), *Cyzicus (Lio.) obliqua* (M.) (Station 14), *Cyclestherioides lenticularis* (M.) (Station 14), and *Cyzicus (Euestheria)* sp. (Station 13). (See "Time-Restricted Occurrences" in a subsequent section.)

Considered by stations, the largest suites of conchostracan species in the Tasch Collection, occurred at Station 5 (ten species), and Station 7 (eight species).

Figure 11 shows that the recurrence of conchostracan species from older to younger beds was spotty. For example, Station 4 has seven species; of these, one species appeared in bed 1, recurred in bed 7, but not in bed 5; *Leaia (H.) mitchelli* Etheridge Jr. appeared in bed 3 and recurred in bed 7. The same sort of selective recurrence by a suite of species is seen at Station 5 with 10 species. Three species of the six which appear in bed 17 of Station 5 recurred in bed 18: *Leaia (H.) collinsi* M., *Leaia (H.) oblonga* M., and *Leaia (L.) oblongoidea* n. sp. At both stations, in the subsequent conchostracan-bearing beds, the species found had not appeared in older beds. This suggests that wind-blown eggs from local dried ponds were brought to these recurrent sites at particular times (cf. Fig. 11 for Station 8, showing vertical succession of species). (See "Recurring Species" in a subsequent section.)

Some *Cyzicus (Lioestheria)* species occur in the Carboniferous and also continue into the Lower Permian (Tasch Station 14; see Fig. 26). With a single exception (Station 13, bed 2c, lateral traverse, site 2g), no cyziciids of the Croudace Bay Formation or of any other of the beds in relevant formations of the Newcastle Coal Measures were laterally traceable over any marked span. That suggests very restricted water bodies ranging from ponds to puddles.

New Species. Four new ribbed species in the Tasch collec-

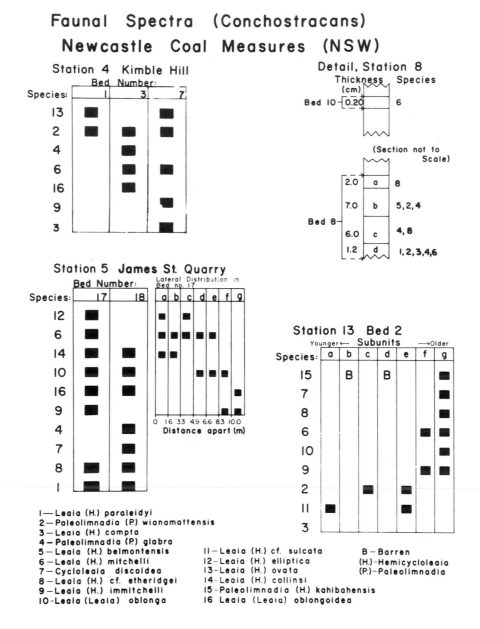

Figure 11. Australia. Vertical and/or horizontal distribution of the Permian (Tartarian) Newcastle Coal Measures conchostracan species at four Tasch Stations.

tion were found in the Newcastle Coal Measures; *Leaia (Leaia) oblongoidea, Leaia (H.) immitchelli, Leaia (H.) magnumelliptica,* and *Leaia (H.) kahibahensis.* (Further discussion of leaiids will be found under a subsequent section, "Evolutionary Trends.")

Associated Biota. (Conchostracans) Throughout the Newcastle Coal Measures there is a strikingly limited diversity of the faunal components of the biota. Conchostracans were the only nonmarine crustaceans and they were widespread geographically

and temporally. (An unconfirmed report of *Limulus* in the Permian Coal Measures was cited by Knight (1950), but no locality data were given.) The complete absence of ostracods is worthy of mention, because they are quite common in conchostracan beds elsewhere.

Nor is there any evidence in this study, or from any previous investigator, of infaunal evidence (burrows, other spoor, etc.) on bedding planes or in the sediment bearing the other biota.

Fish. A single fish was reported in the Permian Coal Measures, *Urosthenes australis* (Feistmantel, 1890, p. 73; also cited by David, 1907, p. 332). Several fossiliferous slices had isolated fish scales (Tasch Station 1, bed 3; Station 5, bed 6; Station 8, units b and c (abundant scales), and Station 10, bed 2). Knight (1950) also noted the presence of isolated fish scales in the insect beds. The general absence of body fossils of fish probably denotes the transport of scales to a given lake margin and/or individual ponds by feeder streams. Mitchell (1925, p. 439) noted that a well-preserved fish, *Elonichthys davidi,* was collected from approximately 91.4 m, below the lowest horizon at which *Leaia* occurs in the Newcastle Coal Measures; i.e., below the Dudley Coal seam.

Insects. Insect fossils were the only other arthropod component that was also widespread. (For map of insect bed, see Knight, 1950; and Fig. 9, this paper.) They were fossilized with elements of the *Glossopteris* flora on either the same bedding plane with, or on lower planes than, those bearing ribbed conchostracan fauna.

The bulk of insect fossils came from the Belmont-Warners' Bay area, east of Lake Macquarie, but in a few places from outcrops at water level or slightly above water level along the existing lake shore. Tasch collections from Stations 4, 5, 7c, 8, and 12A yielded insect fossils. These were associated with leaiids and/or palaeolimnadiid conchostracans. (An extensive collection of the types of the Newcastle Coal Measures insects is on deposit at the Australian Museum, Sydney).

Mollusks. A single pelecypod (fragment only; Fig. 9) was found in rock debris at the base of Tasch Station 1. There were no conchostracans on the same bedding plane; only scattered and sparse, minute plant fragments. The rock debris was derived from bed 4 of Station 1.

Flora. Paleobotanist Mary White of the Australian Museum has supplied a list of plant fossils in the insect beds in the Belmont-Warners' Bay area that she and others had identified and that she had catalogued. In general, the flora came from the area sampled in the present study for fossil conchostracans. Unfortunately, measured-section data are unavailable for the flora which Mrs. White has described from the Australian Museum collection. Selected paleobotanical samples from measured sections have been assembled and sent to the Australian Museum for study and these may help to locate more precisely some floral horizons in the Newcastle Coal Measures.

The *Glossopteris* flora was found at almost all Tasch stations and was dominant in one or more beds; notable were bed 5, Station 1; bed 18, Station 5, and particularly Station 3, Belmont Quarry. Faunal elements, chiefly conchostracans, tended to be sparse to absent in beds dense with plant fossils.

INDIA

In the Spiti District (Himachal Pradesh), the Lipak River section, Himalaya Mountains (see Gansser, 1964 for section), a coquina of mostly flattened and eroded conchostracan valves of variable configuration occurs in the Lower Carboniferous. Only a single slab was available at the Geological Survey of India S.E.M. Laboratory. Study led to the tentative conclusion that a few valves showing polygonal ornamentation were probably *Cyzicus (Euestheria)* sp.

The Upper Permian belt (conchostracan-bearing Raniganj Formation) occurs in the Raniganj and East Bokaro Coal Fields in the Damodar Valley area, southwest and south of Asansol in northern India. The Lower Triassic belt (conchostracan-bearing Panchet Formation) is in direct contact with the Raniganj Formation in the Raniganj Basin (Fig. 13). In the general area of the Pranhita-Godavari Basins (see Fig. 12) the conchostracan-insect beds of Lower Jurassic age crop out (Kotá Formation).

In the Raniganj Coal Field, the end of the *Glossopteris-Schizoneura* flora was taken as the datum separating the Upper Permian Raniganj Formation from the Lower Triassic Panchet Formation. All surface outcrops that were sampled had elevations determined relative to this datum (Fig. 3, Pl. 46). Although searched for, no ribbed conchostracans (*Leaia* and/or related genera) were found in the Upper Permian bed. An Upper Permian *Cyzicus* sp. was found (Fig. 15) in a borehole core slice (Andal area, West Bengal).

The Panchet Formation was explored both in outcrop and in the subsurface cores; both in the Raniganj Coal Field but at different localities. Detailed study of the outcrops disclosed multiple conchostracan-bearing horizons (Figs. 13, 14). Also, from borehole core data (Andal Area), a conchostracan-bearing zone was defined, 40 to 50 m above the Panchet Formation, Raniganj Formation contact (datum) (Fig. 15).

Colleagues from the Geological Survey of India made collections from the East Bokaro Coal Field at Bihar (Fig. 14, right) that yielded *Cornia panchetella* n. sp. and *Estheriella taschi* Ghosh and Shah. The presence of this corniid species in the Panchet Formation of both the Raniganj and East Bokaro basins permitted a correlation between them. The estheriellid species is reassigned and redescribed herein, as *Cornutestheriella* n.g. Tasch. It ties together the evolution of four genera (cf. Tasch, 1963a, and discussion under taxonomy of this new genus, below), listed below.

Cornia panchetella n. sp. was found as noted above, and also was collected from the Mangli site (central India). Besides the ubiquitous genus *Cyzicus* at the three localities (Raniganj and East Bokaro basins and Mangli), the Mangli site yielded *Cyclestherioides*. Previously, only *Cyzicus (Euestheria) mangaliensis* (Jones) was known from there, and were also abundant in this writer's collections. Two genera, *Palaeolimnadia* and *Gabonestheria,* were found in a bore-hole core (RNM-4), in the Andal area (Lower Triassic Panchet Formation). *Gabonestheria* and *Cornia panchetella* n. sp. occur together on the 106 m level of the indicated site (Fig. 15).

The presence of gabonestheriids and corniids in the Triassic of Africa (Lesotho), South America (Brazil), and India would seem to indicate the spread of bioprograms for these taxa between contiguous continents (Gondwanaland), because of the availabil-

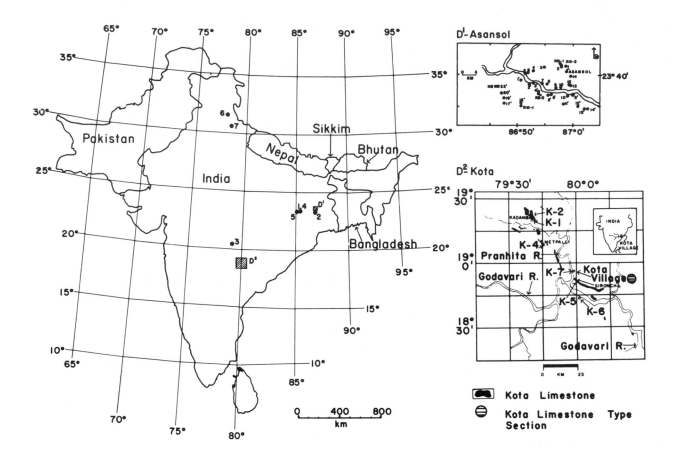

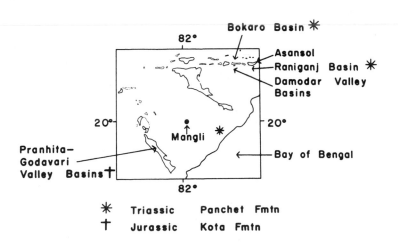

Figure 12. India. Top: Location map of conchostracan-bearing beds, with detail of the Triassic of the Raniganj Basin, Asansol area (D[1]), and the Jurassic Pranhita-Godavari Basin Kotá Formation sites (D[2]). Bottom: Location map of several Triassic and Jurassic conchostracan-bearing sites. Mangli (Triassic). [See Fig. 13, p. 27, for enlarged detail of D'-Asansol area, above.]

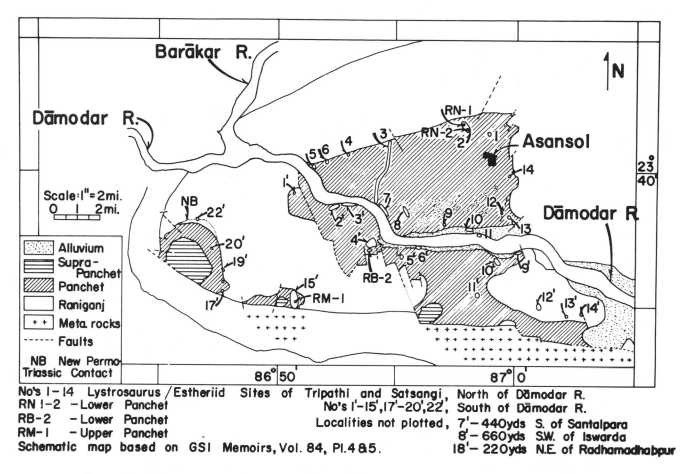

Conchostracan Localities of the Panchet Series
Raniganj Coalfield

Figure 13. India. Panchet Series. Raniganj Coal Field. Geographic location of fossil conchostracans and the *Lystrosaurus* beds. (Geological Survey India, Memoirs, v. 84, pl. 4 and 5; schematic map used to plot localities.)

ity of nonmarine dispersal opportunities. *Cyclestherioides* and *Estheriella* are also found in the Triassic of South America, Africa, and India. To these are now added the presence of a new *Pseudoasmussiata* species in the Mangli section. This genus is also found in South America and Africa.

As with the Lower Triassic beds of India, only a single Lower Jurassic conchostracan species had been known, *Cyzicus (Lioestheria) kotahensis* (Jones, 1862), before research was undertaken in conjunction with colleagues of the Geological Survey of India (Tasch and others, 1975). Six sections were located and measured and systematic fossil collecting undertaken in the Kotá Formation outcrop belt (see Fig. 16). A few conchostracan genera were found, and several new species. The genera recog-

nized are *Pseudoasmussiata, Cyzicus, Palaeolimnadia,* and *Estheriina.* Some species were found at different levels of the same section, others were more restricted.

Biostratigraphic correlation was possible between the southern outcrops of the Kotá Formation area below the 19th Parallel (Sections K-7, K-6, K-5, and K-5A). The new species *Cyzicus (Euestheria) crustabundis* occurs in coquinas at site K-4 in the northern outcrop of the Kotá Formation, and in the southern outcrop belt at K-5, K-5A, K-6, and K-7. This confirms the contemporaneity of the two extremes of the outcrop belt. This correlation is further sustained by the distribution of the insect fauna. Coleoptera and Blattaria are known from Stations K-1, K-2, and K-5, and Blattaria from K-4, whereas Mecoptera occurs

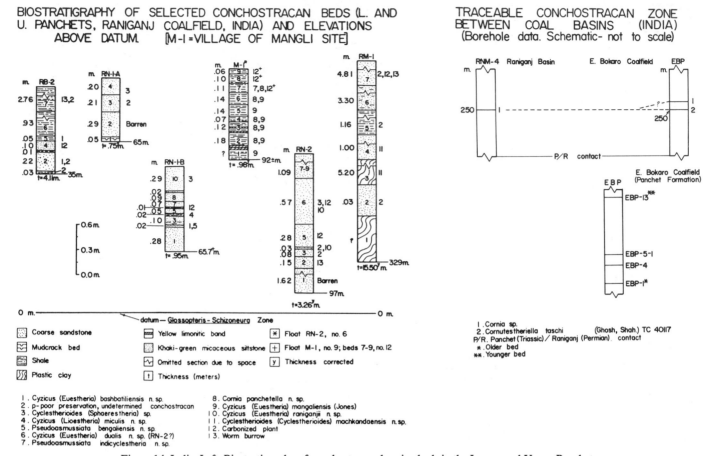

Figure 14. India. Left: Biostratigraphy of conchostracan-bearing beds in the Lower and Upper Panchets, Raniganj Basin, and elevation of sections above the datum (*Glossopteris-Schizoneura* Zone). Right: Correlation between two coal fields (Raniganj and E. Bokaro)* at the 250 m level (Panchet Formation). (*Cornutestheriella taschi* [Ghosh and Shah] was collected from the southern bank of the Dhardharwa nala about 2 km north of Lapanic village (at 23°44′N, long 85°45′E].) Ghosh and Shah (1976, 1977) named the species *Estheriella taschi*. Tasch (herein) named the new genus *Cornutestheriella* based on that species. [10. *"raniganjis" non "raniganjii"*].

at only two stations, one northern (K-2) (above the 19th Parallel) and one southern (K-5) (below the 19th Parallel). Frank Carpenter of Harvard identified insect orders and further study is underway.

The overall view of fossil conchostracans of the Paleozoic and Mesozoic of India indicates low generic diversity. This can be attributed to the limited areal distribution of fossil conchostracans on the subcontinent. In India, Antarctica, and the other southern continents, there is a regional density of collecting sites. Most of the data from India comes from the Raniganj and East Bokaro basins and the Pranhita-Godavari basins (Fig. 12).

In India, as in the other Gondwanaland continents, the estherinids are a prominent part of the conchostracan fauna. In India, *Estheriina* is a dominant genus represented by both subgenera *Estheriina* and *Nudusia,* and several species. The estheriinids were noted above as a prominent constituent of the Australian Triassic; they are also prominent in Brazil, where they occur in five states (Jones [1897a], originally described the

Brazilian collections.) As indicated above, the estheriinids also occur in the Jurassic of Antarctica. Their Gondwana range is throughout the Mesozoic; Triassic (Africa and Australia), Jurassic (Antarctica, India), and Cretaceous (South America). Absence of estheriinids from Africa during the Jurassic and Cretaceous, especially the Cretaceous, because they were then present in South America, may be related to continental drift. (See "Dispersal" below.)

Estheriina (Nudusia) stormpeakensis n. sp. (from this writer's Antarctic Jurassic collection), while differing in several important valve characters and ratios from *Estheriina (Nudusia) indijurassica* n. sp., is similar in several other ratios. These species, with their basic unity of morphological pattern, could have diverged from a common ancestral stock. The spread during Mesozoic time of the estheriinid bioprogram between the above indicated continents sustains the interpretation of Southern Hemisphere dispersal and subsequent divergence and speciation.

BOREHOLES: ANDAL AREA-RANIGANJ COAL FIELD WEST BENGAL, INDIA[+]

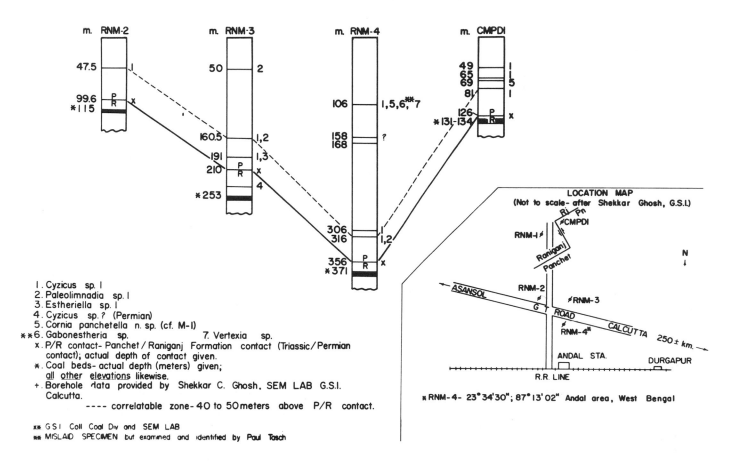

Figure 15. India. Borehole correlation of conchostracan-bearing beds. Andal area, Raniganj Coal Field.

SOUTH AMERICA

Paleozoic and Mesozoic conchostracans from South America are currently known from nine countries; Argentina, Bolivia, Brazil, Chile, Colombia, Guyana, Peru, Uruguay, and Venezuela.

Conchostracan sites occur in greatest density below the equator, as in Africa. The greatest number are in Brazil (Fig. 17), where they range in age from Permian through Upper Cretaceous and occur in 11 states: Bahia (Upper Triassic, Upper Jurassic, Lower Cretaceous), Maranhão (Upper Triassic), Minas Gerais (Lower Cretaceous), Paraíba (Lower Cretaceous), Paraná (Lower and Upper Permian), Pernambuco (Upper Jurassic, Lower Cretaceous), Piauí (Upper Jurassic), Rio Grande do Sul (Upper Triassic), Santa Catarina (Permian), São Paulo (Upper Permian, Lower and Upper Cretaceous), and Sergipe (Upper Jurassic; see Pl. 46).

Argentina (Permian, Triassic) and Chile (Triassic and Jurassic) both have a few sites. The Argentina fossil conchostracan fauna occurs in four provinces: Chubut (Jurassic-Callovian),

Cordoba (Lower Permian), Mendoza (Triassic), and Santa Cruz-Patagonia (Triassic-Liassic). Chilean fossil conchostracans are found in the High Cordillera in the province of Atacama (Jurassic), in the coastal region of the province of Aconagua and Antofagasta (both Upper Triassic; Hillebrandt, 1965; personal commun; Harrington, 1961, as well as in the province of Arauco [Jones, 1897]).

Bolivian conchostracan fossils have been studied by this writer and were found in Gulf Oil Corporation samples from two cores. These were from the Ipaquazu Formation at a depth of 2592 m (Permian-Tartarian), and the Vitiacua Limestone (Permian or Permo-Triassic) of the Aguarague Range, Santa Cruz province. Fossil conchostracans from Colombia are recorded from four localities, two of which yielded fauna studied by this writer and reported here: Bocas Formation, Santander massif (Triassic-Lower Jurassic; Will Maze collection), and an Imperial Oil Company locality G-1818 core; they are also reported from near Montebel (Rhaetic or Lower Jurassic; Bock, 1953). A fourth locality is in the Quebrado de los Indios, Santa Marta Mountains

KOTA   FORMATION   SECTIONS   (NORTH AND SOUTH) LOWER   JURASSIC.
INFERRED   CORRELATIONS – UPPER   FERRUGINOUS   CONCHOSTRACAN   BEDS – DATUM

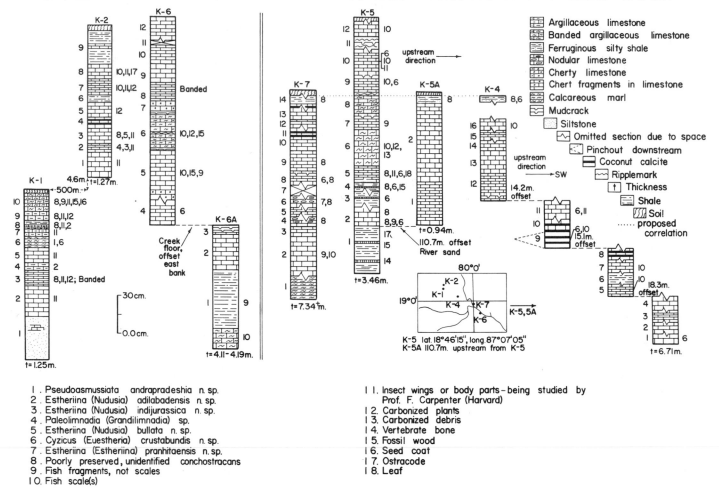

1. Pseudoasmussiata andrapradeshia n. sp.
2. Estheriina (Nudusia) adilabadensis n. sp.
3. Estheriina (Nudusia) indijurassica n. sp.
4. Paleolimnadia (Grandilimnadia) sp.
5. Estheriina (Nudusia) bullata n. sp.
6. Cyzicus (Euestheria) crustabundis n. sp.
7. Estheriina (Estheriina) pranhitaensis n. sp.
8. Poorly preserved, unidentified conchostracans
9. Fish fragments, not scales
10. Fish scale(s)

11. Insect wings or body parts – being studied by
    Prof. F. Carpenter (Harvard)
12. Carbonized plants
13. Carbonized debris
14. Vertebrate bone
15. Fossil wood
16. Seed coat
17. Ostracode
18. Leaf

Figure 16. India. Jurassic Kotá Formation. Biostratigraphy of conchostracan-insect beds and inferred correlation.

(Triassic; Trumpy, 1943). Guyanan (formerly British Guiana) conchostracans come from the Letham Shales (Permian or Triassic). The only Peruvian conchostracan fossils known were described by Jones (1897); the types were restudied for this memoir. These Carboniferous fossils came from Arica, Department of Arequipa.

A conchostracan recently found in Uruguay by an Argentine colleague, Dr. Rafael Herbst, occurs in the Tacuarembó Formation (Jurassic). Venezuelan fossil conchostracans reported in this memoir were collected by Will Maze from the Sierra de Perija, northwest Venezuela, Tinacoa Formation (Lower Jurassic). Other Venezuelan Jurassic conchostracans are from an Imperial Oil Company core near Merida, Sierra de Perija (Tinacoa Formation; Rivas and Benedetto, 1977), and from the La Quinta Formation (Lower Cretaceous).

Argentina

The oldest Argentinian conchostracan fossils are Permian and came from the Tasa Cuna Formation, Sakmarian-Artinskian, Cordoba. These are ribbed leaiids described by Leguizamon (1975) as a new species of *Leaia*, which he compared with *Leaia pruvosti* Reed from Poço Preto, Brazil, and *Leaia gondwanella* Tasch from the Ohio Range, Antarctica. Although not exact time-correlates, as this writer's study indicates, all of these localities are Permian in age which suggests that the leaiid bioprogram was then widespread in the southern continents.

The area of the Mendoza oil fields of Argentina has yielded several Triassic conchostracan fossils as reported by Geinitz, 1876; Philippi, 1887; Rusconi, 1946, 1948a, 1948b; Jones, 1862; and Hedberg, 1964. [For references, see synonomy of *Cyzicus*

(Euestheria) forbesi (Jones), this section.] Geinetz's specimens came from the San Lorenzo District from several Mendoza sites.

Conchostracans occurred throughout the Triassic although reports are limited. Rusconi (1948a) reassigned genus and species of Geinetz (1876) (line drawings only) to *Cyzicus (Euestheria) forbesi* Jones. This writer did not find the original assignment to *"mangaliensis"* or the reassignment warranted. Several of Rusconi's (1946a and 1946b, 1948a) taxonomic placements of his Mendoza species are dubious. His *Pseudestheria* assignments are all questionable and need restudy; his *P. leonense* is the same as his *P. contorta* (1948b); *P. mioprioi* (Lower Triassic), which he compared to *"Estheria" mangaliensis,* is unlike that species. *Cyzicus (Lioestheria) striolatissima* (Rusconi) is a valid species. Finally, Rusconi's *"Estheriopsis bayensis"* is probably a pelecypod (1948a). It is not uncommon to confuse pelecypod and conchostracan valves. The Upper Triassic (Rhaetic) *Cyzicus (Euestheria) forbesi* (Jones) from Cacheuta, Mendoza, is here emended. (See a new diagnosis and measurements of the types as well as new photographs of valve features below.)

The Patagonian region of Argentina has yielded some of the most usable Argentinian species to date, as evidenced by Volkheimer's biostratigraphic collections from Cañadón Asfalto Formation (Chubut Province) (in Tasch and Volkheimer, 1970, p. 12–18) (Lower Jurassic–Callovian), that contain two new species; *Cyzicus (Euestheria) volkheimeri* Tasch and *Cyzicus (Euestheria) patagoniensis* Tasch. There are also three undetermined species of *Cyzicus (Lioestheria)* and one of *Cyzicus (Euestheria)* (Tasch *in* Tasch and Volkheimer, 1970). A Volkheimer collection from Malacara, Santa Cruz Province, directly south of Chubut, contains a new species which is described below.

It is apparent that the Argentine conchostracan fossil record is rather sparse and of low diversity. The Argentine Mesozoic (Triassic and Jurassic) is dominated by a few species of *Cyzicus* and its two subgenera. This is in contrast to the highly diverse Mesozoic conchostracan fauna of Brazil. Although the difference may be attributed to either more intense collecting (and publication) of the Brazilian faunas than in Argentina, it is possible that the sparsity in Argentina reflects the presence of barriers to dispersal.

Bolivia

Only three conchostracan genera are known from Bolivia. These were found in Gulf Oil Corporation samples from the Ipaguazu Formation (Permian-Tartarian), studied by this writer; *Cyzicus, Palaeolimnadia,* and a probable *Palaeolimnadiopsis.* The last two of these genera also occur in the Vitiacua Limestone (Lower or Middle Permian or Permian-Triassic), near Villa Montes.

Moreover, both genera are known widely through Gondwanaland, and no doubt are of great relevance not only to former continental positions but to the dispersal of bioprograms.

Brazil

The broadest conchostracan distribution and diversity on the continent is found in the Brazilian Paleozoic–Mesozoic. They range in age from Permian through Cretaceous and are known from 10 states.

Aside from *Cyzicus,* there are three other genera and/or related genera that have the most extensive distribution; *Palaeolimnadiopsis* (seven states), *Estheriina* (five states), and the estheriellid group (three states).

The following genera are recognized in this memoir: *Cyzicus, Leaia, Estheriina, Palaeolimnadiopsis, Graptoestheriella, Aculestheria, Cornia, Gabonestheria, Acantholeaia, Unicarinatus, Monoleaia, Macrolimnadiopsis, Pseudoasmussia, Asmussia, Pseudoasmussiata,* and *Cyclestherioides.*

Cornia and *Gabonestheria* were recovered by the writer from the Late Permian Rio do Rasto Formation of São Pascoal (Paraná State), and had not been previously reported from South America. They also occur in the Lower Triassic of Angola and in borehole samples of the Panchet Formation from the Lower Triassic, Raniganj Coal Field, India.

The Rio do Rasto Formation (Upper Permian) is the Brazilian Paleozoic unit with several widespread conchostracan genera; *Cyzicus, Leaia, Cornia, Gabonestheria, Monoleaia* (formerly *Monoleiopholus),* and *Palaeolimnadiopsis.* Comparison of this list with those of genera hitherto reported in the literature on the Rio do Rasto (or "Brazilian Permian") would reveal some discrepancies. These will be fully accounted for in the following systematic text.

Conchostracan genera are known from the Jurassic of four states, Bahiá, Pernambuco, Piauí, and Sergipe; *Cornia* (formerly *Echinestheria Cardoso,* 1963), *Pseudoasmussia, Macrolimnadiopsis, Graptoestheriella, Cyzicus, Estheriina, Palaeolimnadiopsis, Palaeolimnadia,* and *Aculestheria.* The last six genera named persisted into the Cretaceous which is found in the first two states named above, Minas Gerais and São Paulo. *Macrolimnadiopsis* and *Graptoestheriella* made an apparent first appearance in the Upper Jurassic of Brazil, and *Estheriina* persisted from the Triassic, and *Cornia* and *Cyzicus* from the Permian.

(Species from Chile, Colombia, Guyana, and Peru are referred to in the catalog and systematics following the Venezuela entry below.)

Venezuela

The Venezuela rock column has yielded Mesozoic conchostracans (Triassic, Jurassic, and Cretaceous) that can be assigned to one genus, *Cyzicus,* and its two subgenera. Apparently the oldest conchostracan fossils came from black laminated shales in a core taken near Merida, *Cyzicus (Lioestheria) olsoni* (Bock) 1953, probably Upper Triassic (Rhaetic) (Bock, 1953). Jurassic (probably Lower Jurassic) cyziciids from the copper mineralization belt, Tinacoa Formation, northwest Venezuela, Sierra de Perija (Fig. 30, upper row, middle) in the Will Maze collection

Figure 17. South America. Location map of conchostracan-bearing sites.

were studied by this writer. They included *Cyzicus (Lioestheria) mazei* n. sp.; *Cyzicus (Lioestheria)* cf *columbianus* (Bock) 1953, and *Cyzicus (Euestheria)* sp. 1 and 2. In addition, from La Quinta formation of probable Lower Cretaceous age, Maze collected *Cyzicus (Lioestheria)* cf *pricei* (Cardoso, 1966).

Some of Bock's (1953) Rhaetic or younger species occur in the Rivas and Benedetto (1977) collection from the Tinacoa Formation or its apparent age equivalent, the Montebel Formation, Montebel (probably Lower Jurassic); *Cyzicus (Euestheria)* aff. *ovata* (Lea.) 1856; *Cyzicus (Lioestheria)* cf *olsoni* (Bock) 1953 and *Cyzicus (Lioestheria) columbianus* (Bock) 1953.

SCHEDULE OF CONCHOSTRACAN-COLLECTING SITES IN SOUTH AMERICA
Numbers correspond to those on the map of South America, Figure 17. Sequence of numbers at a given site denote additional species.

1. Brazil. Rio Grande do Sul, São Pascoal. 800 km southeast of Curitiba.
2. Brazil, Piaui. 16 km northeast of Floriano-Amarante, town of Muzinho.
3. Argentina, Chubut Province. Lat 43°38'S, long 69°10'W about 25 km northwest of Berwyn.
4. Brazil. Santa Catarina. About 1 km from town of Poço Preto, near Rio Timbo.
5. Guyana. Right bank of Takuta River at St. Ignatius Mission.
6. Venezuela. Merida.
7. Colombia. Near Montebel (Loc. 1818, Imperial Oil Co.).
8. Colombia. 177 km north of Bogota.
9. Brazil. Mato Grosso. Road between Jardim and Bela Vista; 51.5 km from Jardim.
10. Argentina. Mendoza. Mendoza Oil Field. Upper Part of Unit B.
11. Brazil. São Paulo. Near town of Igaçuba, region of Rifiano.
12. Brazil. Paraná. Near town of Quatiguá, km 210 from Curitiba to Jacarezinho.
13. Brazil. Bahía. City of Serrinha.
14. Brazil. Bahiá, road between Queimoda Grande and Quererá.
15. Brazil. Bahiá. City of Cacimba do Carvao.
16. Brazil. Sergipe. Riacho dos Pilões.
17. Brazil. São Paulo. City of Serrano, on the Ribeiro–Serrano–Cajurú road.
18. Brazil. São Paulo. Northeast of Rio Claro, km 222 on railroad from Rio Claro to São Carlos.
19. Brazil. São Paulo. Town of Presidente Bernardes (19) and Mirandopolis (59), about 55 km east of Adamantina (60) on south side of Rio Aguapeí. (Bracketed Numbers 59 and 60 are also on the map).
20. Brazil. Rio Grande do Sul. 7.8 km from Santa Maria, north of Arroio dos Tropas.
21. Brazil. Paraná, Santo Antonio de Platiná.
22. Brazil. São Paulo, km 216 +500 m, Rio Claro to São do Pinhal road.
23. Brazil. Northern Sánta Caterina, between Poço Preto and Porto União.
24. Brazil. Paraná. Road from Prudentopolis to Guarapuava at km 109 and 200.
25. Brazil. Paraná. Santo Antonio de Platina; city of Platina, about 1 km along the Rio Boi Pintado.
26. Brazil. Pernambuco. Riacho do Saco do Machado, Petrolândia.
27. Brazil. Pernambuco. Mirandiba, km 555 on National railroad.
28. Brazil. Pernambuco. Várzeado Campinho, Icó.
29. Brazil. Bahía, Riachio São Paulo near Candeias.
30. Brazil. Bahía. Cicero Dantas.
31. Brazil. Bahía. 12–13 km from San Salvador between Periperi and Olaria.
32. Brazil, Bahía. Km 73–74 on railroad from Pitanga to Mato.
33. Brazil. Bahía. At km 82, Pojuca.
34. Brazil. Bahía. Along railroad about 5 km from Salvador.
35. Brazil. Bahía. Along railroad about 83 km from Salvador.
36. Brazil. Bahía. Between Pojuca and São Thiago.
37. Brazil. Bahía. Km 77–78 on railroad from Santo Amaro to Candeias.
38. Brazil. Rio Grande do Sul. 3.9 km from where road Posto Castelinho crosses the Santa Mariz–Camobi road.
39. Brazil. Rio Grande do Sul. 7.8 km from Santa Maria on the São Sepí road.
40. Brazil. Bahía. 5 km from railroad at Santo Amaro.
41. Deleted.
42. Brazil. Maranhão. On road between Floriano and Admirante.
43. Brazil. Maranhão. Rio Tocantins, 5 km from Imperatriz.
44. Brazil. Minas Gerais. Road MG 51 between Patos de Mines and Pirapora.
45. Brazil. Minas Gerais. São Jose near Varjão.
46. Brazil. Minas Gerais. About 1 km from MG 51, near São Gonçalo do Abaeté.
47. Brazil. Minas Gerais. São Gonçalo do Abaeté.
48. Bolivia. Lat 21°10'S, long 63°40'W near Villa Montes on the Aguaque River.
49. Chile. West of the High Cordillera. Lat 27°10'–27°25'S, long 69°20'–69°45'W. Between Quebrado San Pedrito and Yerbos Buenos.
50. Argentina. Mendoza Province. District of San Lorenzo.
51. Argentina. Mendoza Province. Near Challas; at Agua salada west of Challas (51).
52. Chile. Aranca. District of Lebu, on seacoast between lat 33°36'–37°39'S.
53. Tarapaca. City of Arica, lat 18°25'S, long 70°15'W.
54. Brazil. Bahía. 12–13 km from Salvador at Pedra Furada, near Monserrat.
55. Brazil. Bahía. Along railroad from Salvador to San Francisco at km 3.88, 4, and 5.
56. Brazil. Bahía. Km 4–5, Pedra Furada, railroad cutting between Pojuca and São Thiago, 83 km from Salvador.
57. Brazil. Bahía. Between Sargiva and Pojuca.
58. Chile. Antofagasta. Sierra Almeida, town of Pular, between Montiraqui Station and Quebrado de Pajonales.
59. Brazil. Mirandopolis.
60. Brazil. São Paulo. About 55 km east of Adamantina; south side of Rio Aguapei.
61. Chile. Province of Aconcagua. Guaquin and Longotoma districts; 310 km south of Taltal.
62. Argentina. Mendoza Province. Cerro de Cacheuta, south of Challas (lat 33°S).
63. Argentina. Mendoza Province. Agua de la Zorra, Siérra de Uspallata.
64. Argentina. Mendoza Province. 18 km east of city of Mendoza.
65. Argentina. Cordoba Province. Northwest part or the province, lat 30°46'S, long 65°18'W.
66. Brazil. Santa Caterina. Along road BR 116. Serra Espigão, diastem section.
67. Brazil. São Paulo. City of Serrana; on the Ribeirão–Serrana–Cajurá road; 2 km from Serrana.
68. Colombia. Quebrado de los Indies 32 approx. km southeast of Fundación at the southwest of the Santa Marta Mountains.
69. Uruguay. Route 5, km 356, close to city of Tacuarembó, slightly to the north.
70. Argentina. Santa Cruz. Estáncia el Tranguilo, approximately lat 47°50'–48°10'S, long 68°42'W.
71. Brazil. Rio Grande do Sul. From Pôrto Alegre, Santa Maria road, km 81–83.
72. Brazil. São Paulo, 15 km northeast of São Carlos.
73. Deleted.
74. Venezuela. Zulia, Sierra de Períja, near Villa del Rosario.
75. Colombia. Santander, near Bucaramanga.
76. Brazil. Santa Caterina. Along BR 116. Serra Espigão, Estrada Nova Formation.

Systematic Descriptions

Repository of Types. Types in the Tasch fossil collection from the southern continents are on deposit at the U.S. National Museum, Department of Paleobiology, Washington, D.C. Repositories of other types cited or redescribed below are noted where appropriate.

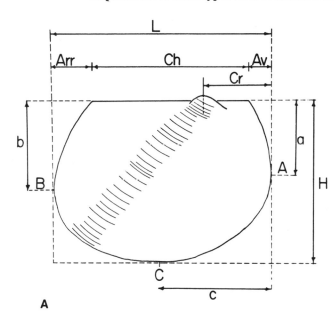

A

MEASUREMENTS AND ANGLES OF RIBBED CONCHOSTRACANS

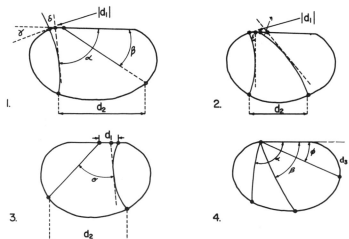

B

Figure 18. Fossil conchostracan measurements. A: Parameters used in this Memoir for ribbed and nonribbed forms. (After Defretin-LeFranc. Measurements in mm. **L** = length of valve; **H** = height of valve; **Ch** = length of hingeline (dorsal margin); **Cr** = distance of beak (or midline of umbo) to most anterior part of valve; **Av** = distance of anterior extremity of the dorsal margin to the most anterior part of the valve; **Arr** = distance of the posterior extremity of the dorsal margin to the most posterior part of the valve; **A** = designates point of the maximum anterior bulge; **B** = designates point of the maximum posterior bulge; **C** = designates point of the maximum ventral bulge; **a** = distance of A to dorsal margin; **b** = distance of B to dorsal margin; and **c** = distance of C to the most anterior part of the valve. From the above parameters, the following ratios are calculated: H/L, Ch/L, Cr/L, Av/L, Arr/L, a/H, b/H, c/L. (These ratios can be compared within a fossil species population and/or between different species of the same genus.) B. Measurements of angles and other parameters for ribbed forms (Tasch). Angles: **Alpha** (α) = measured from dorsal margin to anterior rib; **beta** (β) = measured from dorsal margin to posterior rib; **phi** (ϕ) = measured from dorsal margin to a third rib; **eta** (η) = if both ribs are curved, eta is the angle defined by the tangent to the posterior rib extended until it intercepts the dorsal margin, and the curved rib; **delta** (δ) = the equivalent anterior angle to eta; **sigma** (σ) = the angle measured from point of tangency to the curved rib and the straight rib; **gamma** (γ) = the dorso-anterior angle formed by extension of the dorsal margin and a line to the point of intersection of the anterior rib with that margin.

Other measurements: d_1 = distance between points of intersection of the dorsal margin by both ribs; d_2 = same as d_1, for ventral margin; d_3 = distance between the dorsal margin extended and the point of intersection of the posterior margin by the third rib.

35

AFRICA

Algeria

Only cyziciids have been reported thus far from the Algerian rock column (Pennsylvanian, Triassic, and Cretaceous.) These are known from Columb Bechar-Kenadze Basin, the Guir Basin, and north Constantine, as well as the Algerian Sahara. This writer had the opportunity to study conchostracans in the National Collection of Morocco that also included specimens from Algeria.

A decade ago I described a Late Cretaceous species from Algeria, provided by M. Jean Fabre. This came from a site north of In Salah in the central Algerian Sahara. Because this writer has no knowledge of that paper having ever been published, it is included here with additional measurements, as the specimens are at hand, as are figures of the holotype and paratype. Stratigraphic data supplied by M. Fabre and colleagues are also included.

Pennsylvanian species (Westphalian and Stephanian) occur in the Columb-Bechar and Guir basins. Defretin (1953) described species from north Constantine of Upper Triassic age. Cretaceous forms were collected at Aïn-el-Hadjadj, Algerian Sahara by M. Fabre and colleagues; the bed with a *Cyzicus* species described below (and some ostracods) occurred at the indicated site at the boundary between the Turonian and younger Senonian and that was the proposed new boundary (Fig. 19).

Map Sites

21. *Estheria subsimoni* Deleau 1945. [= *Cyzicus (Euestheria) subsimoni* (Deleau), Columb Bechar-Kenadze Basin]. Lower Westphalian, P. Deleau, 1945, p. 625–632; R. Furon, 1950, p. 107.

21, 26. *Estheria limbata* Goldenberg 1877. [= *Cornia? limbata (Gold.)* Kobayashi, 1954, p. 160.] Same locality as above. Stephanian (also found in the Guir Basin). F. Daguin, 1929; Furon, 1950, p. 107–108.

21. *Estheria simoni* (Pruvost) 1911. [= *Cyzicus (Euestheria) simoni* (Pruvost) 1919.] Westphalian, Kenadze. National Collection of Morocco.

8. *Euestheria minuta* (von Zieten) Raymond 1946. [= *Cyzicus (Euestheria) minuta* (v. Z.), north Constantine, Sid-Al-bou-Krizi.] Upper Triassic. (Defretin and others, 1953, p. 187–188, Pl. 1, Figs. 1–6.)

80. *Estheria (Euestheria) forbesi* (Jones) 1862, Defretin 1953. [= *Cyzicus (Euestheria) forbesi* (Jones). Same locality as "minuta" above.] Upper Triassic. (Defretin and others, 1953, p. 188–189, Pl. 1, Figs. 7, 8.)

Superfamily CYZICOIDEA Stebbing 1910
Family CYZICIDAE Stebbing 1910

Genus *Cyzicus* Audouin 1837

Subgenus *Eustheria* Depéret and Mazeran 1912, emended, Car-apace generally ovate (or subovate), but with wide variation in shape and size. Characterized by a pattern of minute polygons (granules) on growth-bands, and not arranged vertically or in parallel. Type. *Posidonia minuta* von Zieten, 1933.

24. *Cyzicus (Euestheria) lefranci* n. sp.
(Pl. 3, Figs. 3, 4, 6).

Diagnosis: Subovate to subelliptical valves with posterior-dorsal angle greater than antero-dorsal angle; hingeline (dorsal margin) straight; umbo placement variable from subterminal in subovate forms to markedly inset from anterior margin in subelliptical forms; growth band spacing and number variable; bands deeply incised, giving a slightly rugosoid appearance to the valve; double growth lines often occur. Ornamentation mostly obscure, but consists of a fine reticulate pattern on growth bands.

Measurements: (mm) Holotype TC 4055 followed by paratype TC 4006. L-4.0, 3.8; H-2.7?, 2.5; Ch-2.8, 0.7; Cr-0.7, 0.4; Av-0.7, 0.4; Arr-0.5, 1.0; a-1.1, 1.1; b-1.0, 0.8; c-1.7, 1.3; H/L-.675, .658; Cr/L-.175, .105; Av/L-.75, .105; Arr/L-.250, .263; Ch/L-.700, .632; a/H-.407, .440; b/H-.370, .320; c/L-.425, .342; antero-dorsal angle-29°, 41°; postero-dorsal angle -39°, 45°.

Name: Named after Dr. J. P. Le Franc who, in company with Dr. Jean Fabre, collected the described material.

Material: Holotype, TC 4055, internal mold of right valve; paratypes, TC 4006, internal mold of left valve; TC 4007, internal mold of right valve showing reticulate pattern of contiguous granules. (Base of Senonian, Late Cretaceous.)

Locality: Aïn-el-Hadjadj, Plateau du Tademait, 100 km north of In-Salah, Algerian Sahara.

Discussion: The closest species to *C. (Eu.) lefranci* n. sp. is *Cyzicus (Euestheria) sambiensis* (Defretin-Le Franc, 1967, p. 45–47, Pl. 3) from the Congo Basin (at Boma; Cretaceous). Comparing the Congo species postero and antero dorsal angles with those of the new species, one finds that they overlap, *C. (Eu.) lefranci* n. sp. having the lower limit of the range. (These angles were measured on Mme. Le Franc's photographs.) Double growth lines occur on both Congo and Saharan species. Differences include H/L and b/H which are greater, and c/L which is smaller in the Saharan specimens. Morphological features that differ are umbonal placement (subterminal, Sahara; terminal, Congo), greater height posterior (not anterior as in *C. (Eu.) sambiensis*). This last applies only to the subovate forms; subelliptical Saharan forms also differ from the Congo species, all of which are subovate.

These combined differences are sufficient to distinguish the Saharan species from the Congo species. These species also differ in biostratigraphic position: Congo species, Lower Cretaceous (Wealden); Saharan species, Upper Cretaceous, Senonian.

Angola

The Oesterlen collection from the Upper Triassic of northern Angola (Cassanje III Series, Phyllopod Beds, Upper Stage)

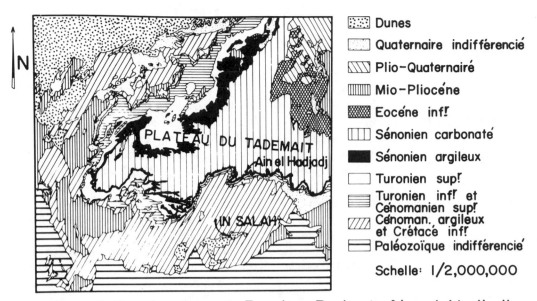

Legend:
- Dunes
- Quaternaire indifférencié
- Plio-Quaternairé
- Mio-Pliocéne
- Eocéne inf.
- Sénonien carbonaté
- Sénonien argileux
- Turonien sup.
- Turonien inf. et Cénomanien sup.
- Cénoman. argileux et Crétacé inf.
- Paléozoïque indifférencié

Schelle: 1/2,000,000

Location of Conchostracan-Bearing Bed at Ain el Hadjadj (Sahara algérien)(100km North of In-Salah). After Dr. J.P.

A Le Franc.

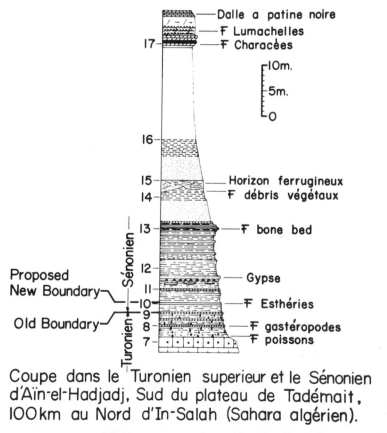

Coupe dans le Turonien superieur et le Sénonien d'Aïn-el-Hadjadj, Sud du plateau de Tadémait, 100km au Nord d'In-Salah (Sahara algérien).

B

Figure 19. Section in the upper Turonian and the Senonian of Aïn-el-Hadjadj, south of the Tademait Plateau, 100 km north of In-Salah (Algerian Sahara). (Bed 7, fish; 8, gastropods; 10, conchostracans; 14, plant fragments; 17, one of the green algae [Characea]; Lumachelles, fossiliferous limestone [probably molluscan].)

yielded some new conchostracan species that are described herein. It also contains established genera which had not previously been reported by other investigators (Passau (1919); Leriche, 1932; Marlière, 1950a) and Defretin, 1967; *Palaeolimnadia, Cornia,* and *Estheriina*). Three previously described species were found in the collection: Marlière's *Estheria (Pemphicyclus) gabonensis* M. (= *Gabonestheria gabonensis* Novojilov, 1958c) from Gabon, *Palaeolimnadia (Palaeolimnadia)* cf. *wianamattensis* (Mitchell), originally described from the Permian and Triassic of New South Wales, Australia, and *Estheriina (Nudusia)* cf. *rewanensis* Tasch, 1979, from the Lower Triassic Bowen basin, Australia (Figs. 26, 27).

The total areal spread of these conchostracan-bearing beds is an estimated 2000 km^2. Oesterlen (in Tasch and Oesterlen, 1977) subdivided the Upper Stage (Phyllopod Beds) of the Cassanje Series into three zones. The conchostracans in his collection are found only in the lower bed of the stage and that is 120 m thick (Fig. 20A; cf. Marlière, 1950a, Table 1).

Conchostracan assignments by Passau (1919), Leriche (1913, 1920, 1932) were modified or extended by Marlière (1950b), and Marlièe's assignments, along with those of Leriche and Passau, were in turn modified by Defretin-LeFranc, because "Estheria", an occupied name, had been employed by Leriche and Marlière. The reassignments of Defretin-LeFranc (1967) involved both Angola and Zaire species and genera.

Estheriella (Estheriella) moutai Leriche, 1913, and *Estheriella (Estheriella) lualabensis* Leriche, 1913, were both placed under the genus *Bairdestheria-Estheriella* Defretin, 1967. The latter was placed under a new subgenus *Estheriella (Lioestheriata)* herein. *Estheria (Euestheria) mangaliensis* var. *angolensis* Leriche (pars) in Marlière 1948 as well as *Estheria (Euestheria) lerichei* Marlière (pars), 1950 were both placed in synonymy of *Pseudestheria lepersonnei* Defretin-Le Franc, 1967. *Pseudestheria* is a synonymy of *Cyzicus* (Tasch, 1969). Finally, Marlière's subgenus *Echinestheria, Estheria (Echinestheria) marimbensis* M., 1950 had earlier been raised to generic status (Kobayashi, 1954).

In summary, the following genera are known from Angola: *Echinestheria, Gabonestheria, Cornia, Estheriella, Palaeolimnadiopsis, Palaeolimnadia,* and *Estheriina.* Of those *Estheriella, Palaeolimnadiopsis,* and *Echinestheria* occur in the Zaire rock column. The presence of species from *Gabonestheria* and *Cornia* in the Oesterlen collection is noteworthy, because these two species also occur in the São Pascoal section of the Brazilian Rio do Rasto Formation, (Appendix, Fig. 28) and in the India Panchet Formation, Raniganj Coal Field subsurface (Fig. 15).

Lower-case letters in this section designate sites on the Angola detail map, Fig. 20A.

"o". *Estheriella moutai* Leriche 1932. [= *Estheriella (Estheriella) moutai* Leriche, Marlière, 1950b.] [= *Bairdestheria – Estheriella* Leriche, Defretin 1967; see below.] [= *Estheriella (Lioestheriata) moutai* Leriche, Tasch, below.] Quela, about 85 km northeast of Malange, Cassanje III, Upper Triassic. (See Marlière, 1950a, p. 37.)

Family ESTHERIELLIDAE Kobayashi 1954
Subfamily ESTHERIELLINAE Kobayashi 1954

Genus *Estheriella* Weiss 1875
Subgenus *Estheriella* Weiss 1875, (Marlière, 1950b)

This subgenus is redefined in keeping with the characteristics of Weiss's original generic type, *Posidonomya nodocostata* (Giebel, 1857). It possesses more than five nodose, radiating costae that may fade out dorsally and are weak on anterior and posterior sides. In contrast to Weiss's concept, Marlière's (1950a) view of the subgenus, while it also had the costae fade out dorsally, had them prominent on both anterior and posterior sides of the valve. Estheriellid species of both investigators have costae that reach the dorsal margin.

Subgenus Lioestheriata **n. subgen.** This subgenus is proposed for forms having more than five to ten or more interrupted, estheriellid, nodose costae that do not reach the ventral margin and are followed ventrad by a *Lioestheria* hachure-type pattern on growth bands.

Type: *Bairdestheria-Estheriella lualabensis* (Leriche) 1913, Defretin, 1967 (Pl. 6, Fig. 6 [cf. text Fig. 9, p. 57] and Pl. 6, Figs. 1, 3, 4, 5.) [= *Estheriella (Lioestheriata) lualabensis* (Leriche) 1913.]
Discussion: *Defretin's (1967)* Figure 6 clearly shows the *Lioestheria*-like growth bands that bear hachure-type markings of the new subgenus. It seems more appropriate to erect a new subgenus for this form than to fuse two generic names as Defretin did; even more so, because *Bairdestheria* appears to be a synonym of *Cyzicus* (subgenus) *Lioestheria* (*non* subgenus *Euestheria*) as in Tasch (1969, p. R151), but *Lioestheria sensu stricto*. (Cf. subsequent discussion about Kozur and others, 1981, below.)

"t". *Estheria (Echinestheria) marimbensis* Marlière 1950. (= *Echinestheria marimbensis* Marlière; Kobayashi, 1954, p. 135. Cassanje III, Marimba, Malange District, Upper Triassic.)

"b". *Estheria achietai* Teixeira 1947. [= *Estheria (Euestheria) achietai* Teixeira, Marlière, 1950.] [= *Cyzicus (Euestheria) achietai* (Teixeira, 1947); Lutoe fish beds, Cassanje I, Upper Permian.]

"a'". *Palaeolimnadiopsis reali* Teixeira 1960. [= *Palaeolimnadiopsis reali* Teix., 1960, *non Pteriograpta reali* (Teixeira); Teixeira, 1961.] Lioestheriid markings on growth bands do not take precedence in generic attribution over the characteristic shape of the valves, as Teixeira thought, although they probably denote an affinity of *Palaeolimnadiopsis* with lioestheriids. The recurved dorso-posterior margin in *P. reali* Teixeira indicates an evolutionary separation between the two genera. Lunda District, at the confluence of the Cassamba and Canquela Rivers. Casasanje III? Upper Triassic. (Teixeira, 1960, p. 82–84, Pls. 1 and 2; Teixeira, 1961, p. 307.)

"a'". *Estheriella cassambensis* Teixeira 1960. [= *Cyzicus (Lioestheria) cassambensis* (Teixeira).] The reassignment is based on Teixeira (1960, Pl. 3, Figs. 3, 4). These figures show a cancellate pattern seen in many lioestheriids (see Uruguay, South Amer-

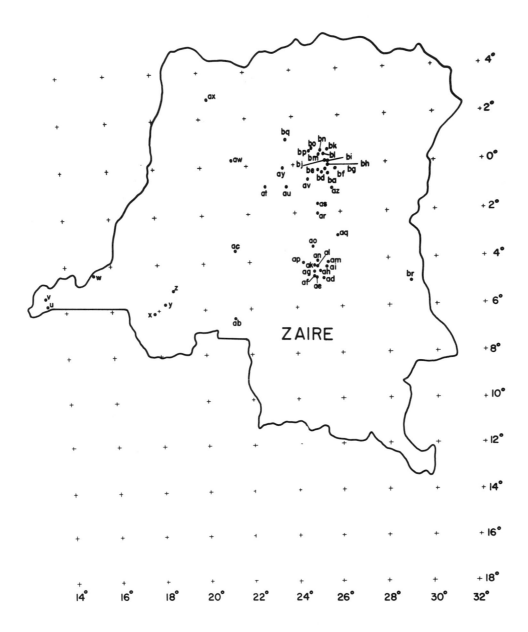

Figure 20. Zaire, detail of fossil conchostracan sites (See also Fig. 22.) **u** = Bomba. Lower Cretaceous; **v** = Kisami Nzambi. Lower Cretaceous; **w** = Leopoldville. Cretaceous **x** = Kitari. Upper Cretaceous; **y** = Falaises Schwetz; **z** = Bumba. Upper Cretaceous **ab** = Pushaluenda. Lower Cretaceous; **ac** = Dekese. Lower Cretaceous; **ad** = headwaters of River Lueki. Upper Jurassic; **ae** = Shema; **af** = River Ole; **ag** = River Lomami between Kimbombo and Lubefu. Upper Triassic/Jurassic; **ah** = Ivunga; **ai** = Limanga; **ak** = Etanga-Shenga. Upper Jurassic; **al** = Combe; **am** = River Lueki between Lubefu and Kibombo; **an** = Wanga; **ao** = Bolaiti; **ap** = Usulunga. Upper Jurassic; **aq** = Bamanga; **ar** = Londo: **as** = Obenge-Benge; **at** = Ndeke. Lower Cretaceous; **au** = Bandu. Lower Cretaceous; **av** = River Lilo. Lower Cretaceous; **aw** = Samba. Lower Cretaceous; **ax** = Katera; Lower Cretaceous; **ay** = Yohila; **az** = Mikaeli; **ba** = Ponthierville. Cretaceous; **bd** = km 25 from Ponthierville. Cretaceous; **be** = Bayanjo; **bf** = Kindu; **bg** = River Oviataku; **bh** = Assengue. Upper Jurassic; **bi** = Songa. Upper Jurassic/Cretaceous; **bj** = Cico-Songa. Cretaceous; **bk** = River Tahopo; **bl** = Cimentstan. Cretaceous; **bm** = CFL. Cretaceous; **bn** = Kisangam (Stanleyville); **bo** = Yangilmo; **bp** = Longa-Longa; **bq** = Bandu; **br** = Makunga.

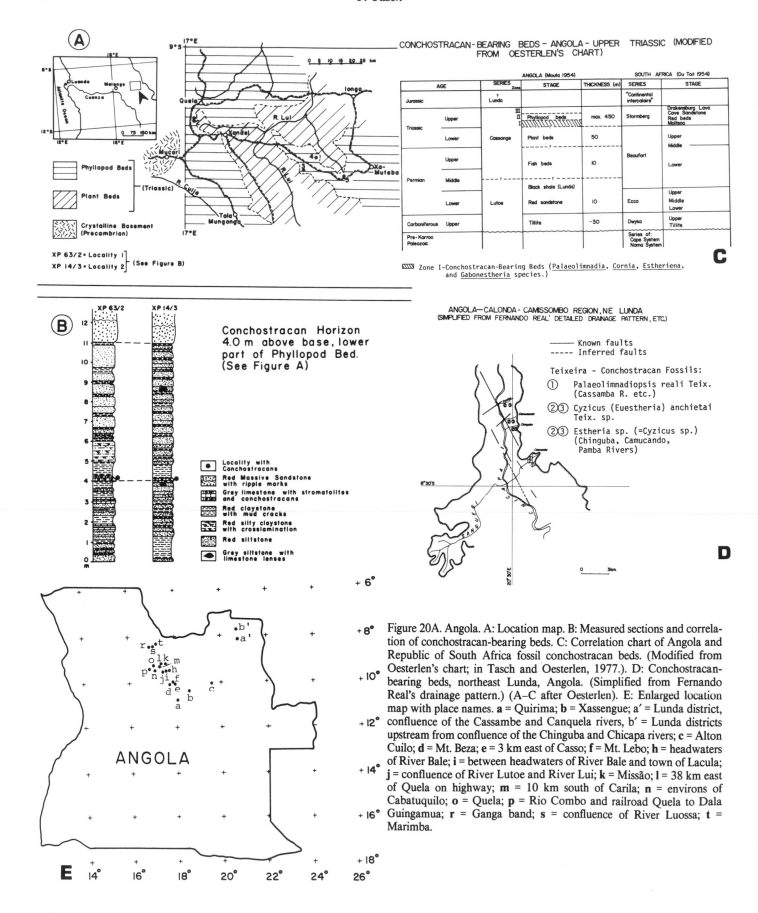

Figure 20A. Angola. A: Location map. B: Measured sections and correlation of conchostracan-bearing beds. C: Correlation chart of Angola and Republic of South Africa fossil conchostracan beds. (Modified from Oesterlen's chart; in Tasch and Oesterlen, 1977.). D: Conchostracan-bearing beds, northeast Lunda, Angola. (Simplified from Fernando Real's drainage pattern.) (A–C after Oesterlen). E: Enlarged location map with place names. a = Quirima; b = Xassengue; a′ = Lunda district, confluence of the Cassambe and Canquela rivers, b′ = Lunda districts upstream from confluence of the Chinguba and Chicapa rivers; c = Alton Cuilo; d = Mt. Beza; e = 3 km east of Casso; f = Mt. Lebo; h = headwaters of River Bale; i = between headwaters of River Bale and town of Lacula; j = confluence of River Lutoe and River Lui; k = Missão; l = 38 km east of Quela on highway; m = 10 km south of Carila; n = environs of Cabatuquilo; o = Quela; p = Rio Combo and railroad Quela to Dala Guingamua; r = Ganga band; s = confluence of River Luossa; t = Marimba.

ica, Pl. 37, Figs. 5, 6). This pattern is unlike the distinctive one of *Estheriella (Estheriella) nodocostata* Weiss with raised, nodose, radiating costae. Nor does it fit other estheriellids. Age uncertain, possibly Cassanje III, Upper Triassic?

"o". *Estheria (Euestheria) mangaliensis* Jones 1862, var. *angolensis,* Leriche 1932. [= *Estheria (Euestheria) lerichei* Marlière 1950.] [= *Cyzicus (Euestheria) lerichei* (Marlière) *pars,* [=*C. (Eu?) leperssonei*], Cassanje III (Upper Triassic).]

"t". *Estheria (Euestheria) malangensis* Marlière 1950 [= *Cyzicus (Lioestheria) malangensis* (Marlière)]. Region of Marimba, Cassanje III, Upper Triassic.

OESTERLEN COLLECTION

(For map numbers for sites 1, 2, and 3, see Fig. 20 A, Angola Map and Sections.)

Superfamily LIMNADIOIDEA Baird, 1849
Family LIMNADIIDAE Baird, 1849
Subfamily ESTHERIININAE Kobayashi, 1954

Genus *Palaeolimnadia* Raymond, 1946

Diagnosis: Relatively long oval carapace, large, smooth umbonal region, and few growth lines (Raymond, 1946, abridged).

Emended herein: Small to large oval carapace with prominent umbonal area of variable size and position on the valve.

Lectotype. *Palaeolimnadia (Palaeolimnadia) wianamattensis* (Mitchell) 1927, Pl. 2, Fig. 8, originally *E. wianamattensis* Mitchell.

Subgenus *Palaeolimnadia* Raymond, 1946

Palaeolimnadia Ray., 1946. Tasch, 1975, p. 101.
Palaeolimnadia Ray., 1946. Tasch *in* Tasch and Jones, 1979, p. 17–18.

Diagnosis: A permanent naupliid carapace, swollen umbonal area occupies a considerable area of the valves; growth bands 3 to 6. Valve form ratio of height to length, approximately 0.82.

Type: Same as for genus.

Subgenus *Grandilimnadia,* Tasch, 1979a

Grandilimnadia, Tasch, 1975, p. 101. Briefly described, no type specified.
Grandilimnadia nov., Tasch *in* Tasch and Jones, 1979a, p. 18.

Diagnosis: Ovate to subovate valves with a comparatively large, smooth, medial to terminal umbo that occupies a much smaller area of the shell than *Palaeolimnadia (Palaeolimnadia).* Growth bands to 20 or more. Valve form ratio, height to length, approx. 0.63.

Type: *Palaeolimnadia (Grandilimnadia) mitchelli* Tasch, 1979a. (= *Estheria wianamattensis* Mitchell, 1927, p. 198, Pl. 2, Fig. 7; non. Fig. 8.)

Palaeolimnadia (Palaeolimnadia) cf wianamattensis (Mitchell)
(Pl. 4, Fig. 4)
(For complete synonymy see Mitchell's types below)

Diagnosis: Valves with a large, smooth umbo that occupies from one-half to three-fourths of the total valve area; three to four growth bands; widely spaced. The wide spacing and sparsity of growth bands indicates a naupliid valve: form ratio of 0.85.

Measurements: (mm) TC 90001. L-1.3, H-1.1; Ch-0.8; Cr-0.4; Av-0.4; Arr-0.2; a-0.5; b-0.5; c-0.6; H/L-.846; Cr/L-.308; Av/L-.308; Arr/L-.154; Ch/L-.615; a/H-.455; b/H-.455; c/L-.462.

Material: TC 90001, 90060, 90011. (From the Oesterlen collection.)

Locality: Northern Angola, Cassanje I Series, Phyllopod Beds Stage, Upper Triassic, XP/63/2 (OEST.). (= Map Site 1, Fig. 20A, upper left.)

Discussion: H/L form ratio of Mitchell's type is .756 (= effectively .800). In this respect, as well as in the size and dominance of the umbo, this Angola species (Upper Triassic) is very close to the Australian type material (Middle or Upper Triassic). It also occurs in the Lower Triassic of the Bowen basin, Australia.

Palaeolimnadia (Grandilimnadia) oesterleni n. sp.
(Pl. 4, Fig. 8)

Diagnosis: Subovate valves with a broad, smooth umbonal area, confined to the antero-dorsal sector, and resulting in a steep descent from the dorsal to the anterior margin; posterior margin rounded. Seven or eight growth bands, deeply incised; H/L-.682.

Measurements: (mm) TC 90010, followed by TC 90009. L-2.2, 2.2; H-1.5, 1.5; Ch-0.9, 1.1; Cr-0.6, 0.9; Av-0.6, 0.6; Arr-0.6, 0.5; a-0.6, 0.5; b-0.8, 0.7; c-1.1, 1.4; H/L-.682, .682; Cr/L-.273, 381; Av/L-.273, .273; Arr/L-.273, .227; Ch/L-.409, .500; a/H-.400, .333; b/H-.500, .467; c/L-.500, .636; h_u*-0.5, 0.5; l_u*-0.7, 0.8; h_u/l_u*-.714, .625 (*h_u and l_u-height and length of umbo, respectively). Growth Bands: 7 (holotype), 8 (paratype).

Material: TC 90010, holotype, internal mold of left valve, paratype, TC 90009, occurs on slab with *Gabonestheria gabonensis* (Marlière); other paratypes, 90715, TC 90725, 90080.

Locality: Northern Angola, Cassanje I Series, Phyllopod Beds, Stage Upper Triassic, XP-72-6 (= Map Site 3, Fig. 20A, upper left).

Name: This species is named for Dr. P. M. Oesterlen who collected the conchostracans and did the biostratigraphy and sedimentology for his doctoral thesis.

Discussion: The closest species to "oesterleni" which has a form ratio of .682 is *Palaeolimnadia (Grandilimnadia) alsatica* Reible 1962, p. 218–219, with a ratio of .690. This species is from the German Triassic. Umbonal placement of the two species is also close but they differ significantly in length of the dorsal margin (which is longer in Reible's species) and in size of the umbo (larger in Reible's species). The new species is quite unlike any of the species assigned to *Limnadia* (= *Palaeolimnadia*) by Novojilov (1970, p. 62) or colleagues, Kapel'ka (1968, Lower Mesozoic) and Zaspelova (1966, Siberia, Mesozoic, central Kazahkstan). It also differs from species assigned to the subgenus by Tasch (*in* Tasch and Jones, 1979a, 1979b, 1979c).

Superfamily VERTEXIOIDEA Kobayashi, 1954
Family VERTEXIIDAE Kobayashi, 1954

Genus *Gabonestheria* Novojilov, 1958

Diagnosis: Large robust spine on initial shell, situated in antero-dorsal sector of the valve; sculpture finely reticulate.
Type: *Estheria (Pemphicyclus) gabonensis* Marlière, 1950.

Gabonestheria gabonensis (Marlière)
(Pl. 4, Fig. 10)

Estheria (Pemphicyclus) gabonensis Marlière 1950, Pl. 3, Figs. 1–3.
Cornia gabonensis (Marilère) Kobayashi, 1954, p. 104–157.
Gabonestheria gabonensis (Marlière) Novojilov 1958a, p. 111, Pl. 2, Figs. 28, 29.

Diagnosis (abridged from Marlière, 1950a, and slightly modified): Elongate, subovate valves; swollen, smooth umbo occupies the upper portion of the antero-dorsal sector and tapers to a prominent node or spine that does not rise above the dorsal margin. Few growth bands, widely spaced ventrad.
Measurements: (mm) L-2.4; H-1.7; Ch-1.1; Cr-0.9; Av-0.3; Arr-0.1; a-0.7; b-0.6; c-1.1; H/L-.708; Cr/L-.375; Av/L-.125; Arr/L-.417; Ch/L-.458; a/H-.412; b/H-.333; c/L-.458.
Material: TC 90009, internal mold of left valve occurring on a slab with *Palaeolimnadia (Grandilimnadia) oesterleni* n. sp.
Locality: Northern Angola, Cassanje III Series, Phyllopod Beds Stage, Upper Triassic, XP/72-6 (= Site 3, about 75 km southeast of Quela).
Discussion: The specimen described above fits Marlière's species in form ratio and conforms to his type in configuration, placement of the umbo, and the prominence and position of the umbonal node. Additional measurements and ratios are included here that are not found in Marlière's original description. Marlière's material came from the Lower Permian beds (d'Agoula Series), Gabon, whereas Oesterlen's collection came from the Upper Triassic (Cassanje I Series, Phyllopod Beds Stage) of northern Angola. Despite this time separation, the Angola specimen(s) are immediately recognizable as belonging to Marlière's species. This suggests stasis in the "gabonensis" bioprogram. Though most common in Permian and Triassic beds, the genus is thought to range into the Cretaceous (Novojilov, 1970, p. 187; Tasch, 1982, p. 45–46).

Superfamily VERTEXIOIDEA Kobayashi, 1954
Family VERTEXIIDAE Kob., 1954

Genus *Cornia* Lutkevich, 1937

Pemphycyclus Raymond, 1946, p. 265, Pl. 3, Fig. 12, Pl. 4, Figs. 1, 2.

Genus *Cornia* Lutkevich, 1937

Emended slightly: Valve shape varying from subovate to subrectangular; umbo, subcentral to anterior in position; small spine or tubercle generally rising from initial valve (but may be flattened on umbo). Ornamentation punctate or with hachure markings on growth bands.
Type: *Cornia papillaria* Lutkevich, 1937.

Cornia angolata n. sp.
Pl. 4, Figs. 3, 6

Diagnosis: Ovate-to-subovate valves with six or more growth bands; a large umbo, slightly or markedly anterior relative to the midpoint of valve length; a prominent node is positioned dorsad on the umbo.
Material: TC 90004, holotype, internal mold of right valve; TC 90742, paratype, internal mold of right valve.
Locality: Northern Angola, Cassanje I Series, Phyllopod Beds Stage, Upper Triassic XP/62/3 (= Site 1).
Measurements: (mm) TC 90004, followed by TC 90742. L-2.5, 1.7?; H-2.0, 1.3; Ch-0.9?, 0.7?; Cr-0.9, 0.4; Av-0.8?, 0.4; Arr-0.7?, 0.6; a-1.0, 0.6; b-0.8, 0.6; c-1.4, --; H/L-.800, .765?; Cr/L-.360, .235; Av/L-.320?, .235; Arr/L-.280?, .353; Ch/L-.360?, .412?; a/H-.500, .462; b/H-.500, .462; c/L-.560, --.
Discussion: The Angola species described above is unlike any of the Permian-Carboniferous or Triassic corniids reported by Novojilov (1970, p. 135–150) and colleagues (Zaspelova, 1965; Molin, 1968, see below) in configuration and position of the umbo on the valve. A Lower Triassic species described by Molin (1968, p. 369, Pl. X, Fig. 3), *Cornia rotunda* from the Vetluga Series, Mezen Basin, USSR, approximates the configuration of *C. angolata* (allowing for eroded sectors on both species). However, the umbonal area is comparatively very small and the node position much lower on the umbo (more anterior - Angola vs. medial - USSR), and the form ratio H/L is greater in Molin's species (.882 vs. .765–.800 Angola).

Superfamily LIMNADIOIDEA Baird, 1849
Family LIMNADIIDAE Baird, 1849
Subfamily ESTHERIININAE Kobayashi, 1954

Genus *Estheriina* Jones, 1897

Diagnosis: Valves more convex for limited area in umbonal region (larval shell) than lower down on the valve (ventrad).

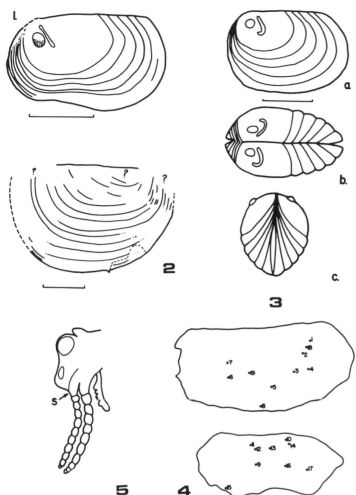

Figure 21. 1: *Lioestheria lallyensis* Depéret and Mazeran 1912. Line drawing of left valve showing features of upwarp of ventral margin, umbonal bulge (node) and the so-called "radial element" (cf. Kozur and others, 1981, text Fig. 1/1). 3a–c (a = lateral, b = dorsal, c = anterior view) *Lioestheria lallyensis* D. & M, emended K.M.P., 1981 (cf. Martens, 1982, text fig. 13). Note how this differs from fig.-1 in lacking ventral upwarp (3a) and in having a "radial element" that is curved or hooked (3b). (Cf. Pl. 44, Figs. 3 and 6 herein.) 5: *Limnadia urukhai* Webb and Bell 1979, a living conchostracan female head in profile, s-scape from which issues the biramous 2d antenna. (Cf. Pl. 44, Figs. 3 and 6.) 2: *"Euestheria" autunensis* Depéret and Mazeran 1912, and Kozur and others, 1981, text-Fig. 1/2. The question mark in the upper anterior sector is unintelligible as Kozur and others Figure 3 of the same species shows an enlargement of that sector, as well as polygonal ornamentation on growth bands (Kozur and others, 1981, p. 1445). (Note also the clearly present growth bands high on the umbo in this enlargement [Pl. 44, Fig. 7 herein].) 4: *Lioestheria lallyensis* Depéret and Mazeran 1912. The original fossiliferous slab showing the fossil arrangement on both sides (Kozur and others, 1981, text-Fig. 2/1, 2). Note that Kozur and others' holotype, i.e., their Number 3 of *Lioestheria lallyensis* Depéret and Mazeran emended, is on the front face of the slab, whereas the figured paratype (Kozur and others, 1981, Pl. 1, Fig. 2), their Number 13, is on the back face of the same slab. It is thus separated in time by "n" seasons (years). That is of interest because the figured paratype (Pl. 44, Fig. 2 herein) has no "radial element" or node. All numbers other than Number 3 have been designated "paratypes." (Note that only one of the many "paratypes" has been figured so that we do not know how many have the so-called "radial element" or what other features of the emended genus they do display, if any.) [See p. 135 for details.]

Emended herein: *Umbo may or may not bear growth bands as does the rest of the valve.*

Type: *Estheriina bresilensis* Jones, 1897.

Subgenus *Estheriina* Jones

Valves with more convex umbo-bearing growth bands (Tasch *in* Tasch and Jones, 1979a, p. 19).

Type: Same as genus.

Subgenus *Nudusia* (Jones)

Valves with swollen umbonal area lacking growth bands (Tasch *in* Tasch and Jones, 1979a, p. 19) present on rest of the valve.

Type: *Estheriina (Nudusia) blina* Tasch, 1979a.

Estheriina (Nudusia) cf *rewanensis* Tasch 1979
(Pl. 4, Fig. 9)

Estheriina (Nudusia) rewanensis Tasch, 1979c, p. 42, Pl. 7, Fig. 7–9.

Diagnosis: Ovate valves with comparatively large subterminal to submedial, irregularly shaped, swollen umbonal area, slightly higher than the dorsal margin and distinct from the topography of the rest of the valve. Below the umbo, successive growth bands are close-spaced, then spacing widens in the middle of the valve and is either more closely spaced or unchanged from there to the ventral margin.

Measurements: (mm) Figures for 90008 followed by those for 90728. L-2.3, 1.7; H-1.6, 1.3; Ch-1.1, 1.0; Cr-0.5, 0.3; Av-0.5, 0.3; Arr-0.7, 0.4; a-0.7, --; b-0.7, 0.5; c- --, 0.8; H/L-.696, .765; Cr/L-.217, .176; Av/L-.217, .176; Arr/L-.304, .235; Ch/L-.478, .588; a/H-.438, --; b/H-.438, .385; c/L- --, .471.

Material: TC 90008, TC 90728 - internal molds of right valves. Other specimens include TC 90731, 90006-A1, 90006, 90007 and 90050.

Locality: Northern Angola, Cassanje I Series, Phyllopod Beds. Stage, Upper Triassic XP/72/6A and XP/72/6B both from Map Site 3.

Discussion: The Angola species differs from estheriinids described by Jones, Kobayashi, and others in either valve configuration, umbonal form, and/or placement on the valve.

ZAIRE − FOSSIL   CONCHOSTRACAN   SITES

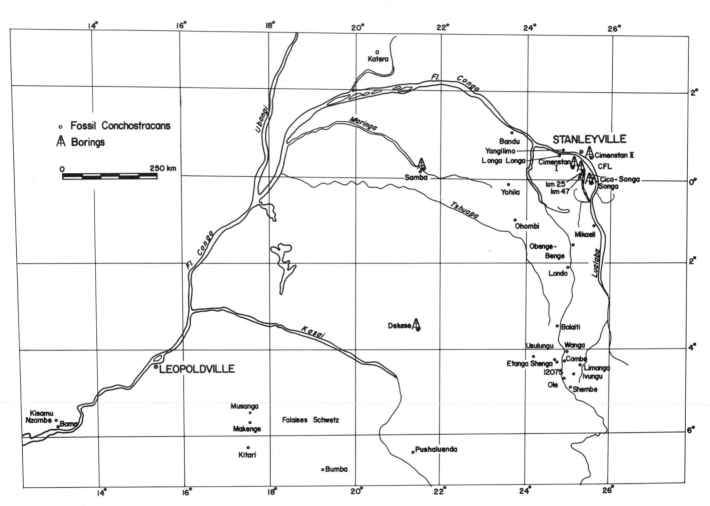

(See systematic taxonomy - text and Africa + Zaire - Angola detail map and schedule of sites)

Figure 22. Zaire. Fossil conchostracan sites. (Data after Mme. Defretin-LeFranc, 1967).

Estheriina (Nudusia) rewanensis Tasch (*in* Tasch and Jones, 1979c) as originally described has a submedial umbo, but in other specimens it is subterminal and fits the Angola species by overlap in H/L and in closeness of the umbonal ratio $u_h/u_1=0.57$ (Angola specimens), 0.55 (Rewan, Australia types).

Palaeolimnadia (Grandilimnadia) africania n. sp.
(Pl. 4, Figs. 5, 7)

Diagnosis: Irregularly ovate valves distinguished by a large, smooth umbo that rises as a convex bulge above the entire dorsal margin, is oblique and tapering from anterior to posterior. Four to six widely spaced growth bands.

Measurements: (mm) Figures for holotype followed by paratype. L-1.9, 1.8; H-1.4, 1.3; Ch-0.8?, 0.6?; Cr-0.4, 0.3; Av-0.4, 0.3; Arr-0.7, 0.9; a-0.7, 0.5; b-0.6, 0.6, c-0.5, 0.6?; H/L-.737, .722; Cr/L-.210, .167; Av/L-.210, .167; Arr/L-.368, .500; Ch/L-.421?, .533?; a/H-.500, .385; b/H-.429, .461; c/L-.263, .333?

Material: Holotype, left valve, TC 90900; paratype, 90729.

Locality: Northern Angola, Cassanje I Series, Phyllopod Beds Stage, Upper Triassic, XP/72/6b (= Map Site 3).

Discussion: This writer chose as the type species of *Grandilimnadia* (Tasch, 1979a, p. 18) one of Mitchell's species, *Estheria wianamattensis* M. (1927, Pl. 2, Fig. 7). This was placed in a new species *Palaeolimnadia (Grandilimnadia) mitchelli* Tasch (Middle Triassic, Sydney Basin, New South Wales). Compared to the new species, "mitchelli" has a smaller smooth umbo that is submedial in position. By contrast *"africania's"* umbo is prominent along the entire dorsal margin above which it arches. The configuration of *"mitchelli"* is subovate (vs. irregularly ovate), and it has a smaller H/L of .666 vs. .722-.737 *"africania"* as well as four growth lines. None of the other Australian species assigned to *P. (Grandilimnadia)* by this writer (Tasch, 1979, *in* Tasch and Jones, 1979a) are any closer to the new species, as they differ in configuration, umbonal size, shape, and placement.

Careful study of the Permian and/or Triassic *Limnadia* and *Limnadia (Palaeolimnadia)* species in Novojilov (1970), all of which belong to *Palaeolimnadia* and most to subgenus *Grandilimnadia,* indicated differences in configuration, umbonal size, shape, and placement on the valve when compared to the new species.

Cameroon

47. *Estheria (Bairdestheria) mawsoni* (Jones) Defretin, 1953. [= *Cyzicus (Euestheria) mawsoni* (Jones), in dark grey shales.] Pont du Mayo Louti, Wealden.

47. *Estheria* cf *dahurica* Chernyshev, Defretin 1953 [= *Cyzicus (Euestheria)* cf *dahurica* (Chernyshev).] Mayo Tafal, in Wealden? shales of Figuil.

47. *Estheriella (Dadaydedeesia) tricostata* Defretin, 1953* [= *Estheriella (Estheriella) tricostata* Defretin 1953.] Non *Afrograpta* (Novojilov, 1958a) (Wealden) Pont du Mayo Tafal. (See below.)

-- *Estheriella (s.s.) camerouni* Defretin 1953* [*Estheriella (Estheriella) camerouni* Defretin] non *Camerounograpta* Nov., 1958.
Pont du Mayo Tafal, Wealden. (See Fig. 23, for map.) (*Novojilov, 1958a, reassigned two of Defretin's species, *Estheriella (Dadaydedeesia) tricostata* D, 1953 and *Estheriella (s.s.) camerouni* D., 1953 each to a new genus: *Afrograpta* and *Camerounograpta,* respectively. The basis for the reassignments was ornamentation. Defretin (1958), however, considered these reassignments unwarranted and thought that conchostracan classification should be based on those valve characteristics which were most prominently and readily observable, and not on fine ornamentation of the valve. This writer reached the same conclusion independently years ago. However, in 1969 this writer accepted *Afrograpta.* (After further consideration, I here reverse that decision.)

-- ? *Estheriella* sp., Dietrich, 1939 (? *Estheriella* sp. Dietrich, 1939), Nord Adamawa. Triassic.

37. *Estheria* sp., Furon, 1950 (= ? *Cyzicus* sp.). Near source of the Cross River in the Mamfe region. Bituminous shales with fish and estheriids, Wealden.

Chad

-- *Estheria* sp., Haughton, 1963 (= ? *Cyzicus* sp.) Léré Series, southwest of the Chad basin, in folded beds of alternating sandstones, green marls, and calcareous beds. Cenomanian(?). (Fig. 23A.)

Gabon

12. *Estheria (Pemphicyclus) gabonensis* Marliére 1950 (= *Gabonestheria gabonensis* Novojilov, 1958c). Red mudstones, Agoula River Series, Lower Permian.

11. *Estheria* sp., *Haughton, 1963.* (= ?*Cyzicus* sp.) Cocobeach Series: Lower-"Estheria", 850-1310 m; Middle-"Estheria," 1380-1675 m. Jurassic.

Ghana

25. *Bairdestheria* sp. [= *Cyzicus (Lioestheria)* sp.] Numerous sites along the coast near the mouth of the Amisa River between Saltpond and Winneba, in pebbly and bouldery argillites. Upper Jurassic–Lower Cretaceous.

Kenya

39. *Estheria greyii* Jones 1878 (nom. imperf.) Latham, 1951 (in 1 Templerlye, 1952. [= *Cyzicus (Euestheria) greyi* Jones. Shales, fish beds, upper Maji ya Chami beds (K5 - Middle Beaufort). Triassic.]

46   P. Tasch

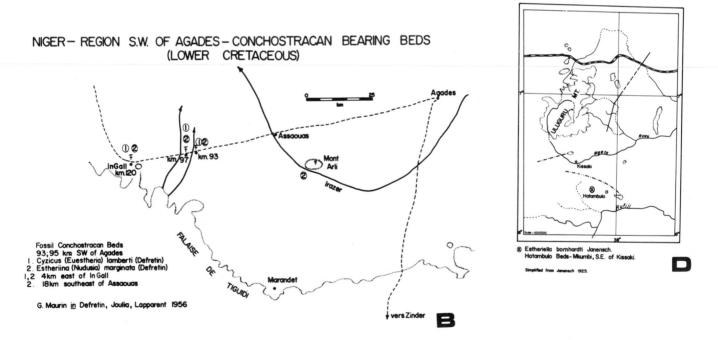

Figure 23. Cameroon, Chad, Niger, and Tanzania. **A:** (Cameroon and Chad.) Lower Cretaceous conchostracan-bearing bed (see Roch, 1953). **B:** (Niger.) Fossil conchostracan sites. **C:** (Tanzania). Jurassic and Triassic fossil conchostracan sites (location sites after the Times Atlas, 1967). **D:** (Tanzania). Triassic site, detail. (Illustrations redrawn and simplified; A, B after Defretin and others (1956) and C. D. after Janensch (1925, 1933). [A. *"dahurica" non "dahurdica"*]

Lesotho

Descriptions of most of the fossils listed below are given in a Festschrift for Haughton (1984, 25, p. 61–85, Pl. 3 and 4).

3. *Cyzicus (Lioestheria) lesothoensis* Tasch n. sp. (Holotype TC 30049B, Thabaning, Mafatene District, Tasch Station 2, beds 1, 10, 11, 18, 20, and 24.) Cave Sandstone, Upper Triassic (Rhaetic).

3. *Cyzicus (Euestheria) thabaningensis* Tasch, n. sp. (Described in this section. Same locality, age, and section as *"lesothoensis"*, Station 2, bed 18, lower B.)

3. *Cornia haughtoni* Tasch, n. sp. (Holotype TC 30067, paratypes: TC 30068, 30090, 30091, and 30093.) Same locality, age, and section as above, Station 2, bed 16, lower; bed 18, lower.

4. *Cyzicus (Euestheria) stockleyi* Tasch n. sp. (Syntypes 30134, 30102, Cave Sandstone, Mofoka's Store, Lesotho, bed 1, layer 7–8). (Also Barkly Pass, R.S.A.; see p. 52 for map and outcrop data.) Upper Triassic (Rhaetic).

3. *Asmussia loockii* Tasch n. sp. (Holotype 30052; paratype TC 30053.) Same locality, age, and section as *"lesothoensis"* above, Station 2, bed 20, unit "c."

3. *Palaeolimnadia (Grandilimnadia)* sp. (Specimen TC 30097) Same locality, age, and section as *"lesothoensis"* above, Station 2, bed 16, "detail, 31-24 mm" (Tasch, 1984, p. 72).

3. *Cyzicus (Euestheria) thabaningensis* n. sp.
(Pl. 3, Fig. 7)

Cyzicus (Euestheria) sp. undet. 1, Tasch, 1984, p. 78, Pl. 4, Fig. 2.

Diagnosis: Small, ovate euestheriids with polygonal ornamentation on growth bands; umbo inset from anterior margin about one-fourth the length of the valve; dorsal margin, slightly curved. Height of anterior third of valve greater than posterior; posterior margin rounded; anterior, more steeply curved. Growth-band number variable: holotype has thirteen widely spaced and strongly expressed concentric bands; the last few are tightly spaced.

Measurements: (mm) Holotype followed by paratypes TC 30037b and TC 30037a. L-2.1, 1.7, 2.0; H-1.5, 1.2, 1.5; Ch-1.1, 0.9, 1.2; Cr-0.7, 0.7, 0.8; Av-0.4, 0.3, 0.2; Arr-0.5, 0.3, 0.3; a-0.6, 0.4, 0.5; b-0.6, 0.5, 0.6; c-1.1, 0.7, 1.0; H/L-.714, .706, .750; Cr/L-.333, .412, .400; Av/L-.190, .176, .100; Arr/L-.238, .176, .150; Ch/L-.524, .529, .600; a/H-.400, .333, .333; b/H-.400, .417, .400; c/L-.524, .412, .500.

Material: Holotype 30037c, paratypes 30037a and 30037b. There are two or three additional fragmented valves in platy shale.

Locality: Thabaning, Lesotho, Cave Sandstone, Tasch Station 2, bed 18, lower B. Upper Triassic.

Discussion: This species differs from *Cyzicus (Lioestheria) draperi* (Jones and Woodward, 1894), in that it has a greater ratio for all parameters except Ch/L and c/L, which are smaller. It differs from *Cyzicus (Lioestheria) lesothoensis* n. sp. from the

Upper Triassic Mafatene District, Lesotho, besides having polygonal vs. hachure markings on growth bands, in having a greater H/L, smaller Av/L, Ch/L, a/H, and b/H.

Malagasy Republic (formerly Madagascar)

10. *Estheria* sp., Besarie, 1952; Haughton, 1963. [= ?*Cyzicus* sp. (Besarie).] Sakamena facies, unit 3, plant-bearing shales; unit 4, shale-sandstone complex with reptiles. Upper Permian. Livret, 1979.

Malawi (formerly Nyasaland)

Estheria (Cyzicus) sp., Haughton, 1963. (= ?*Cyzicus* sp.) Red marls with fish scales, Lower Beaufort (Dixey) or uppermost Ecca (Haughton, 1963, p. 221, cf. Mozambique, p. 220. Southern Malawi, see Haughton, 1963, map, Fig. 28).

Nyasaestheriella nyasana (Newton) Kobayashi, 1954. (= ?*Estheriella nyasana* Newton 1910.) Karoo Beds, Nkana, Malawi (L = approx. 1.0 mm). Permian(?). This is a questionable conchostracan. It may very well be a mytiliform pelecypod. The mode of preservation is also unusual for a conchostracan; it is flattened on bedding plane so that the vertical anterior margins face each other.

Morocco

Termier and Termier (1950) provided a line drawing of what they identified as *Leaia* cf *tricarinata* (Meek and Worthen). All from N de Hamahou, Ida ou Zal, Morocco (Westphalian). Their illustrations appear to represent three species. None of them are *"tricarinata,"* as Kobayashi (1954) pointed out; he referred to them as *Leaia* sp. nov? Apparently these specimens are not in the National Collection of Morocco. From the Termiers (1950) illustrations (Figs. 42–44), three species are represented. (Pl. 5).

Their figures now appear to be assignable as follows: Figure 42 to *Leaia (Hemicycloleaia)* sp. 1; Figure 43 to ?*Estheriella* sp.; Figure 44 to *Leaia (Hemicycloleaia)* sp. 2. A new Stephanian leaiid is described and figured herein. It was found in the National Collection of Morocco among the on-loan conchostracans. Their collection also contained a large number of specimens of *Cyzicus (Euestheria) simoni* (Pruvost), two specimens of which were figured by the Termiers (1950, Figs. 40, 41). These were all from the Westphalian of Djerada, Morocco. This species is redescribed below on the basis of the Moroccan collection.

Superfamily LEAIOIDEA Raymond, 1946
Family LEAIIDAE Raymond, 1946

Genus Leaia Jones, 1862
Subgenus *Leaia* Jones, 1862

Diagnosis: Generally bicarinate leaiids with rectangular, sub-

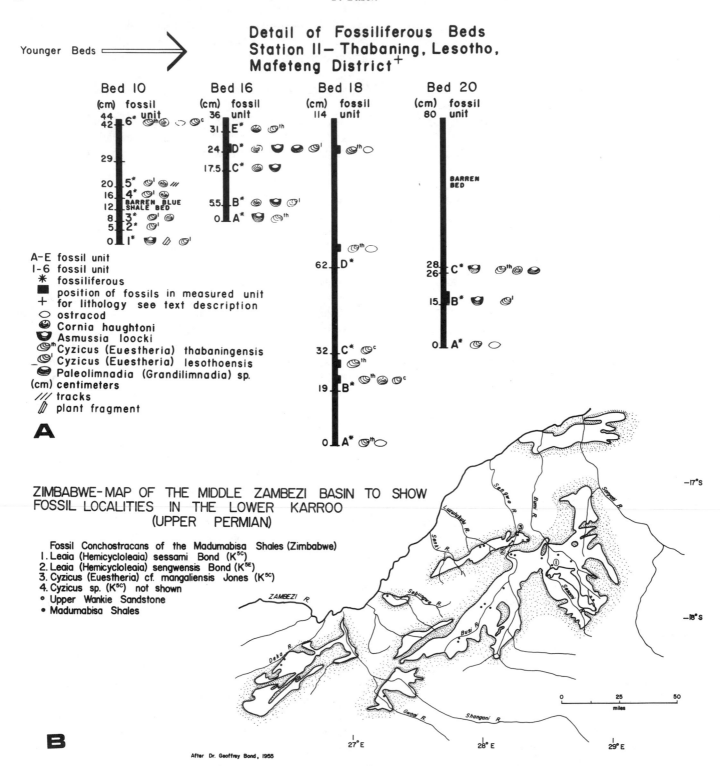

Figure 24. Lesotho and Zimbabwe. **A:** (Lesotho) Thabaning, spectra of fossiliferous beds and conchostracan species distribution. Tasch Station 11. **B:** (Zimbabwe) Map of Middle Zambesi Basin conchostracan-bearing sites (after Geoffrey Bond, 1955, p. 98.) [A. *Cyzicus (Lioesthena) lesothoensis non "(Eustheria)."*]

quadrate, and elliptical forms. (Tasch, *in* Tasch and Jones, 1979c.)

Type: *Hemicycloleaia laevis* Raymond, 1946.

Subgenus *Hemicycloleaia* Raymond, 1946

Diagnosis: *Bicarinate to tricarinate leaiids with circular to subovate forms.* (Tasch *in* Tasch and Jones, 1979.)
Type: *Leaia leidyi* (Lea) 1855.

Leaia (Leaia) bertrandi n. sp.
(Pl. 2, Figs. 1, 2)

Diagnosis: Compact, subquadrate, bicarinate valves, with straight dorsal margin and with middle and posterior sectors approximately equal in size; the anterior is smaller. Two ribs present, anterior rib is slightly curved away from the anterior margin and fades out at the fifth from the last growth band; the posterior rib is slightly curved dorsad, where it crosses the umbo, but otherwise is straight and terminates on the last growth band. Margins: anterior, steep, slightly curved; posterior, gently arched.
Measurements: (mm) Holotype, 1-5.5; H-4.8; Ch-3.5; Cr-0.9; Av-0.9; Arr-1.1; c-2.5; H/L-.872; Cr/L-.729; Av/L-.187; Arr/L-.229; c/L-.454; alpha-85°; beta-37°; sigma-83°; d_1-0.4; d_2-4.3.
Locality: Haute Atlas, Hamalou-Agadir. Morocco, Lower Stephanian.
Name: Named for P. Bertrand, who collected the specimens assigned to this species.
Material: National Collection of Morocco. Holotype, mold of right valve; anterior sector is partially eroded and posterior margin covered (hs q-λ, Slab 1, side B, specimen "a"); paratype, mold of right valve; posterior sector is preserved (Slab 2, side B). Other incomplete specimens of this species on Slabs 1, 2, 3, and 4.
Discussion: The new species is compact and subquadrate, and thus differs from Mitchell's (1925) leaiids from the Newcastle Coal Measures (Upper Permian; see Plates 40–41 below) which are chiefly hemicycloleaiids, i.e., ovate to subovate forms; and a few subelliptical forms. Bond's (1955) Zimbabwe leaiids (Upper Permian) are not compact and subquadrate; rather, they are hemicycloleaiid species. The West Australian Carboniferous (Tasch *in* Tasch and Jones, 1979a) yielded a species *Leaia (Leaia) andersonae* Tasch which differs from the new species in that it is subelliptical. To indicate how this type of species differs from *"bertrandi,"* H/L-0.57 (*"andersonae"*) vs .87 (*"bertrandi"*), alpha-81° vs. 85° (*"bertrandi"*), beta-23.5° vs. 37° (*"bertrandi"*).

Pruvost (1920, Table 1) set subquadrate and subrectangular *Leaia* species from the Carboniferous in two groupings. These leaiids, such as *L. leidyi* (Lea) and *L. williamsoniana* Jones, have a rounded anterior margin and a steep posterior margin; they lack the compact and subquadrate configuration of the new species. Novojilov's (1956a) leaiids (Carboniferous and Permian—USSR and elsewhere) are chiefly hemicycloleaiids, i.e., subcircular to subovate forms or those irregularly ovate. Many of Novojilov's leaiids of eccentric form assigned to separate genera belong under synonymy of *Leaia* (cf. Tasch, 1958). Zaspelova's (1966) leaiids

also are chiefly hemicycloleaiids (Also see Guthörl, 1934.)

Leaia (Leaia) subquadrata Raymond, 1946 based on a Jones species from the Welsh Coal Measures is "irregularly subquadrate." Novojilov (1956a) assigned that species to a new genus *Leaianella Nov.* (which is a synonym of *Leaia;* Tasch, 1958). The new species lacks the obliquity of the anterior margin relative to the rest of the valve that characterizes Raymond's species, and is more compactly subquadrate. Metrically *"bertrandi"* differs from Raymond's species as follows: H/L-.872 vs. .750 *"subquadrata,"* alpha-85° vs. 97° *"subquadrata,"* beta-52° vs. 37° *"subquadrata."*

47. *Cyzicus (Euestheria) simoni* (Pruvost)
(Pl. 2, Figs. 3, 4)

Estheria simoni, Pruvost, 1911, p. 64–68, Pl. 1, Fig 4–8.
Estheria (Euestheria) simoni, Pruvost, 1919, p. 57, Pl. 24, Fig. 29–33, Text-Fig. 15.
Pseudestheria simoni (Pruvost), Raymond, 1946, p. 248.
Euestheria simoni (Pruvost), Termier and Termier, 1950, Pl. 184, Fig 39–41.
Lioestheria simoni (Pruvost), Kobayashi, 1954, p. 55.

Diagnosis (emended): Valves of this species are represented on the same bedding plane by dimorphs. One of these consists of subcircular valves with umbo almost subterminal; a short, straight, dorsal margin, slightly elongate transversely—the *Estheria (Euestheria) simoni"* of Pruvost (1919). The other dimorph is elongate-elliptical, (also with umbo almost subterminal) with a comparatively long dorsal margin for which there is no equivalent in Pruvost (1911, 1919). Ornamentation for both dimorphs consists of small, raised, contiguous granules on growth bands.
Measurements: (mm) Takout slab (short dorsal margin), hw 93-q14, "a" and "b" on opposite sides of slab; numbers represent first "a" and then "b" for each parameter. L-3.7, 3.8; H-2.7, 2.8; Ch-1.5, 1.6; Cr-1.7?, 1.6; Av-0.5, 1.1; Arr-1.4, 0.6; a-1.2, 1.7; b-1.6, 1.1; c- --, 1.5; H/L-.892, .763; Cr/L-.459, .526; Av/L-.216, .289; Arr/L-.378, .342; Ch/L-.405, .342; a/H-.429, .571; b/H-.424, .517; c/L-.838, .911.

Djerada slab (long dorsal margin, hw-92-m84). L-4.6; H-2.8; Ch-2.5; Cr-1.4; Av-0.6; Arr-1.6; a-1.0; b-1.3; c-2.5; H/L-.609; Cr/L-.304; Av/L-.130; Arr/L-.348; Ch/L-.543; a/H-.357; b/H-.464; c/L-.543.

Material and Locality: Ministry of Energy and Mines (Division of General Geology), Rabat, Moroc; hw-91-F378, Oulmes sondage de Sidi Bel Hassam, top of bed A; hw-92-m84, Djerada (see "Africa" map, No. 47); hw-93-q14, Takout, top of bed G; hw-94-F-378, Oulmes sondage de Sidi bel Kessem, top of bed C. There are also some Algerian specimens in the same collection: hw 9-y 17, Kenadze, Algeria. All specimens are labelled *Estheria simoni* Pruvost, Westphalian.
Discussion: The species of *Cyzicus (Euestheria) simoni* (Pruvost) figured in line drawings by the Termiers (1950, Pl. 184,

Fig. 39–40) from the Djerada Collection are closest to Pruvost's subcircular valves that had a transverse slant posteriorly, and a short, straight dorsal margin, *Estheria (Euestheria) simoni* Pruvost, Plate 2, Figure 3 below represents the same dimorph; the elongate-elliptical dimorph with a longer dorsal margin is depicted in Plate 2, Figure 4 below.

One of the Termiers figures displayed hachure-type markings, i.e., longitudinal striae that ended about halfway on the growth bands. This was probably the effect of erosion because this writer and Pruvost originally observed euestheriid ornamentation that Pruvost described as "très finement réticulée," i.e., a network or web which is the polygonal pattern.

The Moroccan and Algerian slabs contain numerous valves in varying conditions of preservation. Few valves on any of the slabs were suitable for photography and measurement. Dimorphs were found on some bedding planes in the Djerada slab as well as on the Takout slab.

22. *Estheria minuta* Alberti, 1832 (Defretin and Fauvelet, 1951). [= *Cyzicus (Euestheria) minuta.*] Region d'Argana-Bigoudine, Haute-Atlas occidental, Morocco. Triassic. Cf. *"Estheria" (Euestheria) minuta* (Defretin, 1950, Pl. VIII, Fig. 1–5; Pl. IX, Fig. 1).

22. *Estheria destombesi* Defretin, 1950. [= *Cyzicus* (subgenus?) *destombesi* (Defretin.] Same site and age as *"minuta"* above.

Mozambique

28. 35. *Estheria borgesi* Teixeira 1943. [= 28:*Cyzicus (?Lioestheria) borgesi* (Teix.).] Emended description herein. Sketch by Teixeira (1943) suggests an asmussid; photograph less clear than sketch. Plant-bearing beds, Tête. See Pl. 5, Fig. 5 below, Upper Permian. (35:below, Lower Shire-Zambesi region. Upper Permian [Haughton, 1963, p. 221].)

9. *Palaeestheria lebombensis* Rennie 1937. [= *Cyzicus (Lioestheria) lebombensis* (Rennie); see below.] Sedimentary intercalations in rhyolites, Lebombo Volcanic Formation, Namahacha, west of Lorenço Marques. Jurassic.

9. *Cyzicus (Lioestheria) lebombensis* (Rennie)
(Pl. 5, Fig. 4)

Palaeestheria lebombensis, Rennie, 1937. p. 20–21, Text Fig. 4.
Euestheria lebombensis (Rennie), Kobayashi, 1954, p. 183, postscript.

Diagnosis: Subovate valves that taper posteriorly; with subterminal umbo and straight dorsal margin; anterior margin broadly rounded; postero-dorsal angle, obtuse; posterior margin, moderately convex; anterior quarter of valve has greatest height; ornamentation, hachure markings on tightly spaced growth bands.

Measurements: (mm) (Rennie's (1937) holotype, additional measurements made on his published figure.) L-6.5, H-4.5, Ch-5.0, H/L-.646, Ch/L-.769.

Material: Holotype, mold of left valve, Rennie's (1937) Slab 11. Another specimen ("a" on Slab 9) shows a complete outline and was described by Rennie. (Unfortunately, the Rennie material has been misplaced or lost.) The Förster Collection from Lebombo, at Munich, is, for the most part, poorly preserved material, but there is one complete valve which furnished the best H/L information; another badly eroded whole valve shows evidence of three growth cycles and clearly displayed lioestheriid hachure ornamentation. Plants, lamellibranchs, and gastropods are associated biota.

Locality: Sedimentary interbeds (flaggy red to purple), Lebombo Volcanic Formation, Namahacha, west of Lourenço Marques. Jurassic (Lias).

Discussion: Rennie provided only a line-drawing of his species and few measurements. He doubted that his "second specimen ("a" on Slab 9) belonged to the same species. However, the form ratio of this specimen is close to the holotype, but no ornamentation was shown.

All three of Förster's specimens are lioestheriids and are very likely identical with, and/or have clear affinities to, Rennie's species. The most reliable H/L measured on one of these specimens is .666, close to Rennie's holotype.

28. *Cyzicus (Lioestheria?) borgesi* (Teixeira)
(Pl. 5, Fig. 5)

Estheria borgesi, Teixeira 1943, p. 71–72, Pl. 3, Figs. 7, 8 (Fig. 7 is a line-drawing reproduced herein, because the photograph [Fig. 8] obscures the chief features). (cf. Borges, 1952.)
Lioestheria borgesi (Teixeira) Kobayashi, 1954, p. 103–104, 154.

Diagnosis: Subovate, small valves with short, straight, dorsal margin and terminal umbo. Steep postero-dorsal angle which continues in a convex, posterior bulge; anterior and ventral margins rounded. Numerous growth-lines. Ornamentation: apparent hachure-like markings on growth bands, as seen in Teixeira's (1943) poorly defined photograph (Fig. 8).

Measurements: (mm) (New measurements [approximate] added on basis of inadequate published figures.) L-1.8; H-1.3; Ch-0.8; Av-0.4; Arr-0.6; a-0.6; b-0.6; c-1.4; H/L-.717; Cr/L-.239; Av/L-.239; Arr/L-.304; Ch/L-.435; a/H-.454; b/H-.484; c/L-.783; postero-dorsal angle-48°.

Material: Left valve (Teixeira's type).

Locality: Plant-bearing, fine sandstone, Karoo Formation, Tête, Mozambique. Permian.

Discussion: The only measurements Teixeira provided were L and H. In light of the limited information available on this species it seems imprudent to make comparisons with other African species assignable to *Cyzicus (Lioestheria)* or the asmussids.

Niger

14. - 19. *Estheria (Euestheria) lamberti* Defretin and others, 1956 [= *Cyzicus (Lioestheria) lamberti* Defretin)]. Kori Ezazel, 95 km

southwest of Agades, on the trail of In Gall at Agades; Kori, 93 km, starting at Agades. (See Fig. 23B.) Southern flank of Mont Arli, 50 km southwest of Agades, 4 km east of Kori Ezazel. (Defretin and others, 1956, Pl. 19, Fig. 1–10; Fig. 10 especially.) Lower Cretaceous. (See Greigert and Pougnet, 1967, p. 112.)

14. - 20. *Estheria (Euestheria) marginata* Defretin, 1956 [= *Estheriina (Nudusia) marginata* (Defretin)]. Same localities and age as "lamberti" with one additional site: 18 km southeast of the well at Assaouas (see Fig. 23B). Defretin, Joulia, Lapparent 1956, Pl. 20, Figs. 1–10, especially Fig. 5 and text-Fig. 2; both major (larger) and minor (smaller) forms of this species have a naked, raised umbo.

--. *Estheria* sp. (= ?*Cyzicus* sp., South of Aïr massif, Irhazer Formation, red lacustrine mudstone; dinosaur beds at Damergou. Lower Cretaceous.) (See Greigert and Pougnet, 1967, p. 112; also discussion in Petters, 1981, p. 139–159, and map, Fig. 3, for Aïr massif and Damergou.)

Republic of South Africa

--. *Cyzicus (Euestheria) stockleyi,* Tasch 1984, 25: p. 74–76, Pl. 3, 4, Barkly Pass, Sheet 3127BP Republic of South Africa, 1st ed., 1968. (See also Mofoka's Store, Lesotho.) Upper Triassic (Rhaetian). (cf. Ellenberger, 1970.)

7. *Estheria anomala* Jones 1901 [= *Cyzicus (Euestheria) anomala* (Jones).] (Redescribed herein.) Conglomerate, basal unit, Uitenhage Series (Cape Folded Belt), about 48 km east of Swellendam District, Republic of South Africa. Lower Cretaceous.

6. *Estheria draperi* Jones and Woodward 1894 [= *Cyzicus (Lioestheria) draperi;* Jones and Woodward, 1984]. (Redescribed herein.) Shale overlying Cave Sandstone. Platberg near Harrismith, eastern side of the Drakensberg, Natal. Upper Triassic (Rhaetian).

2. *Cyzicus (Lioestheria) lesothoensis* Tasch n. sp., Tasch, 1984, p. 77–78, Pl. 4, Fig. 1a, b. Cave Sandstone, Siberia, Republic of South Africa, and Thabaning, Lesotho, various beds, Upper Triassic (Rhaetian).

2. *Afrolimnadia sibiriensis* Tasch n. gen., n. sp. (Described herein.) Cave Sandstone, Siberia, Republic of South Africa, RSA, Station 1A, bed 4. Upper Triassic (Rhaetian).

7. *Estheria greyii* Jones [= *Cyzicus (Euestheria) greyi* (Jones)]. (Restudied herein.) Dark grey shale near Cradock, Cape Province. Upper Permian.

7. *Leaia (Hemicycloleaia) cradockensis* Tasch n. sp. (Described herein.) Near Cradock, Cape Province. Upper Permian.

--. *Leaia* sp., Barnard, 1931. [= *Leaia (Hemicycloleaia* sp.] Approximately 3.1 km south of Mooi River Station, Natal. Lower Beaufort Formation. Upper Permian.

1. *Palaeestheria* sp., Barnard, 1931. (= ?*Cyzicus* sp.) Lacks illustration and formal description. Large valves, L-16.0–23.0 mm; numerous growth bands. Port Elizabeth, Cretaceous (Wealden).

5. *Cyzicus (Estheria)* sp. Roux, 1960. (= ?*Cyzicus* sp.) Appears to have a submedial umbo, H/L-.714, in Roux's Figs. 1, 2. Farm

Leeukuil 81, near Vereeniging, Coal Measures, Transvaal. Plant-bearing. Middle Ecca, Lower Permian. In ferruginous red shale, badly eroded and crushed; poor photograph.

--. *Palaeestheria* sp., Rennie, 1934. [= *Cyzicus (Lioestheria?)* sp.] Quarries near railroad station, Port Alfred, Republic of South Africa. Albany Museum, Grahamstown, No. 4165. Witteberg Series, Cape System. Lower Devonian (see "Africa," Pl. 3, Fig. 5).

Afrolimnadia n. g.

Conchostraca incertae sedis, Tasch, 1984, p. 13, Pl. 4, Fig. 8.
Diagnosis: Large ovate-elongate valves with a comparatively large, prominent, smooth umbo that occupies up to three-quarters of the valve; slightly arched dorsal margin.
Type: *Afrolimnadia sibiriensis* Tasch, n. sp., left valve, TC 30027.

2. Afrolimnadia sibiriensis, n. sp.
(Pl. 3, Figs. 1, 2)

Estheria draperi Jones and Woodward, Stockley, 1947, partim. (Mere mention on p. 41, 42, 45.) Stockley, 1947. Report on the geology of Basutoland. Masero, Basutoland (now Lesotho) p. 30–54. For full synonymy of *Cyzicus (Lioestheria) draperi* (Jones), see subsequent redescription, this section.

Diagnosis: Anterior and posterior margins of the valve are rounded; large subovate umbo, h_u/l_u umbonal ratio varies slightly; dorsal margin slightly arched. Granular ornamentation. Zone of approximately 17 growth bands peripheral to the large umbo, and mostly tightly spaced.

Measurements: (mm)

Holotype	L	H	Ch	Cr	Av	Arr	a	b	c
TC 30027	12.8	8.2	7.0	5.0	2.0	3.2	3.2	3.0	6.0
Paratypes:									
30020	-	7.2	-	-	-	2.5	-	2.2	5.0
30025	14.5	9.5	-	6.5	3.0	3.0	4.0	3.0	6.5
30028	12.7	8.0	8.0	4.6	2.5	2.5	2.7	3.5	5.5

	H/L	Cr/L	Av/L	Arr/L	Ch/L	a/H	b/H	c/L	hu/lu*
TC 30027	.641	.391	.156	.250	.547	.391	.366	.469	.647
30020	-	-	-	-	-	-	-	-	.685
30025	.655	.448	.207	.207	-	.421	.316	.448	.700
30028	.650	.362	.197	.197	.630	.338	.438	.433	.706

*h_u/l_u = height of umbo over length of umbo.

Material: Holotype TC 30027, left valve overlain in postero-dorsal sector by a fragment of right valve. Slight erosion of small area in antero-dorsal sector. Paratypes: TC 30020, 30025, 30028, each less complete than the holotype.
Locality: Siberia, Republic of South Africa, Tasch Station 1A, bed 4. Site and measured section in Tasch, 1984, p. 62–64; Haughton, 1924, p. 326–330.

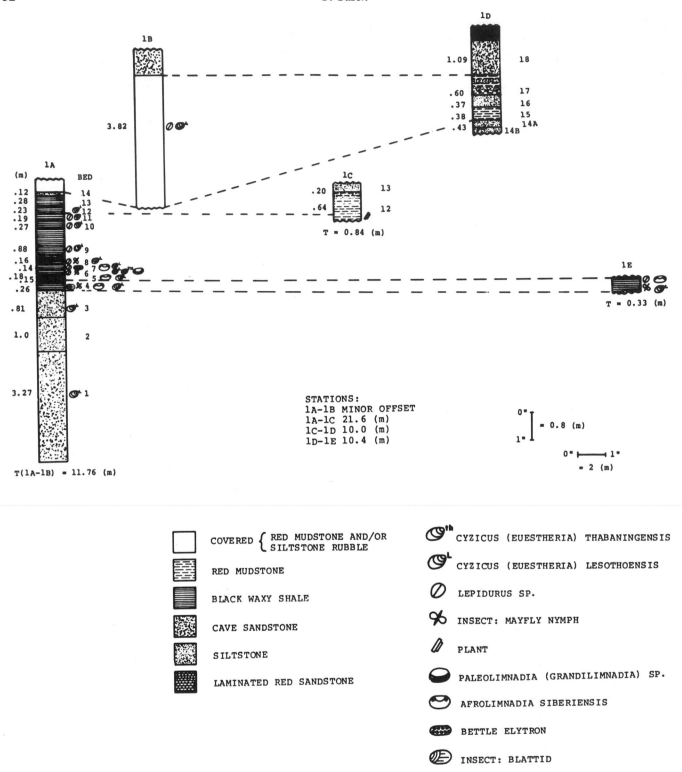

STATIONS:
1A-1B MINOR OFFSET
1A-1C 21.6 (m)
1C-1D 10.0 (m)
1D-1E 10.4 (m)

0" ⌐
   ⎬ = 0.8 (m)
1" ⌐

0" ├——┤ 1"
      = 2 (m)

COVERED { RED MUDSTONE AND/OR SILTSTONE RUBBLE

RED MUDSTONE

BLACK WAXY SHALE

CAVE SANDSTONE

SILTSTONE

LAMINATED RED SANDSTONE

CYZICUS (EUESTHERIA) THABANINGENSIS

CYZICUS (EUESTHERIA) LESOTHOENSIS

LEPIDURUS SP.

INSECT: MAYFLY NYMPH

PLANT

PALEOLIMNADIA (GRANDILIMNADIA) SP.

AFROLIMNADIA SIBERIENSIS

BETTLE ELYTRON

INSECT: BLATTID

(m) METERS

"SIBERIA" SECTION
Conchostracan-Bearing Beds   Cave Sandstone

Figure 25. Siberia, Republic of South Africa. Conchostracan-bearing beds. (Tasch Section; see text.) [Beetle elytron not bettle]. [C. *(Lioestheria) lesothoensis non* (Euestheria)].

Discussion: Several workers (Jones and Woodward, 1894; Stockley, 1947, Haughton, 1924) assigned specimens from Siberia, Republic of South Africa and elsewhere, which belong to this new genus and species, to *Estheria draperi* Jones and Woodward 1894. Study of types of *draperi* indicates that only a single large specimen (British Museum specimen I-3233) was found to be measurable. It has a large, smooth umbonal area which is the major morphological distinction. A faint granular ornamentation on growth bands in the new species contrasts with the punctate pattern found in "draperi." In addition, *C. (Lio.) draperi* has a smaller H/L and Av/L, and a straight dorsal margin.

Furthermore, "sibiriensis," measured on the holotype, has an antero-dorsal angle of 20° ±3° compared to 32° ±4° for *draperi* and a larger postero-dorsal angle, 21° ±3° compared to 15° ±3°, for *draperi*. Equivalent measurements on paratypes of the new species show similar angles.

The nearest previously described limnadiid with a comparable umbo is a Triassic species, *Limnadia tenisseica* Kapel'ka 1968 from eastern Siberia, USSR. Comparable measurements using his photograph indicate that the latter has a greater H/L (.718 vs. .641–.650) than the new species. All other parameters and ratios, with the exception of Av/L, Arr/L, c/L, and h_u/l_u that overlap, are smaller than in "sibiriensis" (Cr/L, Ch/L, a/H, b/H).

That species is more compact-ovate compared to the distinctly elongate-ovate configuration of the new species, and it is one-fourth smaller. The umbo in the Russian species is terminal compared to the new species, with an umbo inset from the anterior margin. In addition, the postero-dorsal angle is 40° compared to 21° ±3° for "sibiriensis".

7. *Cyzicus (Euestheria) greyi* (Jones)
(Pl. 1, Figs. 4, 5)

Estheria greyii Jones (nom. imperf.) Jones, 1878, p. 100, Pl. 3, Fig. 1.
Palaeestheria greyii (Jones). Barnard, 1931, p. 255.
Palaeolimnadia greyii (Jones). Raymond, 1946, p. 265.
Cornia? greyii (Jones), Kobayashi, 1954, p. 103.

Diagnosis: Subelliptical to ovate valves with straight dorsal margin and an apparently smooth umbo that bears obscure growth bands on the holotype, but which are more distinct on the paratype. Umbo occupies anterior third of valve; anterior margin, convex; posterior, rounded; anterior margin join to dorsal margin steeper than that of the posterior margin.

Medial growth lines are paired and more widely spaced than those ventrad to them. (Double and triple growth bands that follow could be illusory and attributed to tight spacing.) The paratype seems to show a minute, granular pattern on a few growth bands.

Measurements: (mm) Figures that follow are for the holotype first and paratype (I-2436a) second. L-3.2, 3.2; H-2.1, 2.2; Ch-2.0, 2.0; Cr-1.0, 1.4; Av-0.5, 0.7; Arr-0.6, 0.3; a-1.0, 1.0; b-1.0, 0.7; c-1.5, 1.3; H/L-.656, .688; Cr/L-.313, .438; Av/L-.156, .219; Arr/L-.188, .094; Ch/L-.625, .625; a/H-.476, .455; b/H-.476, .318; c/L-.469, .406.

Material: Holotype, British Museum (= BM I-2436a.) Right valve, and numerous paratypes on same slab, as well as on Slab I-2436b.

Locality: Hard, dark grey shale, near Cradock, Cape Province, Republic of South Africa. Upper Permian.

Discussion: George Grey (1871) found these fossils, first reported through Jones in print. Jones (1878) misinterpreted the apparently smooth umbo as real, but restudy of all paratypes indicates that growth bands covered the umbo as well as the rest of the valve. That error was the basis for Raymond (1946) assigning "greyii" to *Palaeolimnadia*. Kobayashi (1954) put "greyii" questionably under *Cornia(?)*, where, even if the umbo were smooth, it would not belong, because corniids bear an umbonal node.

8. *Cyzicus (Euestheria) anomala* (Jones)
(Pl. 1, Fig. 3)

Estheria anomala Jones, 1901, p. 353, Figs. 1-4.
Palaeestheria anomala (Jones), Barnard, 1931, p. 255.
Palaeestheria anomala (Jones), Raymond, 1946, p. 237, Pl. 3, Fig. 3.
Lioestheria anomala (Jones), Kobayashi, 1954, p. 105, 153.
Cyzicus (Lioestheria) anomala (Jones), Morris, 1980, p. 36, Fig. 7.

Diagnosis (emended): Elongate, oblong valves with umbo inset from, but close to, anterior margin; dorsal margin straight; anterior margin less steeply convex than the posterior; numerous closely spaced growth bands. Ornamentation consists of minute, close-packed granules on growth bands.

Measurements: (mm) (BM I-6130, holotype). L-4.8; H-2.5; Ch-3.0; Cr-.15; Av-0.8; Arr-0.9; a-1.0; b-1.0; c-2.3; H/L-.521; Cr/L-.313; Av/L-.167; Arr/L-.188; Ch/L-.625; a/H-.400; b/H-.400; c/L-.479.

Material: Two sandstone slabs. Jones specimen 302a-1 (= BM I-6130), here designated holotype, left valve. (New numbers BM-"syntypes" I-6130 and 6131.) Other specimens figured by Jones (paratypes): 302a-2, 302-3 (= I-6131) do not belong to this species. See discussion below and new measurements of Jones' holotype, figured herein (BM I-6130). No syntype can be recognized in light of the above observations.

Locality: About 48 km east of Swellendam District, near Heidelberg; Enon conglomerate unit, base of the Uitenhage Series, Cape Folded Belt (Lower Cretaceous; see Haughton, 1963, p. 273; 1969, p. 421).

Discussion: It is apparent from the following H/L ratios alone, that specimens 302a-2 and 300a-2 (= BM I-6131) cannot belong to *C. (Eu.) anomala* (Jones). Because of poor preservation, H/L is the only parameter determinable in Jones' specimens exclusive of 302a-1 (= BM I-6130). The configuration of I-6131 is entirely different from I-6130; the former has greater height, a shorter dorsal margin, and apparently greater antero-dorsal and postero-dorsal angles (based on Jones' ink drawing). H/L: BM I-6130-.521 (same as Jones; 302a-1, Tasch designated holotype); 302a-1-.533 (measured on Jones [1901] Fig. 1, holotype; it was the first figured but not so identified by Jones); 302a-2-.787 (orbicular form) and 300a-2-.705 (the last two are both included in I-6131).

Flattening and consequent distortion could not produce the resultant configuration seen on the holotype (Jones, 1901, Fig. 1). Jones did seem to recognize that the specimens he assigned to "anomala" were "quasi specific" (i.e., Jones was uncertain if the valves "misshapen by pressure" or "crushed quite flat" as well as being badly eroded, and having variable configurations all belonged to the same species. His doubt was well founded, as re-study of his specimens indicated to this investigator. Kobayashi (1954, p. 105), by inspection of the published drawing of Jones, reached the same conclusion. Accordingly, the holotype BM I-6130 is the sole representative known of the species.

6. *Cyzicus (Lioestheria) draperi* (Jones and Woodward) (Pl. 1, Figs. 1, 2)

Estheria draperi Jones and Woodward 1894, Pl. X, Fig. 1a-c.
Palaeestheria draperi (Jones and Woodward), Barnard, 1931, p. 255.
Orthothemos draperi (Jones) (sic), Raymond, 1946, p. 237, Pl. 2, Fig. 5.
Estherites draperi (Jones) (sic), Kobayashi, 1954, p. 103.
Estheria stowiana (Jones) (sic), Kobayashi, 1954, p. 103.
Cyzicus (Euestheria) draperi (Jones and Woodward). Morris, 1980, Pl. 1, Fig. 7 [Cited in Tasch, 1984, p. 74, an erroneous entry.]

Diagnosis (emended): Suboblong valve with straight, dorsal margin, much broader posteriorly; anterior and posterior margins, both convex, anterior less so than posterior; ventral margin, gently concave upward; anterior margin join with dorsal margin is steeper than the posterior join. Umbo submedial and, on the largest specimen "of the syntypes", crushed and eroded; no growth bands are in evidence. However, on one of the syntypes that has a much smaller valve and the umbo partially buried under the largest specimen, the umbo bears indication of having growth bands, the marginal ones of which retain punctate ornamentation.

Measurements: (mm) (I-3233 only; the large left valve.) L-17.0; H-10.5; Ch-10.5; Cr-6.2; Av-2.5; Arr-4.5; a-3.5; b-4.5; c-8.5; H/L-.618; Cr/L-.365; Av/L-.147; Arr/L-.265; Ch/L-.618; a/H-.333; b/H-.429; c/L-.500; antero-dorsal angle-32° ±4°; postero-dorsal angle-15° ±3°.

Material: Syntypes; British Museum, I-3233. One large left valve overlies the right valve of a smaller specimen. On the same slab a third, still smaller, incomplete right valve occurs. It may or may not belong to this species (formerly *Estheria stowiana* Jones and

Woodward, 1894). Their rendition of this last-named specimen is misleading, because the upper one-third is missing.

Locality: Shale overlying the Cave Sandstone at Platberg near Harrismith (Republic of South Africa), on the eastern side of the Drakensberg, 2438m (8000 ft) above sea level. Grey (1871) reported finding what became the syntype slab restricted to a single horizon.

Discussion: The reconstruction of Jones and Woodward (1894 Pl. 9, Fig. 1a) is a subjective interpretation of the morphology of this species. The photograph provided by Morris (1980, Pl. 1, Fig. 7) and new photographs herein (Pl. 1, Figs. 1 and 2) give better information.

The punctate ornamentation is a characteristic of several lioestheriid cyziciids. Stockley (1947) reported this species at several Cave Sandstone localities, but the conchostracans found at two of these by the writer (Mofoka's Store and Barkly Pass) belong to other species.

Leaia (Hemicycloleaia) sp.

Leaia sp., Barnard 1931, p. 256.

Diagnosis (emended): Bicarinate ovate valves with straight to slightly curved dorsal margin.

Measurements: (mm) (Limited to Barnard's (1931) report because specimens are lost or misplaced.) H-7.5; L-4.0; H/L-0.53; alpha-90°; beta-<45°.

Material: N-13. Geological Survey of South Africa Collection. No other details given.

Locality: 3.2 km south of the Mooi River Station, Natal, Republic of South Africa, Lower Beaufort Beds.

Discussion: Based on Barnard's (1931) limited description, Bond (1955, p. 95) inferred that this species was close to his *Leaia sessami.* Without a photograph or even a sketch of the specimen, such inferred relationship cannot be made with confidence and must remain speculative. All of Bond's five specimens of *L. sessami* had alpha angles less than 90° and beta angles of 30°-39°.

It is not clear from Barnard's description whether the leaiids he reported were all on the same bedding plane or stratigraphically separated. This point arises because he found four specimens with a straight and one with a curved dorsal margin.

Leaia (Hemicycloleaia) cradockensis n. sp. (Pl. 1, Figs. 6, 7; Pl. 2, Fig. 6)

Diagnosis: Ovate, bicarinate valve with gently arched margin, and umbo inset from anterior margin; median sector of valve is largest, the anterior is smallest; anterior rib is slightly concave dorsad, curved more sharply where it crosses the umbo, thickens posteriorly, then tapers toward the last few growth bands and terminates on second growth band above the ventral margin; posterior rib is very slightly curved and anteriorly beaded where it crosses the umbo; it thickens and then tapers before it terminates slightly beyond the third growth band above the ventral margin.

Measurements: (mm) Holotype (I-2436a). L-2.8; H-2.0: Ch-2.0; Cr-1.2; Av-0.5; Arr-0.2; a-0.9; b-0.9; c-1.5; H/L-.714; Cr/L-.429; Av/L-.179; Arr-.171; Ch/L-.714; a/H-.450; b/H-.450; c/L-.536; d_1-0.2; d_2-1.7 ±0.1; alpha-102°; beta-52° ±1.0°; sigma-53° ±1°; eta-4° ±1.0°. Paratype I-2436a is too poorly preserved for most measurements; beta-51° ±3°; sigma-55° ±4°; d_1-0.3.

Material: N-2, holotype, British Museum I-2436a, left valve on black shale slab; occurs with the holotype and multiple specimens of *"Estheria greyii"* Jones. Two cards accompanying this slab read: (1) *Estheriella greyii* "purchased from T. R. Jones. 27 June 1891" and on the reverse side, colld by Dr. George Grey." Card 2 reads: *"Estheria greyii"* and later was modified to *"Estheria? Cornia greyii."* Reverse side of Card 2 similar to reverse of Card 1. Paratype, I-3436b, incomplete left valve.

Locality: Upper part of Karoo Series (hard, dark-grey shale) near Cradock, Cape Province, Republic of South Africa. Middle Beaufort, Upper Permian.

Discussion: Jones overlooked the leaiid described above because it was surrounded by numerous unribbed forms of *"Estheria greyii."*

Although reported to occur in the upper part of the Karoo, which is designated as Triassic, elsewhere in the world no leaiids persisted beyond the Permian. Further, *"cradockensis"* valve characters and general configuration in general terms, fit those of other Permian hemicycloleaiids. Accordingly, an Upper Permian age is indicated.

Several investigators have described leaiids and/or hemicycloleaiids (now a subgenus of *Leaia*), with a slightly curved dorsal margin, flexed downward; for example, Mitchell's (1925) *Leaia pincombei* M. from the Permian (Tartarian) of the Newcastle Coal Measures (see Mitchell's types below). Tasch (*in* Tasch and Jones, 1979c) described a hemicycloleaiid from the Upper Permian Bowen basin of Australia that had a dorsal margin gently flexed downward. Both of the above species have smaller alpha and beta angles than the new species.

Zaspelova's (1966) hemicycloleaiids of subcircular configuration and corresponding curved dorsal margin from the Upper Permian of Kazakhstan have larger or smaller alpha angles than the new species, whereas beta angles are all smaller.

--. *Palaestheria* sp. (sic) Rennie 1934. [= *Cyzicus (Lioestheria?)* sp.] Albany Museum No. 4165, Grahamstown, Republic of Africa; in black carbonaceous mudstone, left valve. H/L-.733. Suboblong valve, subterminal umbo, slightly curved dorsal margin. Witteberg Series (Lower Devonian), Cape System, quarries near railroad station, Port Alfred, Republic of South Africa. (See Pl. 3, Fig. 5 below.)

Rio Muni (formerly Spanish Guinea or Equatorial Guinea)

27. *Estheria* sp. Haughton, 1963. (=? *Cyzicus* sp., Cocobeach Series, Upper Jurassic or Lower Cretaceous.)

Tanzania

Estheria tendagurensis Janensch 1933, has two distinct configurations: Fig. 1, an elongate form, and Fig. 2, a subcircular form. Kobayashi (1954) named the subcircular form *Cyclestherioides janenschi,* and restricted *Euestheria tendagurensis* Jan. to the more elongate form. Novojilov reassigned the latter to *Cyzicus (Lioestheria).* Novojilov (1958c, p. 35) rejected *Cycloesthèrioides* and preferred the modern genus *Cyclestheria* (cf. Novojilov and Kapel'ka, 1960, p. 179–180).

Kobayashi (1954) and Raymond (1946) both chose a more conservative course than Novojilov (1958c) in assigning subcircular or subspherical *Cyclestheria*-like fossil shells to a new genus *Cyclestherioides* Ray. 1946. On the basis of further evidence, this writer concurs. First, Novojilov (1958c) assigned as a new species of *Cyclestheria,* fossil valves of (*Cyclestheria krivickii* Nov.) that did not fit the type of valve of living *Cyclestheria,* because it has an inset umbo (Novojilov, 1958c, p. 35, Pl. 3, Fig. 35) as well as a subovate configuration. Another species originally described as *Cyclestherioides kazachorum* Nov. 1953 was placed by Novojilov under *Cyclestheria* in 1961 (p. 38). This last has a subcentral umbo rising above the dorsal margin. Placement of the umbo becomes important when it is realized that *Cyclestheria* has a terminal umbo that *trends over* the antero-dorsal sector of the valve. Thus, the fit between fossil valves and those of living *Cyclestheria* has not been demonstrated. [Cf. Tasch, 1969; R-150, Fig. 49.1, for the type of the genus *Cyclestheria hislopi* (Baird), 1859.]

Although Janensch attributed the subcircular form to distortion, that is not visible on his photograph (Janensch, 1933, Fig. 2), this writer agrees with Kobayashi that the subcircular forms are best assigned to a separate species, *Cyclestherioides (Cyclestherioides*) janenschi* Kob., 1954. Nevertheless, it may eventually turn out to be a dimorph. (*= A new subgenus, see Antarctica taxonomy under *"Cyclestherioides,"* this Memoir.)

Janensch (1925, p. 136) noted *"Estheriella bornhardti"* as being represented by length groupings: approximately 3.0 mm (specimens 1–10) and 6–7.5 mm (specimens 11–14). In his sketches there is some uncertainty created by what appears to be a ventrad cancellate apron (his Fig. 3) and clear, fine, nodose ribs. Kobayashi (1954) thought that this condition should prohibit assignment to an estheriellid. It is quite possible that the art work is misleading, because where the ribs cross the growth lines small nodes appear in many estheriellids, and the illustrations may over-emphasize the condition.

--. *Estheriella bornhardti* Janensch 1925. [= *Estheriella (Lioestheriata) bornhardti* Jan.] Hatambulo Beds, fourth member above base, in grey and black calcareous shales. (Beaufort Series, Upper Triassic.) Janensch, 1925, Fig. 3, is here designated the holotype. Paratypes include Figures 1 and 2, Janensch, 1925. H/L (1)-.812; (9)-.704; (14)-.526. (Length increases from higher to lower values of the ratios; smaller forms are juveniles. The numbers above represent Janensch's specimens 1, 9, and 14.)

36. *Estheria tendagurensis* Janensch 1933. [= *Cyzicus (Lio-*

estheria) tendagurensis Jan.] Dinosaur beds (Tendaguru Series, Jurassic) northwest of Lindi. The elongate-ovate form (Janensch, 1933, Fig. 1) is the holotype. It has a terminal umbo and H/L-.614.

Zaire (formerly Belgian Congo) (Fig. 20 and 22.)

All Zaire species listed below are on deposit in the *Musée Royal de l'Afrique Central* at Tervuren, Belgium (R. G. nos.). Descriptions and/or synonomies are to be found in Defretin-Le Franc, 1967.

Letters bf, etc., represent conchostracan-collecting stations on the Zaire map. See Fig. 20.

bf *Estheriella lualabensis* Leriche 1913. [= *Estheriella (Estheriella) lualabensis* Leriche.] Holotype Belgium Museum R. G. 2990, left valve. Lualaluba beds at various collection sites along the railroad from Stanleyville, to Ponthierville, in limestones, bituminous mudstones, and shales; also near Lualaba at Bamanga and Kindu, Zaire (Leriche, 1913; see also Defretin Le Franc, 1958). Upper Triassic–Jurassic.

x *Estheriella* Passau 1919. [= *Cyzicus (Lioestheria) kitariensis* Defretin 1967.] Formerly *Bairdestheria kitariensis* Defretin-Le Franc 1967; *Estheria (Euestheria) lerichei* (Marlière) 1950a. Holotype, Belgium Museum R. G. 243, valve no. 1, left valve. Abundant in mudstones of the Lubilash System Kwango Series, beds of the Canyon d l'Inzia, Zaire. Upper Cretaceous.

ab *Estheria (Bairdestheria kasaiensis)* Marlière 1950a [= *Cyzicus (Lioestheria) kasaiensis* (Marlière)], formerly *Bairdestheria kasaiensis* Marlière (Defretin-Le Franc, 1967). (Holotype Belgium Museum R. G. 3109, left valve "a.") Mudstones. Pushaluenda (Kasai) drain. Loia Series, horizon 15, Zaire Wealden.

bd *Estheria (Euestheria) passaui* Marlière 1950a. [= *Cyzicus (Euestheria) passaui* (Marlière.] Formerly *Palaeestheria passaui* (Marlière), Defretin-Le Franc, 1967. Holotype, Belgium Museum R. G. 4475, Fragment 1, valve "a." In arenaceous, fissile shale at 25 km along the Stanleyville to Ponthierville railroad. Stanleyville Series, horizon 2-4?. Upper Jurassic (Kimmeridgian).

y;z *Pseudestheria* lepersonnei* Defretin-Le Franc 1967. [= *Cyzicus (Euestheria?) lepersonnei* Defretin, Le Franc 1967).] Holotype, Belgium Museum R. G. 4773, right valve no. 12. Falaises Schwetz, right bank of Kwenge River; also Bumba between Kitwambari and Kikwit. Kwango Series; beds of l'Inzia, km 5 from Kitwambari. Upper Cretaceous. [*The genus is so broad as defined by Raymond (1946), that it embraces forms which are difficult to place in any other known taxa. Accordingly, it is not useful in any meaningful taxonomy.]

ag;am;ad *Pseudestheria malangensis* (Marlière) emend. Defretin 1967. [= *Cyzicus (Lioestheria?) malangensis* (Marlière).] Formerly *Estheria (Euestheria) malangensis* Marlière 1950, based on Angola material. Holotype on deposit at Instituto Tecnico Superior, Lisboa, Portugal (Mouta Collection from Angola). Cotype (=

syntype) and topotype on deposit in the Belgium Museum: R. G. 12074 and R. G. 4765, respectively. Haute Lueki Series, Cassanje Stage, Cassanje III (cf. Angola entry). Along various rivers in Zaire; rivers Lueki, Lomani, Ole, Sungusungu, Kelekele. Specimens from these sites on deposit at the Belgium Museum. Upper Triassic or Jurassic (pre-Oxfordian).

u;w *Euestheria sambaensis* Defretin-Le Franc 1967 (= *Cyzicus (Euestheria) sambaensis* Defretin-Le Franc 1967). Holotype, Belgium Museum R. G. 12.463. Samba boring, depth 333.09–336.70 m; Leopardville; environs of Boma, sequence of sublittoral sands; Boma-Lukule route at km 14.4 (Kisamu Nzambe) Zaire. Bokungu Series, Lower Cretaceous (post-Wealden).

ab;bg;aw *Bairdestheria kasaiensis* Marlière. (Defretin-Le Franc, 1967). See previous entry for this species. Besides Pushaluenda (Kasai), it also occurs in Samba boring samples, depth 660.99–840.24m, at the source of Rivers Eyano and Ohombi, and at Bandu upstream from Elizabeth, Zaire.

x *Bairdestheria kitariensis* Defretin-Le Franc 1967. See previous entry for *Estheriella* Passau 1919, and this species *"kitariensis."* Also found at Bumba, right bank of Kwenga River, and between Kitwambari and Kitwit.

bn, bj *Bairdestheria caheni* Defretin-Le Franc 1967. = *Cyzicus (Lioestheria)* caheni (Defretin). Holotype, Belgium Museum R. G. 12.102, fragment IV, left valve no. 3. Environs of Songa on the Maosaosa River. Stanleyville Series, horizon 2, Kimmeridgian. Other specimens from region of Stanleyville; Cico-Songa borings, borings C.F.L., and those of Samba.

bn *Bairdestheria*-Estheriella lualabensis* (Leriche) Defretin-Le Franc, 1967. [= *Estheriella (Lioestheriata) lualabensis* (Leriche.)] Holotype, Belgium Museum R. G. 2990, left valve. (See *"Estheriella"* under Angola entry, this section, for new subgenus.) Along railroad from Stanleyville to Ponthierville, km 31.9; Stanleyville Series, Central Basin, Kimmeridgian. (For other localities, see Defretin, 1967, p. 59-62.)

bn *Bairdestheria-Estheriella evrardi* Defretin-Le Franc 1967. [= Estheriella (Lioestheriata) evrardi (Defretin)]. Always found associated with *"lualabensis"*; hence of the same Series (Stanleyville) and age (Kimmeridgian). The two species holotypes came from different nearby sites; "evrardi" from Cimestan Boring IV. Holotype, 11.368, fragment III, right valve No. 4.

am *Bairdestheria-Estheriella moutai* Leriche. Defretin-Le Franc, 1967. [= *Estheriella (Estheriella)* moutai* Leriche]. Holotype at Instituto Tecnico Superior de Lisboa, Portugal, Belgium Museum R. G. 4.762, environs of Quela, Angola, at 85 km to east-northeast of Malange. Note that this subgenus does not fit Defretin-Le Franc's description of *"Bairdestheria-Estheriella"*. There is no fringe margin on the dominant estheriellid radials of the valve. Rather, the radials are only visible close to the ventral border [Defretin-Le Franc, 1967, Pl. 7, Fig. 5]. Accordingly, they are assigned to subgenus *Estheriella*. See Angola entry, this section.) Zaire occurrences; along Rivers Lueki Sungusungu, Lomani, Luhuhu, Ole, Wudji (near Combe), and Kelekele (near Shembe). Between Upper Triassic and Jurassic (pre-Oxfordian).

bn;bj;bm;aw *Polygrapta biaroensis* Defretin-Le Franc 1967. [= *Cyzicus (Lioestheria) biaroensis* (Defretin).] Holotype, Belgium Museum R. G. 3096, fragment 1, right valve, no. 1. Songa near Wanie Rakula, Biaro Quarry, Stanleyville Series, horizon 2, Kimmeridgian. (Other localities in Stanleyville Region, and borings Cic-Songa, C.F.L., and Samba.) (See Fig. 22.)

bj;bh *Estheria corneti* Marlière (*nomen nudem,* Defretin-Le Franc, 1967, p. 71). (= *Glyptoasmussia corneti* (Marlière), Defretin-Le Franc, 1967. = Holotype, Belgium Museum R. G. 11.955, right valve no. 2. Cico-Songa boring 0, core slice 126/131. Stanleyville Series, horizon 3^1, Kimmeridgian.) (Other localities, along various rivers [Central Basin]; Mekombi, Loango, Assengwe, and Maosaosa [and see Defretin-Le Franc, 1967, p. 74–75, for listings of samples from borings] Cimenstan, Cico-Songa, Samba, also Lomani and Bilima.)

af;ad *Glyptoasmussia luekiensis* Defretin-Le Franc 1967. Type, Belgium Museum R. G. 12.118, fragment II, left valve no. 2. Haute Lueki Series, Zaire (equivalent to beds of Cassange III, Angola). Luhuhu River, upstream from confluence with the Lomani River, near Wanga. (Other sites along various rivers [Haute Lueki Region], Lueki, Ole, Ivunga, Sungasunga, Wudji, Kelekele, and head of source near Limanga.) Upper Triassic or Jurassic (pre-Oxfordian).

aw;at *Pseudoasmussia ndekeensis* Defretin-Le Franc 1967 [= *Pseudoasmussiata ndekeensis* (Defretin)]. (See "South America" below for details on the designation of a new name, *Pseudoasmussiata,* for the preoccupied name of *Pseudoasmussia* Novojilov, 1954c). (Holotype, Belgium Museum R. G. 12.480, fragment 1, valve 1, and its mold, fragment 2, valve 1. Samba boring, series 4, depth 652.35–654.01 m. Loia Series, Upper Wealden. Other locality at Ndeke on River Tshuapa.)

ba;ay;aw *Pseudoasmussia banduensis* Defretin-Le Franc 1967 [= *Pseudoasmussiata banduensis* (Defretin)]. Holotype, Belgium Museum R. G. 12.134, fragment II, left valve, and print of right valve (fragment 1). Bandu, site 21, upstream from Elizabeth; Loia Series, old horizon 15, Wealden. (Other sites; boring, Samba, series 4, depth 652.35–654.01 m., core sample 48/31. Central Basin; Yohila, River Makanga, and source of River Ryano at Ohombi.)

b1;bm;aw *Pseudoasmussia duboisi* (Marlière) 1953, Defretin-Le Franc, 1967. [= *Pseudoasmussiata duboisi* (Marlière).] Holotype, Belgium Museum R. G. 2173, fragment II, left valve "f". Confluence of the Londo and Lomani. Stanleyville Series, horizon 6, Kimmeridgian. (Other sites; borings, Cimenstan, C.F.L., Samba. Also along rivers Assengwe and Lockwaye near Yangilimo.)

aw *Asmussia dekeseensis* Defretin-Le Franc 1967. Holotype, Belgium Museum R. G. 12.473, fragment 1, left valve no. 1, and its print; fragment II, valve 1. From a boring at Samba, series 3, depth 508.34–513.68 m. Bokunga Series, Lower Cretaceous (post-Wealden). (Other samples and sites; Samba, depth 418.79–632.21 m; Dekese, depth 326.04–380.04 m.)

aw *Asmussia ubangiensis* Defretin-Le Franc 1967. Holotype, Belgium Museum R. G. 12.461, fragment III, left valve and its impression, fragment IV. Boring; Samba, series 2, depth 194.69–195.84 m. Bokunga Series, Lower Cretaceous (post-Wealden). (Other localities and samples; Ubangi, source of Barabara River at Katera; Samba boring, 222.83–223.36 m.)

am;al;ae *Echinestheria marimbensis* Marlière. Type, sample no. 35, Instituto Technico Superior de Lisboa (Angola type). Cotype (Angola) at the Belgium Museum, R. G. 4.765, Marimba. In addition, other Zaire specimens are Belgium Museum R. G. 12.120, 12.122, and 12.123. Haute Lueki Series (Zaire) is equivalent to Cassanje III, Upper Triassic or Jurassic (pre-Oxfordian). (Haute Lueki Region; River Sungusungu, Wudji [near Combe], River Kelekele [near Shembe].)

am;ag *Palaeolimnadiopsis lubefuensis* Defretin-Le Franc 1967. Type, Belgium Museum R. G. 12.074, fragment VIII, right valve no. 2. Haute Lueki region, at the crossing of River Lueki by the route of Kibombo to Lubefu. Equivalent to beds of Cassanje III, Angola. Haute Lueki Series. Upper Triassic or Jurassic (pre-Oxfordian). Haute-Lueki Region; River Lomani, Sungusungu, source of River Wudji (near Combe).

w *Palaeolimnadiopsis lombardi* Defretin-Le Franc 1967. Holotype, Belgium Museum 12.225, left valve, Ndolo bore. Leopardville. Stanleyville Series, horizons 2 and 4, Kimmeridgian. (Other specimens are R. G. 12.223–12.227.)

b1 *Palaeolimnadiopsis* sp. Holotype, Belgium Museum R. G. 11.348, fragment VI, right valve. Cimenstan, bore 1, No. 51, core sample 219. Stanleyville Series, horizon 5a, Upper Jurassic.

To set the record straight after reassignments and corrections, the Zaire rock column yields species of the following genera: *Cyzicus, Estheriella, Glyptoasmussia, Pseudoasmussiata* n.g., *Asmussia, Echinestheria,* and *Palaeolimnadiopsis.*

Zambia (formerly Northern Rhodesia)

46. *Pseudestheria welleri* Bond 1964. [= *Cyzicus (Lioestheria) welleri* (Bond) herein.] Holotype and topotypes unnumbered, National Museum Zimbabwe. Holotype figured in ink sketch; H/L-.750. Bond (1964, p. 3) reported "no reticulate structure" and "irregular punctation." Luano Valley, lat 14°53'S, long 29°26'E, Zambia (Pl. 5, Fig. 6 herein). In bedded, grey, micaceous siltstone above the Escarpment Grit (near base of Upper Karoo). Upper Permian or Lower Triassic.

--. *Isaura* cf. *mangliensis* (Jones), Bond 1955 (= *Cyzicus* sp.). Chongola Coal area, Guembe District, Upper Matabola Beds. Northern Rhodesia (Zambia). Cited in Bond, 1955, p. 96. Upper Permian.

Zimbabwe

29, 30, 31. *Leaia (Hemicycloleaia) sengwensis (Bond)* (Pl. 2, Fig. 7)

Hemicycloleaia cf. *haynesi,* Bond 1952, p. 209–216.
Hemicycloleaia sengwensis Bond 1955, p. 93–94, Pl. 13, Fig. 3.

Diagnosis (emended): Bicarinate, robustly ovate valves, with straight dorsal margin. Holotype with ribs covered and/or obscure ventrally; paratype indicates that ribs terminated 3 to 4 growth bands above the ventral margin; both ribs reach, but do not completely cross umbo; anterior rib gently curved dorsad. Median sector of valve largest.

Measurements: (mm) (Holotype.) L-6.0; H-4.0; Ch-2.5; Cr-2.5; Av-1.0; Arr-0.5; a-1.7; b-1.8; c-2.5; H/L-.667; Cr/L-.417; Av/L-.167; Arr/L-.083; Ch/L-.750; a/H-.425; b/H-.450; c/L-.416; alpha-99° ±2°; beta-26° ±2°; sigma-64° ±3°; d_1-0.8 ±1; d_2-4.9 ±2. (Paratype, incomplete specimen; d_1-0.6 ±1; alpha-100°; beta-27°; sigma-62° ±3°. Range of Bond's measurements for four specimens; alpha-100°–105°; beta-40°–48°; H/L [for three specimens] -.554–.675.)

Material: Holotype, external mold of left valve, National Museum Rhodesia (NMR) 7148, paratype, internal mold of left valve. Two small, incomplete, nonribbed conchostracans overlay the holotype, one in the antero-ventral sector; the other overlies the posterior rib at its extremity.

This writer had the opportunity to study rubber replicas of the types provided by M. P. Cooper, Curator (NMR), which allowed a new series of measurements as well as new photographs.

Locality: According to Cooper, the holotype is from a cliff section through the Upper Madumabisa Mudstone (bed K^{5c}) to the east of the Sengwe River and to the north of Neyunka's Kraal. The stratigraphic level is high *Cistecephalus* Zone, very late Permian; lat 17°25′S, long 28°16′E. Cooper noted that although the paratype is from the same locality, it is not specifically located. (Fig. 24.)

Discussion: When Bond (1955) described his leaiid species, he noted the Gondwanaland distribution of this genus did not include Antarctica or India. A decade later, this writer described a new species, *Leaia (Hemicycloleaia) gondwanella* Tasch in the Doumani Collection (Doumani and Tasch 1965) from Antarctica (Ohio Range; for Tasch collection, same site, see Tasch, 1970a, and below under "Antarctica"). Search of the Permian beds of India yielded no leaiids, nor have any *Leaia* species been reported (see Tasch and others, 1975). Significance of the apparent absence of leaiids from the Permian of India will be considered further in a subsequent section of this study.

34. *Leaia (Hemicycloleaia) sessami* (Bond)
(Pl. 2, Fig. 5)

Hemicycloleaia cf. *normalis,* Bond 1952, p. 215.
Hemicycloleaia sessami Bond 1955, p. 94, 95, Pl. XIII, Figs. 1, 2.

Diagnosis (emended): Subovate, bicarinate valves with straight dorsal margin. Medial sector of valve, largest, defined by ribs, anterior sector, smallest. Both ribs somewhat eroded in umbonal area, apparently crossed the umbo and reached the ventral margin; the posterior rib is not visible on the holotype but both ribs

are visible on the paratype. Anterior rib, as noted by Bond, is slightly curved; posterior rib is straight.

Measurements: (mm) (Measured from Bond's published photographs of the types and later checked on the rubber replicas provided by M. P. Cooper of the National Museum of Rhodesia (Zimbabwe). Holotype followed by paratype; a dash denotes that the item could not be measured due to erosion or missing portion.) L-7.0 (incomplete), 8.33; H-5.9, 5.0; Ch- --, 4.7; c-2.6, 2.9; Av-1.5, 1.7; Arr- --, 1.7; a-1.8, 2.7; b- --, 2.1; c-4.3, 3.6; H/L- --, .602; Cr/L- --, .349: Av/L- --, .204; Arr/L- --, .204; Ch/L- --, .566; a/H-.305, 325; b/H- --, .253; c/L- --, .434; d_1-1.07, 1.06; d_2-6.15, 6.06; alpha-86° ±3°, 86° ±1°; beta-32° ±3°, 33° ±0°; sigma-53° ±1°, 54.5° ±1°. (The angles measured on five specimens by Bond were alpha-80°–87° and beta-30°–39°.)

Material: Geological Collection of the National Museum of Rhodesia (Zimbabwe): Holotype 7196 (A), right valve, posterior quarter missing; paratype 7196 (B), right valve, ventral piece of valve eroded.

Locality: Cliff 180 m downstream from drift across Sessami River on road from Gekwe to Binga's, near Madziwadzide's Kraal. Shales and calcareous concretions in shales. Middle Madumabisa Shales, bed K^{5e}, Lower Beaufort, Upper Permian.

Discussion: *Leaia (H.) sessami* is distinct from *Leaia (H.) sengwensis* in all parameters such as smaller alpha and sigma angles, larger beta angle, and smaller H/L, Cr/L, Ch/L, a/H, and b/H.

45. *Cyzicus* sp. Haughton 1927 (in McGregor, 1927). Mafungabusi Plateau of southern Zimbabwe. Forest Sandstone, 12 m below base of Karoo Basalts. Haughton Appendix to McGregor, 1927 (see also Haughton, 1963, 1969). *Geol. Surv. S. Rhod. Short Report No. 20.* Cited in Bond, 1964, p. 15.

29, 30, 31. *Isaura* cf. *mangaliensis* Jones, 1862, Bond, 1955 (= *Cyzicus* sp.) National Museum of Rhodesia (Zimbabwe), No. 6951, 7147, 7148, and 7159. Upper Madumabisa Shales, Zimbabwe, Upper Permian. Bond, 1955, p. 95–96.

--. *"Estheria"* Lightfoot 1914, 1929. (= ?*Cyzicus* sp.) Wankie District. Gorges of Chibondo Range. Bedded, arenaceous limestones. Lower and Middle Madumabisa Shales. Upper Permian. Lightfoot 1929, p. 38, and cited in Bond, 1955, p. 79.

ANTARCTICA (Middle to Upper Devonian)

Lashly Mountains

Cyzicus (Euestheria) lashlyensis n. sp.
(Pl. 6, Figs. 2, 3)

Diagnosis: Subovate valve with a short, dorsal margin and subcentral umbo; anterior margin, which is covered by a portion of the left valve, forms a convex bulge in contrast to the posterior margin, which is uniformly rounded. More than 12 growth bands; five or six bands are more tightly spaced than preceding. Ornamentation consists of contiguous granules only microscopically visible on growth bands.

Measurements: (mm) (Specimen GS 7399/8.) L-3.2; H-2.7; Cr-1.0; a- --; b-1.6; c-1.5; H/L-.844; Cr/L-.312; Av/L-.094; b/H-.593; c/L-.469. All of the other fragments of the new species, other than GS 7399/8, are so incomplete that they cannot be usefully measured.

Material: Syntype, New Zealand Geological Survey, GS 7399/8. The syntype consists of the anterior section of the left valve overlying the interior face of the right valve, and incomplete valve fragments, black with high gloss; GS 7399/4, two crushed and eroded, very poorly preserved, incomplete valves; GS 7399/11, three fragments with growth bands.

Locality: McMurdo Sound/6. Transantarctic Expedition localities 388 and 432, Beacon Sandstone, laminated, dark-grey siltstone. Lat 77°54'S; long 159°35'E, Lashly Mountains, Antarctica. Middle to Upper Devonian.

Discussion: On the same slab as GS 7399/8 there is an arctolepid plate. The other slabs also have fish fragments. The dolerite intrusion into the Beacon Sandstone during Jurassic–Cretaceous time was the likely source of igneous heat that carbonized these conchostracan valves (Tasch, 1982b). For the paleoecology and food web during the Devonian, see Tasch, 1977b. The only other Antarctic Devonian conchostracan comes from the Aztec Siltstone and is described directly below. It differs from *lashlyensis* in various parameters and morphological features (see below).

Portal Mountain

2. *Cyzicus (Euestheria) ritchiei* n. sp.
(Pl. 6, Figs. 4, 5)

Diagnosis: Robust, ovate valves with a straight dorsal margin and subterminal umbo; posterior and anterior margins, rounded; multiple growth bands with closer spacing ventrad. Ornamentation consists of minute contiguous granules on growth bands.

Measurements: (mm) (Figures for holotype followed by those for paratype for each parameter and ratio.) L-3.8, 3.6; H-2.9, 3.2; Ch-2.0, 2.0; Cr-1.4, 1.2; Av-1.2, --; Arr-0.5, 0.7?; a-1.5, 1.5; b-1.4, 1.5; c-1.8, 1.5; H/L-.763, .888; Cr/L-.368, .333?; Av/L-.316, --; Arr/L-.132, .194; Ch/L-.526, .555; a/H-.448, .469; b/H-.483, .468; c/L-.474, .417.

Material: Australian Museum, Sydney, Australia: Slab 6, F 54956f, part of left valve underlain by internal mold, holotype, F 54956c, paratype, also slabs 1-5; F 54955a, F 54055b, F 54957, F 54958, and F 54959 (a coquina of valves). All of these have valves occurring one one or more bedding planes. Numerous valves are incomplete, eroded, and/or partially covered. Erosion of general white enamel, outer layer sometimes reveals a carbonized underlayer of the valve. This is seen on all the slabs of finely laminated, hard, dark-grey siltstone.

Locality: East face of the Portal Mountain, South Victoria Land, Antarctica, Aztec Siltstone (probably Upper Devonian).

Name: After Alex Ritchie, Australian Museum, Sydney, Australia, who collected these specimens.

Discussion: This new species differs from *Cyzicus (Euestheria) lashlyensis* n. sp. in that it has a broadly ovate configuration, longer dorsal margin, differences in curvature of the anterior margin, and it differs in the following ratios: Cr/L, Av/L, and b/H. There is an overlap in H/L between the two species.

Despite the poor state of preservation, the coquina of valves permits reconstruction of the species characteristics.

It is possible that the Lashly Mountains and Portal Mountain conchostracan-bearing beds are actual, or near, time correlates, as the distance apart of the two sites is only about 24 km.

No other Devonian species known (e.g., Lutkevich, 1929, Novojilov, 1954a, 1955, 1961; Novojilov and Varentsov, 1956, 1958), corresponds to either of the new Antarctic species.

ANTARCTICA (Permian)

Ohio Range (Mercer Ridge)

3. *Leaia gondwanella* Tasch, 1965 [= *Leaia (Hemicycloleaia) gondwanella* Tasch]. Holotype, AD 6114, complete right valve; paratypes, AD 6118-1, right valve, and AD 6130, left valve. Repository for types, Institute of Polar Studies, Ohio State University and United States National Museum. Mt. Glossopteris Formation (*Leaia* Zone), Middle to Upper Permian. Doumani and Tasch, 1965, p. 237–238, Table 1, Plate 14, Fig. 10–11. *Emended diagnosis* is based on new material in the Tasch Collection; angle measurements have been improved and additional measurements used throughout this memoir have been included. Topotypes (USNM) N=15; angles: alpha-75° to 103° (nine specimens with angle of 90°); beta-38° to 60° (eight specimens with angle of 45°); sigma-38° to 48° (ten specimens with angles of 40° to 45°); eta-0°, 0°; delta-0.0° to 3.0° (eight specimens with 0.0°); d_1 (mm)-0.0–1.0; d_2 (mm)-3.2–5.6.

4. *Cyzicus (Lioestheria) doumanii* Tasch 1965. Holotype, AD 6130, complete right valve; paratypes, AD 6115-1 and AD 6131-1. Repository, formation, zone, and age are the same as "3" above. (Doumani and Tasch, 1965, p. 238, Plate 14, Fig. 12, and Tasch, 1971.)

ANTARCTICA (Jurassic)

Brimstone Peak

5. *Cyzicus (Lioestheria) rickeri* n. sp.
(Pl. 6, Fig. 1)

Name: Species is named for J. F. Ricker, who collected the specimens.

Diagnosis: Robust and laterally ovate valves with straight dorsal margin; umbo inset from the anterior margin about one-fourth the length of the valve. Anterior and posterior margins rounded. Ornamentation: hachure markings on growth bands; early growth bands widely spaced, last eleven more closely spaced.

Measurements: (mm) L-3.2; H-2.1; Ch-1.3; Cr-1.3; Av-0.7; Arr-0.9; a-1.0; b-0.9; c-1.2; H/L-.656; Cr/L-.406; Av/L-.219; Arr/L-.281; Ch/L-.406; a/H-.476; b/H-.429; c/L-.375.

Material: Syntype, British Museum of Natural History (BMNH L-3452). Two large slices bearing numerous fragments of this species and one small chip with the specimens described here. Specimens are all carbonized and/or fragmented. Lower Jurassic.

Locality: "MS 116, Mawson-Priestley Glacier, Southern Victoria Land, Brimstone Peak, lat 75°38'S, long 158°33'E. Southwest ridge of highest nunatak in the far western mountains between Mawson and David Glaciers (Jurassic). The fossiliferous interbed is about 151 m above the bottom of the exposed section, and above the eighth basaltic flow. Summit of the Peak is approximately 477 m above the base. The sedimentary deposits at one fossiliferous horizon only occur as lenses of siltstone-fireclay overlying scoriaceous lava and below other basaltic flows." (Data of Ricker; note accompanying B.M. fossils, L-3452.)

Discussion: The closest form to *C. (Lio.) rickeri* n. sp. is *Bairdestheria caheni* Defretin-Le Franc, 1967 [= *Cyzicus (Lioestheria)* from the Middle–Upper Jurassic of Zaire]. The Zaire species configuration varies from subovate with umbo inset about one-fourth the length of the valve, to either compactly ovate or more broadly ovate, both with a submedial umbo (Defretin-Le Franc, 1967, Pl. 5, Figs. 5 and 7). The last two specimens are not from the same site as the holotype and may not be conspecific. The holotype of *"caheni"* has an obtuse join of the dorsal and posterior margin that is lacking in *"rickeri"*. Other differences are; Cr/L is smaller in *"caheni"* (.210–.290 vs. .406 for *"rickeri"*), and c/L is greater (.440–.540 vs. .375 *"rickeri"*). H/L, a/H, and b/H overlap. (Other ratios not given.) The indicated differences suffice to distinguish the two species.

The Carapace Nunatak lioestheriids (cf. this Memoir under that heading) are somewhat vertically ovate and hence commonly have a more abbreviated dorsal margin than the new species. The placement of the umbo also differs, in accord with the contrasting configurations.

Mauger Nunatak

6. *Cyzicus (Euestheria) juravariabalis* n. sp.
(Pl. 7, Figs. 6–8)

Diagnosis: Ovate valves with umbo inset from anterior margin somewhat less than one-third the length of the valve; dorsal margin behind umbo comparatively short; more strongly curved join of dorsal with anterior and posterior margin; ventral margin gently curved. Growth bands, about 15. Ornamentation: contiguous granules on growth bands.

Measurements: (mm) (TC 8027d, followed by TC 8027c.) L-2.1, 3.4; H-1.6, 2.6; Ch-0.8, 1.7?; Cr-0.6, 0.8?; Av-0.6, 0.8?; Arr-0.7, 0.9?; a-0.8, 1.0; b-0.6, --; c-0.8, 1.7; H/L-.762, .765; Cr/L-.286, .235?; Av/L-.286, .235?; Arr/L-.313, .265?; Ch/L-.381, .520?; a/H-.500, .385; b/H-.375, --; c/L-.381, .500.

Antero-dorsal angle approx. 30°, 30°; postero-dorsal angle approx. 20°, --.

Material: Syntype, TC 8027a–d and TC 8039a and b. Fragment of inner layer of left valve underlain by the internal mold (TC 8027a–d); TC 8014 (site 62-10-1B). Other specimens: TC 8003a (62-10-1A) and TC 8311 (ornamentation displayed).

Locality: Mauger Nunatak, (MN) lat 85°44'S, long 176°44'E, David Elliot site, MN-62-10-BC (for TC 8027a-d and TC 8039a); also 62-10-1B (TC 8014) and 62-10-1?A (TC 8003a). Other specimens of this species came from sites 62-10-1B (TC 8023), 62-10-2B (TC 8042), and 62.3 (TC 8054a). For diagram of MN sites, see Fig. 3A. Lower Jurassic.

Discussion: The closest species to *"juravariabalis"* is *Cyzicus (Euestheria) ellioti* n. sp., herein, from Blizzard Heights. The latter differs in having a more variable configuration from subovate to broadly ovate, a smaller H/L (.717–.759 vs. .762–.765 *"juravariabalis"*) and larger b/H (.424–.558 vs. .375 *"juravariabalis"*); other ratios overlap but *"juravariabalis"* has both lower and higher range limits for Cr/L, a/H, and Av/L than *"ellioti."* The overlap may reflect derivation from a common gene pool. (Mauger Nunatak and Blizzard Heights are separated by more than 172 km.)

7. *Cyzicus (Lioestheria) maugerensis* n. sp.
(Pl. 7, Figs. 1–4)

Diagnosis: Irregularly subovate valves with a straight dorsal margin; umbo inset one-fourth the length of the valve; anterior and posterior margins rounded; double growth lines. Ornamentation of hachure markings on growth bands with one pseudo-cancellate specimen due to close spacing of growth bands (Pl. 7, Fig. 3).

Measurements: (mm) (TC 8316a followed by TC 310b, plus appended ratio ranges for 8015a, c, and e.) L-3.3, 2.9; H-2.6, 2.1?; Ch-1.2?, 1.2?; Cr-0.5, 0.7; Av-.788, .724; H/L-.788, .724; Cr/L-.151, .241; Av/L-.151, .241; Arr/L-.485, .310; Ch/L-.364, .414; a/H-.462, .286; b/H-.146, .381; c/L-.424, .414. (Ranges of ratios for TC 8015a, c, and e: H/L-.643?–.741; Cr/L-.185–.396; Av/L-.385–.324; Arr/L-.235–.302; Ch/L-.415–.441; a/H-.425–.588; b/H-.375–.529; c/L-.370–.453.) **Note:** Variation in valve parameters above can be accounted for by flattening, partial embedding in substrate, slight erosion on margins, and occasional slight cover at one or more places along valve margins.

Material: Syntypes, TC 8316a, internal mold of left valve overlain by portion of inner layer of original valve. TC 8313, fragment showing common ornamentation. TC 8310b, portion of left valve overlying internal mold. In addition to the syntypes there are numerous fragmented, incomplete and/or eroded valves: TC 8020c, TC 8010b, TC 8022a, TC 8301b, TC 8000, TC 8039, TC 8350, and TC 8022. TC 8015a, c, and e is the most complete of these.

Locality: Mauger Nunatak (MN), Queen Maud Mountain, Antarctica, coordinates as in *"juravariabalis"* above. David

Elliot's (D.E.) site 62-10-1A, 62-10-B, 62-10-1BC, and 62-10-2B yielded the MN fossils. Lower Jurassic.

Discussion: Antarctic Jurassic lioestheriids differ from the new species *"maugerensis"* as follows. From Blizzard Heights; *Cyzicus (Lio.) brevis* (Shen), in Chang and others, 1976, from the Chinese Mesozoic; it has a steeper join of the dorsal and anterior margins, larger H/L, b/H, and smaller Cr/L, and Av/L. *Cyzicus (Lio.) antarctis* n. sp. is laterally to broadly ovate, with a comparatively short dorsal margin behind the umbo. *Cyzicus (Lio.) longulus* n. sp. has an elongate elliptical valve, smaller H/L, Arr/L, Ch/L, and c/L. *Cyzicus (Lio.) malangensis* (Marlière) has more robust ovate valves, an absence of double growth lines, and a lesser inset of the umbo from the anterior margin. From Storm Peak: *Cyzicus (Lio.) disgregaris* n. sp., dimorph 1 with its laterally to vertically ovate configuration, generally smaller H/L, greater b/H, and occurrence of irregular hachure markings on growth bands not found in *"maugerensis."*

Cyzicus (Lio.) mawsoni (Jones, 1897b), from the Cretaceous of Brazil, displays the cancellate pattern of ornamentation seen on specimen TC 8301 of the new species. A Lower Cretaceous species of the Chinese rock column, *Orthoestheria youghangensis* Chen (in Chang and others, 1976, Pl. 71, Figs. 1 and 3, and Pl. 72, Figs. 6–8), shares some valve characteristics with the new species (which is Jurassic); subovate to irregularly subovate configuration, double growth lines, ornamentation, and hachure markings on growth bands. The Chinese species has a smaller H/L and a subterminal umbo.

Palaeolimnadia (Palaeolimnadia) sp.

Diagnosis: Incomplete valve; very large smooth umbo with dorsal portion missing; antero-ventral sector with three growth bands. No useful measurements possible due to incompleteness of the valve.
Material: One fragment, mold of left valve, TC 8308.
Locality: Mauger Nunatak, D.E. site 62-10-1A? (Jurassic).

8. *Palaeolimnadia (Grandilimnadia)* sp. 1

Diagnosis: Minute ovate valves with comparatively large smooth umbo that trends towards anterior end; six growth bands.
Measurements: (mm) (Due to erosion, measurements are necessarily incomplete.) L-1.9; H-1.3; H/L-.684; l_u~0.9±, h_u~0.6, h_u/l_u-.667? (u =umbo.)
Material: TC 8055a, internal mold of left valve eroded on posterior and anterior margins. The umbo is split vertically.
Locality: Mauger Nunatak; see species "6" above for coordinates. D.E. collection site 62.4. Lower Jurassic.
Discussion: Despite the incompleteness of this specimen, it is recorded and figured here because palaeolimnadiids are rare in the *Cyzicus*-dominated fauna of the beds at Mauger Nunatak. On the reverse side of this slab (separated by "n" seasons) are indeterminate cyziciid valves.

9. *Cyzicus (Lioestheria) longacardinis* n. sp.
(Pl. 6, Figs. 6, 7; Pl. 7, Fig. 5)

Diagnosis: Elongate, subovate valves, with a comparatively long, straight dorsal margin; umbo inset about one-third the length of the valve; valves taper posteriorly. Double growth lines. (Each growth band has an upper and lower margin, each margin reflected as a line on the valve. For successive growth bands on the valve, the upper growth line of any succeeding band will overlap and cover the lower growth line of the immediately preceding growth band. The last, lower growth line of a valve is also its ventral margin. When double growth lines occur they are separated by a small space and recur during the growth cycle.) Antero-dorsal angle 40°, postero-dorsal angle 15°. Ornamentation: hachure markings consisting of vertical alignment of three or four granules each on growth bands. There are approximately 36 growth bands noted.
Measurements: (mm) (TC 8033a followed by TC 8026a.) L-4.2, 6.1; H-2.9, 3.9; Ch-1.7?, 2.5?; Cr-0.8, 1.7?; Av-0.8, 1.7; Arr-1.7, 2.0; a-1.1, 2.0?; b-0.9, 1.6; c-1.3, 2.5; H/L-.690, .639; Cr/L-.190, .279; Av/L-.190, .279; Arr/L-.405, .328; Ch/L-.405, .410; a/H-.379, .513; b/H-.310, .410; c/L-.310, .410. (TC 8026a corresponds to TC 8033a in antero and postero dorsal angles.)
Material: Syntype TC 8033a, internal mold of left valve eroded along dorsal margin, under remnants of the inner layer of the original valve; TC 8026a, the same as above but eroded behind the umbo and along the dorsal margin; TC 8030b, internal mold underlying inner portion of the right valve; and TC 8056b, an incomplete specimen showing parts of both valves.
Locality: Mauger Nunatak, see *"juravariabalis"* above for coordinates. D.E. site 62-10-BC and site 62.4 (for TC 8056b only). Lower Jurassic.
Discussion: The above species is characterized by its elongate-subovate valve, umbonal position (inset one-third the length of the valve vs. one-fourth, *"maugerensis,"* and comparatively large size. It differs from *Cyzicus (Lio.) maugerensis* n. sp. in the above features as well as in having a smaller H/L, c/L, and longer dorsal margin (Ch. 1.7–2.5 vs. 1.2 *"maugerensis"*).

Blizzard Heights and Environs

10. *Cyzicus (Lioestheria) malangensis* (Marlière) emend.
Defretin-Le Franc, 1967
(Pl. 9, Figs. 1, 2, 3; also Pl. 8, 11)

Estheria (Euestheria) malangensis Marlière 1950a, p. 27, Pl. 2, Figs. 3, 4 text-Fig. la–d, f.
Pseudestheria malangensis (Marlière) Defretin-Le Franc, 1967, p. 41., Pl. 2, Fig. 1–6, Text-Fig. 4.

Diagnosis: Robust subovate valves with straight dorsal margin and umbo slightly inset from anterior margin. Margins: anterior, rounded; posterior, convex; ventral, gently curved. Hachure markings on growth bands. Growth-band number variable;

widely spaced for one-third to two-thirds of the valve and then more closely spaced to the ventral margin.

Measurements: (mm) (TC 11214A followed by TC 11233B and TC 11184c.) L-4.6, 4.5, 5.5; H-3.5, 3.2, 4.5; Ch-2.5, 1.7?, 2.0; Cr-0.9, 1.3, 1.2; Av-0.9, 1.3, 1.2; Arr-1.3, 1.4, 2.3; a-1.8, 1.3, 2.3; b-1.3, 1.6, 2.3; c-2.1, 1.8, 2.6; H/L-.761, .711, .719; Cr/L-.196, .289, .218; Av/L-.196, .289, .218; Arr/L-.283, .311, .418; Ch/L-.543, .378, .364; a/H-.514, .406, .511; b/H-.371, .500, .511; c/L-.435, .400, .473. (TC 11109A [Station 1, bed 6] overlaps *"malangensis"* in all parameters except Arr/L, which is greater, and c/L, which is smaller.)

Material: TC 11214A, fragmentary portion of right valve overlying internal mold of same; TC 11233B, internal mold of left valve. Additional specimens: TC 11171A, badly eroded internal mold of left valve; TC 11184C, internal mold of left valve; TC 11109A, internal mold of right valve. (Two transported, fossiliferous slabs, TC 11328A, mold and overlying fragments of left valve and TC 11328B, internal mold of right valve; and TC 11329A-C, found by David Elliot in a Jamesway hut at Mt. Falla, were located biostratigraphically [i.e., same horizon] by comparison with Tasch's Blizzard Heights collections. On these slabs are specimens that overlap *"malangensis"* in all ratios except c/L for TC 11328A and B in which c/L is greater by less than 0.1.)

Locality: Blizzard Heights, Marshall Mountains, lat 84°37′S, long 163°53′E, Queen Alexandra Range. TC 11214A and TC 11233B (Tasch Station 1, bed 7). TC 11184c (Station 1); TC 11109A (Station 1, bed 2b); TC 11171 (Station 1, bed 6). (Sta. 1, bed 2A [Jamesway hut specimens]: TC 11328, TC 11329, see Fig. 3B and Appendix for measured sections; Fig. 4C and Fig. 2 for coordinates.) Lower Jurassic.

Discussion: The larger variation in parameters and ratios in the B.H. specimens can be attributed to flattening. Lioestheriid hachure markings can be seen on slab TC 11184c and TC 11214 on the last few growth bands. The specimens described above are morphologically close to and overlap the ratio ranges given by Defretin-Le Franc (1967, p. 43) for *Pseudestheria malangensis* (Marlière).

	Zaire*	Antarctica**
H/L	.690–.780	.711–.761
Cr/L	.250–.300	.196–.289
Av/L	?	.196–.289
Arr/L	.210–.320	.283–.418
Ch/L	.390–.530	.364–.543
a/H	.380–.500	.406–.514
b/H	.380–.520	.371–.511
c/L	.390–.530	.400–.473

*Pseudestheria malangensis (Marlière), emended Defretin-Le Franc, 1967 (Zaire).
**Cyzicus (Lioestheria) malangensis (Marlière). Blizzard Heights (Antarctica).

Raymond's *Pseudestheria* species was placed under synonomy of *Cyzicus (Lioestheria)* (Tasch, 1969). Defretin-Le Franc's specimens from Zaire were placed in the time frame of Late Triassic to Jurassic (pre-Oxfordian). The overlap in valve characters and measurements in the Antarctic specimens with those of the Zaire specimens, suggests a Jurassic age for the Zaire beds.

The significance of this hitherto wholly African species occurring in the Antarctic Jurassic will be discussed in a subsequent overall evaluation of the evidence.

11. *Cyzicus (Lioestheria) antarctis* n. sp.
(Pl. 9, Figs. 5, 7; Pl. 12, Fig. 3)

Diagnosis: Valves laterally ovate (greater length) to broadly ovate (greater height) with comparatively short, straight dorsal margin behind the umbo; umbo inset from the anterior margin about one-third the length of the valve; anterior and posterior margins, rounded; ventral margin gently or broadly curved upward. Ornamentation: hachure markings that may be nonparallel as well as parallel on growth bands.

Measurements: (mm) Holotype (TC 11313) followed by paratype (TC 11232). L-5.5, 3.8; H-3.7, 2.5; Ch-1.4?, 1.8; Cr-1.4?, 0.9; Av-1.4, 0.9; Arr-2.7, 1.1; a-1.2, 1.2; b-1.6, 1.0; c-3.2, 2.8; H/L-.672, .658; Cr/L-.254, .237; Av/L-.254, .237; Arr/L-.491, .289; Ch/L-.254, .474; a/H-.324, .480; b/H-.432, .400; c/L-.582, .474. (Some ranges are extended for the species when one includes specimens from younger beds TC 11117A and B: H/L-.654–.785; Cr/L-.145–.237; Av/L-.145–.254; Arr/L-.291–.491; Ch/L-.254–.564; a/H-.324–.500; b/H-.378–.549; c/L-.418–.600. Ratios for TC 11305 overlap those cited above except for H/L, which it extends to .820.)

Material: Holotype, TC 11313, internal mold of right valve; paratype TC 11232, inner layer of left valve and underlying mold; paratypes: 11117A and B, both left valves; fragments of inner layer of valve overlies internal mold.

Locality: Blizzard Heights (see species "10" above for coordinates), Station 1, bed 4 (TC 11313 and TC 11232); Station 1, bed 7 (TC 1117A and B, TC 11305, TC 11315, and TC 11288). Lower Jurassic.

Discussion: This species differs from other Jurassic species described by Novojilov, Defretin-Le Franc, Kobayashi, and others from northern areas in its laterally ovate configuration and umbo position. It is unlike *Cyzicus (Lioestheria) longulus* n. sp., that species being more elongate-subelliptical, although many parameters are close. The new species differs from *Cyzicus (Lioestheria) disgregaris* from Storm Peak as follows: from dimorph 2, which is circular but of similar configuration to dimorph 1; in both dimorphs umbonal placement may vary, from slightly inset to subterminal (dimorph 1), subterminal to submedial (dimorph 2) vs. inset one-third the valve length *("antarctis")*; absence of double growth lines in *"antarctis"* but present in *"disgregaris."* At Storm Peak the new species has a smaller H/L (.658–.672 vs. .703–.727 *"disgregaris"*). The last named species occurs at Agate Peak and Carapace Nunatak also (described below), and H/L is larger than at Agate Peak, and overlaps the ratio of Carapace Nunatak specimens.

The Blizzard Heights species *"antarctis"* differs as noted above, yet seems closest to dimorph 1. There is still a qualification on this similarity as dorsal margin length (CH) is 1.4–1.8 mm, *"antarctis"* overlaps the value at Storm Peak (1.7–2.2 mm, dimorph 2). The latter, however, has a circular configuration very different than that of *"antarctis."* Dimorph 1, which is closest to the new species of the two dimorphs of *"disgregaris,"* by contrast has a *(Ch)* at Storm Peak of 2.5–3.8? mm, and 2.0–2.6 mm at Carapace Nunatak.

The new species differs from *Cyzicus (Lio.) longacardinis* n. sp. in that it has a short vs. long dorsal margin (1.4–1.8 mm vs. 1.7–2.5 mm); its valve does not taper posteriorly as in *"longacardinis"* and c/1 is greater (.474–.583 vs. .310–.410, *"longacardinis"*); other ratios overlap.

12. *Cyzicus (Lioestheria) longulus* n. sp.
(Pl. 10, Fig. 5)

Diagnosis: Elongate, subelliptical valves with a long, straight dorsal margin; umbo inset from the anterior margin about one-third the valve length; anterior and posterior margins rounded. Ornamentation: hachure markings on growth bands. Growth bands widely spaced, except last few.
Measurements: (mm) L-5.0; H-3.5; Ch-2.4; Cr-1.1; Av-1.1; Arr-1.5; a--; b-1.5; c-1.8; H/L-.700; Cr/L-.220; Av/L-.220; Arr/L-.300; Ch/L-.480; a/H--; b/H-.429; c/L-.360.
Material: Holotype TC 11097, flattened and eroded inner layer of right valve and its underlying internal mold.
Locality: Blizzard Heights. (See species "10" above for coordinates.) Tasch Station 0, bed 3. Lower Jurassic.
Discussion: The elongate-subelliptical configuration differentiates this species from the other Antarctic Jurassic lioestheriids described herein. The closest Jurassic species in terms of a long dorsal margin is an Upper Jurassic species, *Sinokontikia youngi* Novojilov, 1958e, from the Turfan Basin, Sin-Kiang, western China. This Chinese species is markedly elongate with a straight dorsal margin and subovate valves, and has the following parameters: L-6.9 mm, H-4.1 mm, H/L-.594 (vs. .700, *"longulus"*), Ch 4.7 mm (vs. 2.4 mm, *"longulus"*). The large, smooth? umbo constitutes 1/4 to 1/5 of the valve, unlike *"longulus"* in which the umbo is small and bears growth bands.

13. *Cyzicus (Lioestheria?)* cf. *brevis* (Shen 1976)
(Pl. 9, Fig. 8)

Diaplexa? brevis Shen 1976, p. 139, Pl. 27, Fig. 3-7. (in Chang and others, 1976).

Diagnosis: Ovate valves with short, straight dorsal margin behind the subterminal umbo; the anterior margin is steep and forms an angle of ~87° with the dorsal margin; the posterior angle is rounded and forms an angle of ~37° with the dorsal margin.
Measurements: (mm) L-5.7; H-4.3; Ch-3.1; Cr-1.0; Av-1.0; Arr-1.6; a-2.8; b-1.7; c-2.5; H/L-.754; Cr/L-.175; Av/L-.175; Arr/L-.281; Ch/L-.544; a/H-.651; b/H-.372; c/L-.439.

Material: Fragment of inner layer of right valve overlying its mold, TC 11074.
Locality: Blizzard Heights (see species "10" above, for coordinates). Tasch Station 0, bed 4. Lower Jurassic.
Discussion: The morphologically closest species to the Blizzard Heights specimen described above is Shen's *Diaplexa? brevis* (1976, Pl. 27, Fig. 3 in Chang and others, 1976) from the Chinese Mesozoic (Triassic), which has a similar configuration; comparative ratios: H/L (.814 vs. .754), b/H (.400 vs. .372), c/L (.465 vs. .439), Cr/L (.209 vs. .175), and Av/L (.209 vs. .175) as measured on Shen's photograph. (Shen's species is distinct from Duan's [1982], *Eosestheria brevis* D.)

The new species differs from *C. (Lio.) longulus* in its ovate configuration (vs. elongate-subelliptical), short (vs.) long dorsal margin, subterminal umbo (vs. inset 1/3 valve length), and generally larger ratios: Cr/L, Av/L, Arr/L, b/H, other ratios are smaller.

14. *Cyzicus (Euestheria) minuta* (von Zieten)
(Pl. 11, Fig. 7)

Euestheria minuta (von Zieten) Defretin-Le Franc 1969, p. 126, Pl. 1, Figs. 1 and 2. (For complete synonymy, see Defretin-Le Franc, 1969, cf. Defretin 1950, and discussion below for a resume.)

Diagnosis (slightly modified): Broadly ovate valves with straight and comparatively short dorsal margin. Margins: anterior, convex, posterior, tapered and rounded; ventral, gently curved. Umbo, subterminal; umbonal terminus rises above dorsal margin. Ornamentation: minute contiguous granules on growth bands.
Measurements: (mm) (Figures for TC 11161 will be followed by those for TC 11170.) L-3.4, 3.3; H-2.5, 2.3?; Ch-1.5, 1.3; Cr-0.9, 0.8; Av-0.9, 0.8; Arr-0.9, 1.1; a-1.0, 0.7; b-1.1., 1.0; c-1.6, --; H/L-.735, .697; Cr/L-.265, .242; Av/L-.265, .242; Arr/L-.261, .333; Ch/L-.441, .393; a/H-.412, .304; b/H-.440, .435; c/L-.471, --.
Material: TC 11115, TC 11293, TC 11161, and TC 11170, eroded internal molds of left valves overlain by remnants of inner shell layer; TC 11243, internal mold of left valve.
Locality: Blizzard Heights. (See species "10" above for coordinates.) Tasch Station 1, beds 4, 6, 7, and 8. Also, cf. TC 9082, *C. (Eu.)* cf. *minuta*, Storm Peak, Upper Flow, bed 6. Lower Jurassic.
Discussion: *Posidonia minuta* Alberti was originally credited to that investigator by De la Beche, 1832, in a mention (a *nomen nudem*). Alberti (1832) attributed the genus and species to Goldfuss, a mere mention (a *nomen nudem*). In 1833, von Zieten attributed the species to Alberti and provided the first valid "indication" (i.e., a figure of the named species, but no description). Not until Raymond (1946, p. 239) was the correct attribution given, *Euestheria minuta* (von Zieten). Jones (1862, p. 42, Pl. 1, Figs. 28-30) used two collections for his description of *"minuta Alberti"*, one from the Triassic of Sinsheim, Germany, and one

from Pendock, Worcestershire, England (Jones, 1862, Pl. 2, Figs. 1–3).

Defretin-Le Franc (1969, p. 127, Pl. 1, Figs. 3, 4) described two specimens; that were closest to Jones Sinsheim material (1862, Pl. 1, Fig. 28) and that had a shorter dorsal margin, *Euestheria minuta* (von Zieten). Her specimens of the species were described, and measurements and ratios given. The latter overlap or, in one instance, are close to the Blizzard Heights specimens. (Ratios for east Greenland specimens are followed by those for Blizzard Heights, Antarctica: H/L-.570–.750; .697–.735; Cr/L-.190–.280; .242–.265; a/H-.430–.560; .304–.412, b/H-.390–.510; .435–.440; c/L-.410–.520; .471– --.)

15. *Cyzicus (Euestheria) beardmorensis* n. sp.
(Pl. 10, Figs. 4, 6)

Diagnosis: Ovate valves that taper posteriorly with straight, comparatively long dorsal margin; subterminal umbo that rises slightly above dorsal margin; anterior margin slightly convex, posterior margin rounded; ventral margin gently curved. Ornamentation of closely spaced granules on growth bands. Growth bands: on umbo more closely spaced, below umbo widely spaced, except last few.

Measurements: (mm) Figures for holotype followed by paratype 1 (Slab 1). L-5.1, 4.9; H-3.5, 3.4; Ch-2.5, 2.4; Cr-1.0, 1.0; Av-1.0, 1.0; Arr-1.5, 1.3; a-1.6, 1.7; b-1.5, 1.6; c-1.9, 2.2; H/L-.686, .694; Cr/L-.196, .204; Av/L-.196, .204; Arr/L-.294, .265; Ch/L-.490 .490; a/H-.457, .500; b/H-.428, .470; c/L-.372, .449.

Material: David Elliot Collection (D.E.). D.E 23.59AA (= Tasch Station 0, bed 2), holotype TC 11506a, slab 1,a; interior mold of left valve. Paratypes 1 and 2, slab 1 (D.E 23.59AA = TC 11506), and slab 2 c—left valve molds from a coquina of valves on same bedding plane. Specimens on slab 2 c were distorted by flattening. Paratype (TC 11503), Station 0, bed 2b. Other specimens; TC 11257, TC 11267.

Locality: Blizzard Heights environs. (See species "10" for coordinates.) D.E site 23.59AA (= Tasch site BH Station 0, beds 2a and 2b). Lower Jurassic.

Discussion: The closest approach to the species described above is *Cyzicus (Euestheria) minuta* (von Zieten) from the German Triassic. However, the posterior tapering of the valve is more strongly accented in the Blizzard Heights species, and the dorsal margin is comparatively longer.

Compared to the Blizzard Heights *"minuta"* von Zieten (see preceding entry on this species and discussion), the new species *"beardmorensis"* has a greater Ch/L (.490 vs. .393–.441), a/H, and smaller H/L (.686–.694 vs. .697–.735), Cr/L (.196–.204 vs. .242–.265), Av/L, and c/L; Arr/L overlaps.

16. *Cyzicus (Euestheria) ellioti* n. sp.
Pl. 11, Figs. 1, 8

Diagnosis: Subovate to broadly ovate valves with comparatively short, straight dorsal margin; umbo variably inset from the ante-

rior margin, less than one-third the length of the valve to subterminal; anterior and posterior margins rounded; antero-dorsal angle greater than postero-dorsal. Ornamentation: fine, contiguous granules on growth bands.

Measurements: (mm) (Holotype followed by paratypes 1 and 2, respectively.) L-6.0, 5.4, 4.4; H-4.3, 4.1, 3.3; Ch-1.4, 2.9, 1.9; Cr-1.5, 1.3, 1.2; Av-1.5, 1.3, 1.2; Arr-3.1?, 1.3, 1.3; a-1.9, 1.8, 1.6; b-2.4, 2.0, 1.4; c-3.4, 2.5, 1.8; H/L-.717, .759, .750; Cr/L-.250, .241, .272; Av/L-.250, .241, .272; Arr/L-.517?, .241, .295; Ch/L-.233, .537, .431; a/H-.442, .439, .485; b/H-.558, .488, .424; c/L-.567, .463, .409.

Material: Holotype TC 11010, internal mold of right valve. Paratypes: TC 11039, fragment of inner layers of right valve and underlying internal mold; TC 11500, inner portion of left valve and its internal mold.

Locality: Blizzard Heights. (See *"malangensis"*, species "10" above for coordinates.) Tasch Station 0, bed 2, holotype TC 11010. Paratypes: 1, TC 11039; 2, TC 11500; D. E. site 23.57A (= .59AA corrected, = Tasch Station 0, bed 2, slab 5). Lower Jurassic.

Name: Named for Ohio State University colleague David H. Elliot, who collected the specimens described herein, and many others.

Discussion: The new species differs from other Blizzard Heights euestheriids in the following: configuration, character of the umbo and/or elevation or lack of it above the dorsal margin, and in several ratios. Taking these in seriatim: *C. (Eu.) minuta* (von Zieten) emend. Defretin-Le Franc tapers posteriorly, umbo subterminal and raised slightly above dorsal margin; *C. (Eu.) juracircularis* n. sp., subcircular configuration, greater H/L, and umbo subterminal; *C. (Eu.) beardmorensis* n. sp., comparatively long dorsal margin, umbo subterminal, and smaller H/L; *C. (Eu.) formavariabalis* n. sp., valves subovate to subelliptical, posterior sector of valve has a greater height than anterior.

17. *Cyzicus (Euestheria) juracircularis* n. sp.
(Pl. 18, Fig. 2)

Diagnosis: Small, subcircular valves with short, straight dorsal margin and subterminal umbo; anterior and posterior margins, rounded; ventral margin concave upward. Ornamentation: minute contiguous granules on growth bands.

Measurements: (mm) (Figures for holotype followed by those for paratype.) L-3.7, 3.5; H-3.0, 3.1; Ch-1.5; Cr-0.8, 1.0; Av-0.8, 0.7; Arr-1.4, 1.2; a-1.4?, 1.0?; b-1.6, 1.6; c-1.9, 1.9; H/L-.811, .886; Cr/L-.216, .286; Av/L-.216, .200; Arr/L-.378, .343; Ch/L-.405, .429; a/H-.467?, .323; b/H-.533, .516; c/L-.513, .543.

Material: Holotype TC 11270-01, carbonized left valve; paratype TC 11260, carbonized right valve; both specimens have slightly eroded anterior margin.

Locality: Blizzard Heights (see species "10" above for coordinates). Tasch Station 1, bed 6. Lower Jurassic.

Discussion: The species described above differs from *C. (Euestheria) minuta* (von Zieten) (Defretin-Le Franc, 1969, Pl. 1, Figs. 1, 2) in its subcircular configuration and a markedly larger H/L. Other ratios that are larger than *"minuta"* include Arr/L, b/H, and c/L. Also, the umbonal terminus does not rise above the dorsal margin as in *"minuta."* The remaining ratios overlap.

18. *Cyzicus (Euestheria) formavariabalis* n. sp.
(Pl. 8, Figs. 2, 4–6; also Pls. 10, 12)

Diagnosis: Valves of variable configuration from subovate to subelliptical; umbonal area close to but inset from anterior margin; posterior sector higher than anterior; anterior margin convex; posterior margin round; dorsal margin straight or slightly curved. Ornamentation: either contiguous granules on growth bands, or a reticulate mesh where more deeply eroded and granules no longer present.

Measurements: (mm) (TC 11331c, followed by 11331d and 11331a.) L-5.3, 5.0, 6.2?; H-3.1, 4.0, 4.3; Ch-1.3?, 2.0?, 2.4; Cr-1.7?, 1.4?, 1.7; Av-1.7?, --, 1.7?; Arr-2.3, 1.6?, 2.1; a-1.3, 2.1, 2.2; b-1.2, 2.4, --; c-2.8, 2.6, 2.8; H/L-.585, .800, .694?; Cr/L-.321?, .280, .274; Av/L-.321?, --, .274?; Arr/L-.434, .320?, .399; Ch/L-.245?, .400?, .387; a/H-.419, .525, .512; b/H-.387, .600, --; c/L-.528, .520, .452.

Material: Syntype TC 11331c, remnant of shell of right valve and underlying interior mold. Syntypes TC 11331a, b, and d. Other specimens: TC 11149, TC 11189A and B, and TC 11319.

Locality: Blizzard Heights (see species "10" above for coordinates). Tasch Station 1, bed 4. Other specimens Station 1, bed 2b. Lower Jurassic.

Discussion: The new species differs from *C. (Eu.) beardmorensis* in having a greater H/L and also differs in most other parameters. Similarly, the new species differs from *C. (Eu.) minuta* (von Zieten) in having a larger H/L, Cr/L, Av/L, and c/L. It differs from *C. (Eu.) ellioti* in most measured parameters, though it overlaps in H/L. It has a greater Cr/L, Av/L, Arr/L, and a smaller Ch/L, a/H, and b/H. Finally, the new species differs from *C. (Eu.) juracircularis* n. sp. in having a smaller lower limit of the range for H/L, smaller Ch/L and b/H, and has a greater Cr/L and Av/L.

Family CYCLESTHERIIDAE Sars, 1899

Carapace laterally compressed; few and indistinct growth lines. Shell almost circular with beak [umbo] far forward.)

Genus Cyclestherioides Raymond, 1946

Diagnosis: Cyclestheriidae with a beak (umbo) a bit farther back than in modern *Cyclestheria*.

Type: *Estheria lenticularis* Mitchell, 1927.

Discussion: The genus here subdivided into two subgenera based on two distinct groups of fossils: *Cyclestherioides (Cyclestheri-*

oides) Raymond, 1946, and *Cyclestherioides (Sphaerestheria)* Novojilov, 1954c.

Subgenus *Cyclestherioides* Raymond, 1946

Diagnosis (emended): Minute, subcircular valves with subround, subterminal umbo directed anteriorly, distinctly unequal height and length; few widely spaced growth bands.

Type: *Estheria lenticularis* Mitchell, 1927.

Subgenus *Sphaerestheria* Novojilov, 1954c

Diagnosis: Trigonal to subcircular valves with generic characters and height and length almost or actually coequal.

Type Species: *"Estheria" koreana* Ozawa and Watanabe, 1923.

19. *Cyclestherioides (Sphaerestheria)* cf. *aldanensis* Novojilov, 1959 (in Molin and Novojilov, 1965).
(Pl. 12, Fig. 4)

Sphaerestheria aldanensis Novojilov, 1959, Fig. 29 *in* Molin and Novojilov, 1965, p. 63–64, Text-Fig. 75.

Diagnosis: Subcircular valves, height and length unequal; straight dorsal margin; umbonal area close to, but inset from the anterior margin, 21 or more growth bands.

Measurements: (mm) L-2.7; H-2.5; Ch-1.3?, Cr-0.7; Av-0.7; Arr-0.7; a-1.3; b-0.9; c-0.8; H/L-.852; Cr/L-.259; Av/L-.259; Arr/L-.259; Ch/L-.481; a/H-.565; b/H-.391; c/L-.296?.

Material: TC 11280b, external mold of right valve with dorsal margin and some adjacent area partially eroded.

Locality: Blizzard Heights. (See species "10" for coordinates.) Tasch Station 1, bed 5. Lower Jurassic.

Discussion: The species closest to the Antarctic species described above is *Sphaerestheria aldanensis* Novojilov, 1959 *in* Molin and Novojilov, 1965, from the Lower Triassic of northern USSR, which has an H/L-0.85–1.0, a straight dorsal margin, and steep anterior and postero-dorsal angles; however, the Antarctic form has a steeper antero-dorsal angle and is more compactly subcircular, with an umbo not subterminal.

Family ASMUSSIDAE Kobayashi, 1954
Subfamily ORTHOTHEMOSINAE Defretin, 1967
(*non* 1965)

Valves relatively short with regularly rounded margins.

Genus *Glyptoasmussia* Novojilov and Varentsov, 1956

Novojilov and Varentsov, 1956, p. 670–673.

Emended. Anterior and posterior margins narrowly or broadly rounded; the dorsal margin in front of the umbo is two-thirds or more the length of the margin behind the umbo, except where umbo is submedial. Valve height less than or almost equal to the length. Ornamentation fine and alveolar (i.e., with small depressions or pits).

Type: *Glyptoasmussia kuluzunensis* Nov. and Varentsov, 1956.

20. *Glyptoasmussia* sp. 1
(Pl. 8, Fig. 7)

Diagnosis: Ovate valves with straight dorsal margin; submedial umbo (flattened by postmortem deformation and slightly elevated above the dorsal margin); anterior and posterior margins steeply curved; valves with a slight axial obliquity posteriorly, i.e., axis through midpoint of umbo to ventral margin. Ornamentation: contiguous granules on growth bands.

Measurements: (mm) (Based on estimates of the reconstructed valve.) L-3.4; H-2.5?; Ch-2.8; Cr-1.5; Av- --; Arr-0.8; a-0.8; b-1.6; c-1.6; H/L-~.659; Cr/L-.441; Av/L --; Arr/L-.235; Ch/L-.823; a/H-.320?; b/H-.640?; c/L-.382.

Material: A single, incomplete right valve, the lower quarter of which is partially eroded and covered, and a small area of the antero-dorsal sector covered. TC 11502, slab 7.

Locality: Blizzard Heights environs, D. E. site 23 =TC 11502 (a correlate of Tasch Station 1). Lower Jurassic.

Discussion: This species has closest affinities to the Mesozoic *Glyptoasmussia* species described by Defretin-Le Franc (1967) from the Congo basin: *Glyptoasmussia corneti* (Marlière) and *Glyptoasmussia luekiensis* Defretin. It differs from the latter in having a longer and more closely equivalent portion of the dorsal margin on either side of the submedial umbo. This last feature is characteristic of *Asmussia* species but the obliquity of the valve rules out assignment to that genus. A new species seems indicated but incomplete preservation precludes assignment of a name.

21. *Glyptoasmussia prominarma* n. sp.
(Pl. 10, Fig. 8)

Diagnosis: Broadly subquadrate (from dorsal margin to midway on the valve) to ovate (from midway on the valve to ventral margin); umbo inset from anterior margin almost two-thirds the length of the valve; umbo rises slightly above the long, straight dorsal margin, which makes a markedly curved join to the anterior margin. Valve has a slight axial obliquity posteriorly. Growth bands: 25 or more; ornamentation consists of contiguous granules on growth bands.

Measurements: (mm) L-3.5; H-2.9; Ch-1.5?; Cr-1.0; Av-1.2; Arr-0.9; a-1.2?; b-1.1; c-2.0?; H/L-.828; Cr/L-.286; Av/L-.286; Arr/L-.257; Ch/L-.428; a/H-.414; b/H-.379; c/L-.517.

Material: Holotype 11127.01 c, inner layer of left valve depressed into substrate; partly carbonized, with outer layer white enameloid; covered marginally along postero-ventral margin.

Locality: Blizzard Heights. (See species "10" above for coordinates.) Tasch Station 1, bed 5. Lower Jurassic.

Discussion: The above species differs from *Glyptoasmussia* sp. 1, described above, in lacking a submedial umbo, and hence differs

in configuration and relevant parameters. It differs from *Glyptoasmussia meridionalis* n. sp. in being subquadrate to ovate; it has a larger H/L, smaller Arr/L, b/H, and c/L. It differs from *Glyptoasmussia* sp. 2, Storm Peak (Upper Flow, bed 4), in overall configuration, umbonal position, and greater H/L, and Av/L, and smaller Arr/L, Ch/L, and b/H; c/L is similar.

The new species differs in configuration and umbonal placement from Mesozoic glyptoasmussids from the Congo Basin (Defretin-Le Franc, 1967); *G. corneti* and *G. luekiensis.* Compared with the latter Congo species *G. prominarma,* it has a larger H/L and a smaller Cr/L, Ch/L, and b/H, but overlaps in Arr/L, a/H, and c/L. *G. corneti* overlaps the new species in several parameters and ratios: H/L, Ch/L, a/H, b/H, and c/L. However, the new species differs from *"corneti"* in having a smaller Cr/L, and Av/L, and a longer dorsal margin, reflected in its subquadrate to ovate configuration (see diagnosis).

22. *Glytpoasmussia meridionalis* n. sp.
(Pl. 10, Fig. 7; Pl. 12, 18)

Diagnosis: Laterally subovate valves obliquely oriented posteriorly (i.e., axis through midpoint of the umbo to ventral margin has a posterior slant) with straight, comparatively short dorsal margin; umbo is close to anterior margin; height of anterior sector less than that of the posterior sector of the valve; double growth lines. Ornamentation: oblique to parallel hachure markings on growth bands (i.e., hachure markings consist of minute granules vertically arranged and generally parallel to each other; sometimes they fuse and/or slant).

Measurements: (mm) (Parameters and ratios for holotype followed by paratype 1 and paratype 2.) L-4.5, 5.4, 5.6; H-3.9, 4.6, 4.2; Ch-2.4, 2.4?, 1.9; Cr-0.7, 1.0, 1.5; Av-0.7, 1.0, 1.5; Arr-1.4, 2.0, 2.2; a-1.2, 2.0, 1.8; b-1.3, 2.0, 2.0; c-1.7, 2.4, 2.7; H/L-.867, .852, .750; Cr/L-.156, .185, .268; Arr/L-.311, .370, .393; Ch/L-.533, .444, .399; a/H-.308, .435, .429; b/H-.333, .435, .476; c/L-.378, .444, .482.

(If the Storm Peak specimens are included in the range of the different ratios, the species-ratio ranges would become: H/L-.742-.867; Cr/L-.156-.292; Av/L-.156-.292; Arr/L-.246-.375; Ch/L-.393-.533; a/H-.308-.540; b/H-.333-.460; c/L-.378-.565. The Storm Peak specimens share all major characteristics of the Blizzard Heights specimens of this species.)

Material: Holotype TC 11245A, fragment of inner layer of left valve and its imprint. Paratype 1, TC 11245G, possible dimorph; shows remnants of inner portion of right valve and its imprint. Paratype 2, TC 11245B, remnants of inner portion of left valve and underlying mold. Both paratypes eroded; also TC 11312 (Station 1, bed 4).

Locality: Blizzard Heights. (See species "10" above for coordinates.) Tasch Station 0, bed 4; also found in bed 1 and at Storm Peak, Upper Flow debris (TC 9065b). Lower Jurassic.

Discussion: The new species differs from the Mesozoic glyptoasmussids of Defretin-Le Franc in that it has a more laterally subovate configuration than *G. corneti* (Marlière) or *G. luekiensis*

Defretin, both of which are more vertically subovate, and in its umbonal position, which is closer to the anterior margin than in either of the Congo species (Defretin-Le Franc, 1967, p. 71–78, Pls. 9 and 10; Novojilov, 1961, p. 61–71, Text-Figs. 25–33. See discussion for Species 21, above).

Family VERTEXIIDAE Kobayashi, 1954

Genus *Cornia* Lutkevich, 1937

Diagnosis (emended): Valve shape varies from subovate to subrectangular; beak (umbo) position subcentral to anterior; small spine or tubercle (node) that may taper to a spine rises from center of the initial valve; sculpture punctate or with hachure markings on growth bands.
Type: *Cornia papillaria* Lutkevich, 1937.

23. *Cornia* sp. 1.
(Pl. 8, Fig. 8)

Diagnosis: A corniid displaying a large umbonal node that tapers to a point (= spine) and is curved downward; inferred configuration irregularly ovate; about six growth bands can be seen counting from the ventral margin.
Measurements: (mm) Only a few measurements are possible given this incomplete material. Greatest height and length of the fragment (measured on mold) and length of the umbo (1_u): L_f-3.1; H_f-2.3; 1_u~0.5 (f = fragment).
Material: TC 10014A, internal mold of left valve; TC 10014B, external mold of same valve. The dorsal margin is missing and an oblique, eroded portion of the valve occupies the postero-dorsal sector up to and above the umbonal node; few growth bands.
Locality: Blizzard Heights. (See species "10" for coordinates.) Tasch Station 0, bed 3. Lower Jurassic.
Discussion: This species is reported here for completeness because it is distinctly different in its umbonal tapered node and size of the umbo from *Cornia* sp. 2 (see below). The umbonal node suffices to identify the genus.

In corniid development (Permian–Jurassic) distinct types of umbonal nodes and/or spines appeared: a mere pimple-size rise (node) with no tapered terminus, a minute tapered node or a node with a tapered terminus (spine) that bent downward, in addition to many variants (in shape and size) of node and spine. In Permian time (Tasch, 1969, p. R161), and extending into Triassic time (Novojilov, 1970, p. 150), a markedly curved or looped spine arose directly from the initial valve in species of genus *Curvacornutus* Tasch, 1961. Clearly genetic loci governing conchostracan umbonal development were active sites for evolutionary change.

The Antarctic Jurassic corniids are of interest in that each of the two species have different development (see below).

24. *Cornia* sp. 2.
(Pl. 11, Fig. 6)

Diagnosis: Broadly ovate valve with slightly arched dorsal margin, submedial, smooth, ovate umbo that dorsad bears a minute, tapered node, the terminus of which is a spine. Growth bands greater than 15, counting from the last band near the ventral margin.
Measurements: (mm) L-2.7?; H-2.3; Ch-1.3?; Cr-0.6; Av-0.6; Arr-0.7; a-0.9; b-0.9; c-1.5; H/L-.853?; Cr/L-.222; Av/L-.222; Arr/L-.259; Ch/L-.481?; a/H-.391; b/H-.391; c/L-.556.
Material: TC 11420, internal mold of left valve with remnants of inner layer of original shell; the spine base visible in inner layer although valve eroded along anterior and ventral margins.
Locality: Blizzard Heights environs. (See species "10" for coordinates, and Fig. 4 for site 23.) D. E. site 23.58c (equivalent to Tasch Station 1). Lower Jurassic.
Discussion: This species differs from *Cornia* sp. 1 in size and nature of the spine. It is minute and rises directly from the umbo in *Cornia* sp. 2, whereas it is comparatively larger and broadly based (i.e., tapered node) in *Cornia* sp. 1. In the latter, the spine is also curved under. Of interest is the occurrence of these two corniid species at two different sites; however, both are related to approximately the same time interval.

25. *Estheriina (Nudusia)* sp. 1.
(Pl. 19, Fig. 2)

Diagnosis: Estheriniids with small, subovate valves; umbo inset from the anterior margin about one-third the length of the valve; not raised above dorsal margin; a smooth, ovate umbonal terminus is followed by numerous growth lines; short, straight dorsal margin behind umbo; anterior margin arcuate; posterior margin less steep, rounded; ventral margin slightly eroded but apparently gently curved. Growth bands, 14–15; double growth lines.
Measurements: (mm) L-2.3; H-1.6; Ch-1.1; Cr-0.7; Av-0.7; Arr-0.6; a-0.7; b-0.8; c-0.9; H/L-.696; Cr/L-.304; Av/L-.304; Arr/L-.261; Ch/L-.478; a/H-.437; b/H-.500; c/L-.391.
Material: TC 11134, internal mold of left valve.
Locality: Blizzard Heights. (See species "10" for coordinates.) Tasch Station 1, debris (probably eroded from bed 8). Lower Jurassic.
Discussion: The species described above differs from *Estheriina (Nudusia)* sp. 2 in being subovate, with an inset (not subterminal) umbo that apparently is not raised above the dorsal margin; also, most ratios are larger, although a few are close: H/L-.696 vs. .711; b/H-.500 vs. .513.

26. *Estheriina (Estheriina)* sp. 1.
(Pl. 9, Fig. 6)

Diagnosis: Eccentrically ovate valves with slightly arched dorsal margin; umbonal area inset from the anterior margin one-third the length of the valve; umbonal nucleus more convex and prominent than rest of valve; anterior and posterior margins steeply convex; round joins to the dorsal margin.

Measurements: (mm) L-3.5; H-2.8; Ch-1.3?; Cr-1.1; Av-1.1; Arr-1.1?; a-1.2; b- --; c-1.9; H/L-.800?; Cr/L-.314?; Av/L-.314?; Arr/L-.314?; Ch/L-.371?; a/H-.429; b/H- --; c/L-.543?.

Material: TC 11319-05A (BH, Station 1, bed 4) portion of right valve overlies the internal mold; postero-dorsal sector partially covered; also TC 11134 (Station 1, debris); TC 11404 (D. E. 23.58c = Station 1, bed 4).

Locality: Blizzard Heights. (See species 10 above for coordinates.) Tasch Station 1, bed 4. Lower Jurassic.

Discussion: There is a small amount of cover over the postero-dorsal sector of the valve which accounts for the uncertain ratios.

The above species has no near-equivalents in the Blizzard Heights sections, nor does it fit any described species of *Estheriina* in the literature of the Mesozoic.

27. *Estheriina (Nudusia)* sp. 2.
(Pl. 10, Fig. 2)

Diagnosis: Ovate valves with small, smooth, subterminal, elliptical umbo that rises above the dorsal margin, the latter being short behind the umbo, and ending with an oblique join to the otherwise rounded posterior margin; anterior margin is more steeply arched. Growth bands: 16 or more.

Measurements: (mm) L-5.2; H-3.7, Ch-2.0; Cr-1.3; Av-1.3; Arr-1.9; a-1.6; b-1.9; c-2.6; H/L-.711; Cr/L-.250; Av/L-.250; Arr/L-.365; Ch/L-.385; a/H-.432; b/H-.513; c/L-.500.

Material: TC 11400A, external mold of left valve, retains a small portion of the original shell.

Locality: Blizzard Heights environs. (See species "10" above for coordinates.) D. E. site 23, bed 23.59b (equivalent to Tasch Station 1). Lower Jurassic.

Discussion: This species differs from *Estheriina (Nudusia) brevimargina* n. sp. in that it has ovate valves, a longer dorsal margin and, in most parameters, this species has a smaller H/L, and greater Cr/L, Av/L, Arr/L, Ch/L, a/H, and b/H. It also differs significantly from the estheriinids this writer described from the Australian Triassic (Tasch *in* Tasch and Jones, 1979a, 1979c): *Estheriina (Nudusia) blina* Tasch, from the Lower Triassic Blina Shale, subovate valves with a large elliptical umbo, and larger H/L; *Estheriina (Nudusia) circula* Tasch, Lower Triassic Rewan Group, with circular valves, a conspicuous elliptical umbo, shorter dorsal margin, and a greater H/L; *Estheriina (Nudusia) rewanensis* Tasch from the Lower Triassic Rewan Group, with a comparatively large, submedial umbo.

28. *Estheriina (Nudusia?) brevimargina* n. sp.
(Pl. 11, Fig. 5)

Diagnosis: Eccentrically subovate to subcircular valves with a small, smooth, subterminal umbo which is elevated above a short,

straight dorsal margin; anterior margin arched, posterior margin rounded. Antero-dorsal angle 40°; postero-dorsal angle 15°. Growth bands, 15 to 16.

Measurements: (mm) (Holotype figures precede those for paratype.) L-3.8, 4.8; H-3.3?, 4.2?; Ch-1.8, 2.5?; Cr-0.8, 1.0; Av-0.8, 1.0; Arr-1.3, 1.4; a-1.0, 1.3; b-1.0, 1.5; c --; H/L-.868, .875?; Cr/L-.210, .208; Av/L-.210, .208; Arr/L-.342, .292; Ch/L-.474, .521?; a/H-.303, .310; b/H-.303, .357; c/L --.

Material: Holotype TC 11405A, inner layers of flattened left valve; Paratype TC 11405B; this valve is crushed.

Locality: Blizzard Heights environs. D. E. site 33, bed 33.56A (Blizzard Peak). Lower Jurassic.

Discussion: The closest species to *"brevimargina"* is *Estheriina (Nudusia) rewanensis* Tasch, from the Lower Triassic, Bowen Basin, Australia (Tasch *in* Tasch and Jones, 1979c, p. 42, Fig. 7) which has a smaller H/L (.65–.80 vs. .87), larger postero-dorsal angle (22° vs. 15°); the antero-dorsal angle is 40°, the same as *"brevimargina."*

Storm Peak

29. *Glyptoasmussia australospecialis* n. sp.
(Pl. 13, Fig. 7)

Diagnosis: Eccentrically subovate valve with a posterior trend and a short dorsal margin; umbo nearly central but inset slightly farther from the anterior margin than the posterior; posterior margin rounded; anterior margin arched. Growth bands: 16 or more.

Measurements: (mm) L-3.0; H-2.4; Ch-1.0; Cr-1.0; Av-1.0; Arr-1.0; a-1.1; b-1.0; c-1.6; H/L-.800; Cr/L-.333; Av/L-.333; Arr/L-.333; Ch/L-.333; a/H-.458; b/H-.417; c/L-.533.

Material: Holotype TC 9205, internal mold of right valve. The valve is flattened; an oblique crack extends from below the posterior end of the dorsal margin to the sixth from the last growth band; growth bands below the crack are continuous. Restoration to the original position prior to flattening does not affect the length of the dorsal margin because the crack started below it.

Locality: Storm Peak, lat 84°35′S, long 163°55′E. Upper Flow, bed 11, middle. Lower Jurassic.

Discussion: The Blizzard Heights glyptoasmussids differ from Storm Peak *"australospecialis"* as follows: *Glyptoasmussia* sp. 1 has an ovate valve (vs. eccentrically ovate); a long dorsal margin (vs. short); a smaller H/L, Ch/L, and b/H, and a greater Cr/L, Arr/L, a/H, and c/L; *Glyptoasmussia prominarma* n. sp. has a broadly subquadrate valve, long dorsal margin, smaller Cr/L, Av/L, Arr/L, Ch/L, and a/H; H/L and b/H overlap. Storm Peak *Glyptoasmussia* cf *luekiensis* Defretin, in its subovate valves, height of anterior margin less than height of posterior, double growth lines, and smaller H/L, Cr/L, Av/L, Arr/L, and greater Ch/L differs from *"australospecialis";* a/H and b/H are similar.

30. *Glyptoasmussia* cf. *luekiensis* Defretin-Le Franc, 1967
(Pl. 15, Fig. 1)

Diagnosis: (Abbreviated and emended slightly.) Subround valves, with umbo inset from anterior margin about one-third the length of the valve; dorsal margin straight; umbo slightly raised above dorsal margin; valves have a posterior slant. Growth bands: 11 or more. Ornamentation: contiguous granules on growth bands.

Measurements: (mm) Right valve, a' only: L-4.0; H-2.9?; Ch-2.2?; Cr-1.1; Av-1.1; Arr-0.8; a-1.3; b-1.2; c--; H/L-.725?; Cr/L-.275; Av/L-.275; Arr/L-.200; Ch/L-.550; a/H-.448?; b/H-.414; c/L--.

Material: TC 9090a,a', a pair of carbonized valves (internal molds). In the left valve, "a," the umbonal terminus, is somewhat eroded but remains above dorsal margin; ventral-posterior sector covered. The right valve, "a'," also has an eroded umbonal terminus but less so than the left valve.

Locality: Storm Peak. (See species "29" for coordinates.) Upper Flow, loose piece. Lower Jurassic.

Discussion: This Antarctic specimen is closest to Defretin-Le Franc's species from the Haute Lueki Series, Upper Triassic or Jurassic (pre-Oxfordian), Congo Basin. The Storm Peak species overlaps the Congo species in H/L, Arr/L, a/H, and b/H; it has a smaller Cr/L, Ch/L, and a larger Av/L. It shares several of the main morphological features, subround valves and umbo inset one-third the length of the valve. Defretin-Le Franc's species differs from the Blizzard Heights *Glyptoasmussia* sp. 1 in that it lacks a submedial umbo, is subround, and has larger H/L, Av/L, a/H, and smaller Cr/L, b/H, and c/L; from *Glyptoasmussia prominarma* n. sp., it differs in being subround (vs. subovate), with a smaller inset of the umbo from the anterior margin (1/3 vs. 2/3), and in a smaller H/L, Cr/L, Arr/L, and Av/L, and a larger a/H, b/H, and Ch/L; from *Glyptoasmussia meridionalis* n. sp. it differs in being subround, with height of anterior and posterior coequal, and in its larger Ch/L, Cr/L, and Av/L, and smaller H/L, and Arr/L; b/H overlaps.

Family CYCLESTHERIIDAE Sars, 1899
Subfamily CYCLESTHERIOIDES Raymond, 1946

Diagnosis: Ovate valves with terminal umbo and distinctly unequal height and length.
Type: *Estheria lenticularis* Mitchell, 1927.

31. *Cyclestherioides (Cyclestherioides) alexandriae* n. sp.
(Pl. 14, Figs. 7, 8; also Pls. 12, 18)

Diagnosis: Robust, subcircular valves with short, straight, or gently arched dorsal margin; umbo subterminal to submedial; antero-dorsal join steeper than postero-dorsal. Ornamentation: pits and/or nonparallel hachure markings on growth bands.
Measurements: (mm) Holotype (TC 9095-01), followed by paratype (TC 9132c) and paratype (TC 9151). L-4.2, 4.0, 4.7; H-2.5, 3.1, 3.7; Ch-1.9?, 1.9, 2.0?; Cr-0.9, 1.2, 1.0; Av-0.9, 1.2, 1.0; Arr-1.3, 1.6, 1.7; a-1.3?, 1.7, 1.5; b-1.3?, 1.7, 1.7; c-1.9, 2.5, 2.0; H/L-.833, .761, .787; Cr/L-.214, .261, .213; Av/L-.214, .261, .218; Arr/L-.310, .348, .362; Ch/L-.412, .413, .426; a/H-.429, .486, .459; b/H-.371, .486, .459; c/L-.452, .543, .426.

The ratios for the best-preserved Blizzard Heights specimen are: H/L-.766; Cr/L-.276; Arr/L-.282; Av/L-.276; Ch/L-.340; a/H-.500; b/H-.472; c/L-.425.

Material: Holotype TC 9095-01, partly eroded, inner layers of right valve, flattened peripherally around posterior margin. Paratype TC 9132c, inner layer of original right valve impressed into substrate; paratype TC 9151, crushed inner layer of left valve.

Locality: Storm Peak. (See species "29" for coordinates.) Upper Flow, bed 10, and float (four specimens). Also, TC 11152B, Upper Flow, bed 10. Lower Jurassic.

Discussion: The closest species to "*alexandriae*" is *Cyclestherioides (Cyclestherioides) janenschi* Kobayashi, 1954, from the Upper Jurassic of Tanzania. The two species share a subcircular configuration, subterminal umbo, and hachure markings on growth bands, but differ in that "janenschi" has a distinctly larger H/L (.945 vs. .761–.833) and an umbo that rises above the dorsal margin.

Blizzard Heights specimens (TC 11153A through C. BH Station 1, bed 2) overlap "*alexandriae*" in H/L, Arr/L, and b/H, and share *all* of this species morphology with the exception of a submedial umbo. In the other parameters, the differences are generally in the second decimal place.

Another Blizzard Heights specimen assignable to "*alexandriae*" is TC 11225C (BH, Station 0, bed 6) because it fits the detailed morphology of that species, as does TC 11156 (BH, Station 1, bed 9A).

32. *Estheriina (Nudusia)* sp. 3.
(Pl. 15, Fig. 5)

Diagnosis: An apparently ovate valve with prominent, swollen, naked, bubble-like umbo set off from the rest of the valve which displays successive growth bands.
Measurements: (mm) l_u-0.7; h_u-0.5; h_u/l_u-.714 (u = umbo). No other meaningful measurements possible.
Material: TC 9097b, internal mold of incomplete right valve.
Locality: Storm Peak. (See species "29" above for coordinates.) Upper Flow, loose. Lower Jurassic.
Discussion: The characteristic of the umbo is distinctive relative to other Antarctic estheriinids, which accounts for its being recorded here.

33. *Estheriina (Estheriina)* sp. 2
(Pl. 17, Fig. 8)

Diagnosis: Subovate valves with subround, subterminal umbo rising above dorsal margin; terminal segment of umbo elliptical, with growth bands; umbo inset from anterior margin approximately one-third the length of the valve. Growth bands widely spaced; the last six bands are widest apart.

Measurements: L-4.8; H-3.6; Ch-2.4?; Cr-1.0; Av-1.0; Arr-1.3; a-2.2?; b-1.3; c-1.6; H/L-.750; Cr/L-.208; Av/L-.208; Arr/L-.271; Ch/L-.500? a/H-.611?; b/H-.361; c/L-.333.

Material: External mold of right valve, TC 11431.

Locality: Storm Peak. (See species "29" above for coordinates.) Lower Flow, Station 0, bed 2. Lower Jurassic.

Discussion: The species described above is closest to *Eshteriina (Estheriina)* sp. 3 (no. 34 below): both species have an umbo that rises above the dorsal margin and are close in H/L (.750 n. sp. vs. .737), but several ratios are smaller in the new species (Cr/L, Av/L, b/H (.361 vs. .464), c/L (.330 vs. .421); a few are larger, Ch/L (.500? vs. .342), and a/H (.611? vs. .500); umbonal shape differs (subround n. sp. vs. tapered) as does umbonal position (inset 1/3 vs. 3/7).

34. *Estheriina (Estheriina)* sp. 3
(Pl. 15, Fig. 3)

Diagnosis: Compact, irregular subovate valves with a small, tapered umbo with growth bands; umbo raised above the short dorsal margin and presents a low pinnacle aspect; the umbo is anteriorly inset three-sevenths of the length of the valve. The anterior margin is steeper than the posterior and joins the dorsal margin at an angle of 30°; the posterior margin is rounded and joins the dorsal margin at an angle of 15°. Growth bands: first fourteen bands are widely spaced and the last nine are more tightly spaced.

Measurements: (mm) L-3.8; H-2.8; Ch-1.3; Cr-1.1; Av-1.1; Arr-1.3; a-1.4; b-1.3; c-1.6; H/L-.737; Cr/L-.289; Av/L-.289; Arr/L-.342; Ch/L-.342; a/H-.500; b/H-.464; c/L-.421.

Material: TC 9185B, left valve, internal mold; upper third, including umbo, carbonized.

Locality: Storm Peak. (See species "29" above for coordinates.) Upper Flow, bed 5. Lower Jurassic.

Discussion: The closest species to the one described above is *E. (E.)* sp. 2 (see species "33" above for discussion). It is of interest that Estheriina *(Estheriina)* sp. 3 occurs in the Upper Flow interbed, while Estheriina (Estheriina) sp. 2 occurs in the Lower Flow interbed, both at Storm Peak (Fig. 5). These interbeds are approximately 273 m apart.

35. *Estheriina (Nudusia) stormpeakensis* n. sp.
(Pl. 17, Fig. 2; also Pls. 10, Fig. 3; 13, Figs. 1–4)

Diagnosis: Dimorphic valves: smaller valves (less than 2.0 mm in length), with subovate to subelliptical valves; large subterminal, smooth and tapered, umbo of irregular shape raised above a gently arched dorsal margin. Larger valves (4.0 to 5.0 mm in length) are elongate-subovate and have a proportionately smaller umbo which is more prominent than in the smaller dimorph, and also raised above the dorsal margin; umbo is inset from, but close

to, the anterior margin. Both large and small valves have a steeply arched anterior margin and a more rounded posterior margin.

Measurements: (mm) (Figures for TC 9300, then TC 10121c, followed by dimorphs TC 9015 and TC 9119c.) L-1.9, 1.7, 5.0, 4.0; H-1.4, 1.0?, 3.0, 2.4; Ch-1.1, 1.3, 2.0, 1.5; Cr-0.4, 0.4, 1.3, 1.0; Av-0.4, 0.2, 1.3, 1.0; Arr-0.4, 0.2, 1.3, 1.0; a-0.6, 0.3, 1.7, 1.2; b-0.4, 0.3, 1.3, 1.1; c-0.7, 0.6, --, --; H/L-.719, .687, .588, .600; Cr/L-.211, .118, .260, .250; Av/L-.211, .118, .260, .250; Arr/L-.211, .138, .340, .375; Ch/L-.579?, .765, .560, .375; a/H-.429, .350, .567, .520; b/H-.286, .300, .286, .458; C/L-.368, .353, ---, ---,.

Material: Syntype: TC 9300, internal mold of left valve; postero-ventral margin is slightly eroded and slightly covered along antero-dorsal margin. Syntype: TC 10121c, external mold of right valve, antero-ventral margin and a small part of ventral margin, respectively, covered and eroded. Syntype TC 9015, internal mold of right valve; portion of ventral margin and terminus of umbo eroded. Syntype: TC 9119C, internal mold of left valve, eroded along part of ventral margin and with limited cover behind flattened umbo. Also, syntypes TC 9196f and TC 9071A, float.

Locality: Storm Peak. (See species "29" for coordinates.) Upper Flow. (TC 9300), Upper Flow, bed 6, lower half: TC9015, same; TC 10121C, Upper Flow, bed 1, lower; TC 9119c, Upper Flow, bed 10. (Also found at Blizzard Heights, Station 1, bed 5 [TC 11256], bed 9A [TC 11169]; Sta. 0, bed 3 [TC 11109b], bed 7 [TC 10012], bed 8 [TC 11082]. Other: D. E. 23.59b, [TC 11066-10].) Lower Jurassic.

Discussion: This writer would reassign all of Kapel'ka's (1968) (Upper Triassic, Siberia) limnadiids to *Palaeolimnadia* with the exceptions of *Limnadia* sp. undet. Kapel'ka, 1968 (*in* Novojilov and Kapel'ka, 1968, Fig. 8) and *Eulimnadia ishikawai* Kapel'ka (1968 Pl. A, Fig. 9; = *Estheriina* Kobayashi, 1975). The undetermined "limnadiid" noted above had an umbo that rises well above the dorsal margin and is set off from the rest of the shell that bears growth bands. Furthermore, it is devoid of growth bands and hence exhibits the early larval shell condition. Accordingly, it should be placed under *Estheriina (Nudusia)* sp.

It is to this reassigned species that *E. (H.) stormpeakensis* n. sp. can be compared. Both have similar H/L ratios (0.60, Storm Peak, larger dimorph; 0.66, Siberian form) and both have a smooth umbo that rises above the dorsal margin. However, they differ in that the Siberian form expands posteriorly, the opposite of the new Storm Peak species.

Another of Kapel'ka's species, *Eulimnadia ioanessica* (1968, Text-Fig. 4, p. 124) is also an estheriinid and reassigned here to *Estheriina (Nudusia),* but the umbonal area which rises above the dorsal margin is even more broadly convex than the undetermined limnadiid species discussed above, and its configuration is more compactly ovate and elongate.

A Korean (Rhaetic–Liassic) species that Novojilov assigned (1954c) to *Estheriina kawasakii* (Ozawa and Watanabe) (1923, p. 41-42, Fig. 4b) is comparatively four times the new Antarctic species in size, and has a small, convex umbo which is barely raised above the dorsal margin. It is reminiscent of the larger dimorphs of the new species, however, in its elongate-oval configuration. *Estheriina (Nudusia) indijurassica* n. sp. from the Lower Jurassic Kota Formation of India is close to *"stormpeakensis"* in many valve characteristics; for example, the umbo is also smooth, tapered, and rises above the dorsal margin, and a/H and b/H overlap *"stormpeakensis."* Differences include a smaller H/L than in the larger dimorph, and a larger H/L than in the smaller dimorph of the new species, and there are other metrical differences; Cr/L, Av/L, Arr/L, and c/L are either larger than either dimorph or smaller than both.

36. *Cyzicus (Lioestheria) disgregaris* n. sp.
(Pl. 16, Figs. 1, 2, 4; also Pls. 17–20)

Diagnosis: Dimorphic valves. Dimorph 1 with laterally-ovate to vertically ovate valves, umbo usually close to, but inset from anterior margin, but in some specimens more subterminal; steeply arched to rounded anterior margin. Growth bands variable, approximately 29. Dimorph 2 with subcircular valves, umbonal position variable from subterminal to submedial. Growth-bands approximately 33. For both dimorphs, double growth lines may or may not be clearly defined. Ornamentation: vertical alignment of granules creating the hachure markings. On some growth bands some vertical markings may be irregular; i.e., nonvertical or slanted at an angle to the dominant trend. These markings cannot be attributed to shell flattening or distortion, but possibly are due to genetic errors.

Material: Syntype: TC 10006. Six reference specimens plus several others too poorly preserved for measurements. Of the six specimens, four are usable for measurements. TC 10006d, fragment of inner layer of left valve overlying its mold; 10006c, internal mold of left valve; 10006f, fragment of left valve and underlying mold; 10006e, internal mold of right valve, medial-anterior ventral margin partly eroded. Syntype: TC 9102d, internal mold of right valve; TC 9072 and TC 9045b, both internal molds of left valve; also TC 9894, left valve and underlying mold. Other specimens from same bed that overlap in a suite of measurements of the larger of the dimorph valves include: TC 9532, 9199, 9203, 9140, 9141, 9543, 9202, 9296, and 9093 (all from SP, Upper Flow, bed 11, lower).

Locality: Storm Peak. (See species "29" above for coordinates.) Upper Flow, bed 6, lower, and bed 11, lower. (See Agate Peak and Carapace Nunatak entries for more localities bearing the above species.) Lower Jurassic.

Measurements: Dimorph, D-1, laterally ovate to vertically ovate valves: TC 10006d followed by three other specimens of D-1; TC 10006c, e, and f, in that order: L-6.4, 6.6, 6.8, 6.3; H-4.5, 4.8, 4.7?, 4.5; Ch-2.5, 3.8?, 3.0, 2.8; Cr-1.6, 0.9?, 1.9, 1.7; Av-1.6, 0.9?, 1.9, 1.7; Arr-2.3, 2.1, 1.9, 1.8; a-2.1, 1.6, 2.4, 2.3; b-2.1, 2.0, 2.2, 1.9; c-3.0, 2.6, --, 3.1; H/L-.703, .727, .691, .714; Cr/L-.250, .136?, .279, .279; Av/L-.250, .136, .279, .279; Arr/L-.359, .318, .279, .286; Ch/L-.391, .576, .441, .444; a/H-.467, .333, .511, .444; b/H-.467, .417, .468, .442; c/L-.469, .394, --, .492. Figures

for dimorph D-2, the subcircular valves, are TC 9102d, 9072, and 9045b, in that order: L-5.1, 5.0, 4.8; H-3.7, 4.0, 3.8; Ch-1.7, 2.0?, 2.2; Cr-1.4, 1.1, 1.3; Av-1.4, 1.1, 1.3; Arr-2.0, 1.9, 1.3; a-1.7, 1.8, 1.8; b-1.7, 1.4?, 1.3; c-2.4, 2.0, 1.7; H/L-.725, .800, .792; Cr/L-.275, .220, .271; Av/L-.275, .220, .271; Arr/L-.392, .380, .271; Ch/L-.333, .220?, .458; a/H-.459, .450, .474; b/H-.459, .350, .342; c/L-.471, .400, .354.

Comparative Ratios of Specimens Assignable to
Cyzicus (Lioestheria) disgregaris n. sp.
at three Antarctic Sites

	Dimorph 1 Laterally Ovate to Vertically Ovate Valves		
	Storm Peak	Agate Peak*	Carapace Nunatak*
H/L	.703–.727	.596–.621	.656–.745
Cr/L	.136–.279	.228–.310	.306–.344
Av/L	.136–.279	.228–.310	.315–.344
Arr/L	.279–.359	.316–.383	.333–.367
Ch/L	.391–.576	.345–.456	.313–.370
a/H	.333–.511	.441–.472	.405–.575
b/H	.417–.468	.353–.472	.406–.514
c/L	.394–.492	.351– ---	.375–.540

	Dimorph 2 Subcircular Valves		
	Storm Peak	Agate Peak	Carapace Nunatak
H/L	.725–.800	.735–.764	.740–.800(?)
Cr/L	.220–.275	.218–.383	.250–.340
Av/L	.220–.275	.218–.388	--- ---
Arr/L	.271–.392	.286–.382	.320–.450
Ch/L	.220–.458	.327–.400	.239–.340
a/H	.450–.474	--- .555	.405–.575
b/H	.342–.459	.417–.548	.406–.514
c/L	.354–.471	.306–.455	.385–.540

*For further data at these two localities, see entries under site names.

Discussion: The lioestheriid closest to the above species Dimorph 1 in general configuration is *Eosestheria brevis* Duan Wei-wu (1982, Pl. 2, Figs. 1, 2) from the Chinese Mesozoic (Upper Jurassic). However, the latter has a greater antero-dorsal, and a smaller postero-dorsal angle, smaller ratio H/L, and larger a/H and b/H; c/L overlaps. The difference suffices to distinguish the above species.

Although the Storm Peak, Agate Peak, and Carapace Nunatak samples were studied separately, review of valve characteristics, measured parameters, and their ratios indicated that the same species occurred at all three sites. Locality and measurement data for the last two sites named above will be given separately under their respective site names, while the comparative ratio ranges are given above. The degree of overlap of valve characters and ratios, with few exceptions, at these three widely separated sites is extraordinary. [See note added in proof, p. 149.]

37. *Cyzicus (Euestheria) ichthystromatos* n. sp.
(Pl. 13, Figs. 5, 6; Pl. 18)

Diagnosis: Subovate valves with comparatively straight dorsal margin and small, elliptical umbo, inset close to anterior margin, that arches slightly above the dorsal margin; anterior margin steeply arched; posterior margin rounded. Growth bands: 17 or more. Ornamentation: minute contiguous granules on growth bands.

Material: Syntype: TC 9143A, external mold of left valve; posterior fifth of valve missing. Syntype: TC 9143B, internal mold of right valve; valve flattened and a small part of posterior margin, as well as the more anterior part of the umbo, eroded. Syntype: TC 9200A and B, inner shell layers of left valves, flattened and eroded slightly. Fish fossils found on same bedding plane. Additional incomplete and distorted valves in a coquina included TC 9200C, inner portion of shell of left valve.

Locality: Storm Peak. (See species "29" above for coordinates.) Tasch Station 2, bed 4 (TC 9143A and reverse side of this slab, 9143B); Tasch Station 0, bed 3 (Lower Flow). This last bed is equivalent to D. E. bed 27.85 (TC 9200-Fish Bed). Beds at both stations represent the same Lower Fish Horizon. Lower Jurassic.

Measurements: (mm) (TC 9200A followed by TC 9143A and 9143B.) L-3.3, 5.7, 5.5; H-2.3, 4.3, 4.1; Ch-1.5, 2.6, --; Cr-0.8, 1.4, 1.7; Av-0.8, 1.4, 1.7; Arr-1.0, 1.7, --; a-1.0, 2.0, 1.7; b-1.0, 2.0, 1.8; c-1.4, 2.0, 1.8; H/L-.697, .754, .745; Cr/L-.242, .246, .309; Av/L-.242, .246, .309; Arr/L-.303, .298, --; Ch/L-.455, .456, --; a/H-.435, .465, .415; b/H-.435, .465, .439; c/L-.424, .507, .545?.

Discussion: On slab TC 9143B there are a tail fin and a few connected fish vertebrae on the same bedding plane as the conchostracan fossils. On slab TC 9200, approximately one-half of a fish body with scales and fragments of a fin occurs on the same bedding plane as some of the conchostracan fossils.

A comparison of the fish-bed conchostracan species described above with those of the Mesozoic elsewhere yielded no close fit. For example, *Eosestheria lingyuensis* Chen from the Chinese Upper Jurassic, shares a few valve characteristics: H/L (.710 vs. .697–.754 Storm Peak species), umbo inset and close to anterior margin, and ornamentation of contiguous granules. However, the anterior margin is rounded and the posterior margin is convex; the dorsal margin join to the posterior margin is steeper than in the Antarctic species. The Chinese species also has double growth lines lacking in *"ichthystromatos."* Other described species differed in some of the above-cited features as well as in length of dorsal margin, shape and size of umbo, and various ratios.

38. *Cyzicus (Euestheria) crustapatulis* n. sp.
(Pl. 15, Figs. 2, 4; Pl. 14, 17)

Diagnosis: Dimorphic valves. Dimorph 1. Expansive, robust, subovate valves with a comparatively short, dorsal margin behind the umbo; umbo inset one-third the length of the valve from the anterior margin; terminus of dorsal margin joins the posterior margin in a gentle arcuate curve, whereas the anterior-dorsal join is steeper; anterior margin is arcuate, the posterior is rounded. Growth bands, 18 or more; ornamentation, contiguous granules. Dimorph 2. Ovate to subcircular valves with umbo inset less than

one-fourth the length of the valve from the anterior margin; height of anterior sector of valve greater than posterior. Growth bands, 20 or more. Ornamentation, reticulate markings. (Note: these markings may be viewed as the pattern left by contiguous granules of varying shapes and sizes.)

Measurements: (mm) (Syntype D-1, TC 10122b followed by syntype TC 9077A.) L-5.5, 5.2; H-3.9?, 3.9; Ch-2.0, 2.0; Cr-1.5, 1.5; Av-1.5, 1.5; Arr-2.0, 1.7; a-1.9, 1.5; b-1.9, 1.5; c-2.5, 2.2; H/L-.709, .750; Cr/L-.273, .288; Av/L-.273, .288; Arr/L-.364, .327; Ch/L-.364, .385; a/H-.487, .385; b/H-.487, .385; c/L-.455, .423.

	L	H	Ch	Cr	Av	Arr	a	b	c
					Syntype D2, TC 10016a-f				
10016a	6.3	4.9	2.8	1.6	1.6	2.3	2.8	2.0	3.0
10016b	5.4	3.9	1.9	1.5	1.3	2.3	1.4	2.0	2.8
10016c	5.0	4.1	3.0	?	1.3	0.8	2.1	2.0	2.7
10016d	6.0	4.4	3.8	?	1.4	1.4	2.2	1.7	2.6
10016e	6.1	4.3	3.0	1.5	1.5	2.0	1.8	2.5	3.4
10016f	5.4	4.0	2.1	1.3	1.3	1.8	1.8	1.7	2.4

	H/L	Cr/L	Av/L	Arr/L	Ch/L	a/H	b/H	c/L
10016a	.778	.254	.254	.365	.444	.571	.408	.476
10016b	.722	.278	.241	.426	.362	.359	.513	.519
10016c	.820	?	.260	.160	.600	.512	.458	.540
10016d	.733	?	.233	.213	.550	.500	.386	.433
10016e	.705	.246	.346	.328	.492	.419	.581	.557
10016f	.741	.241	.241	.333	.389	.452	.425	.444

Ranges of some ratios have a higher and/or lower upper limit when the D-1 types and the D-2 dimorphs are added: H/L-.680–.820; Arr/L-.160?–.426; Ch/L-.350–.600; a/H-.319–.574; b/H-.385–.581; c/L-.423–.557. This can be considered the species range.

Material: Syntype: dimorph, D-1, TC 10122b, internal mold of left valve. Syntype: TC 9077A, mold of left valve with remnant of inner portion of original shell. Syntype: TC 9071B, inner carbonized portion of original right valve impressed into substrate; part of dorsal margin and postero-dorsal margin missing. Dimorph, D-2 (TC 10016a-f), a coquina of completely or partially carbonized valves; in addition, juvenile valves include: TC 9053A, 9535, 9075, 9139A, and 9071.

Locality: Storm Peak. (See species "29" above for coordinates.) Upper Flow, float, assigned to Upper Flow, bed 5; bed 5, lower (TC 10016a-f, and bed 11. Lower Flow, Station 0, bed 2 = D.E site 27.85, TC 9077A. Lower Jurassic.

Discussion: The closest species is one from the Algerian Sahara Cretaceous, *Cyzicus (Euestheria) lefranci* n. sp. (see "Algeria" under Africa, this memoir). It shares some characteristics of the new species, for example one of its two configurations, subovate (which corresponds to Dimorph 1), H/L, Arr/L overlap the new species in the lower part of its range, and a/H also overlaps. Differences between the two species include double growth lines (absent in the new species), another configuration, subelliptical (vs. ovate to subcircular, Dimorph 2); and b/H, and c/L are greater than in *"crustapatulus"*.

The two types of carapace configuration appear to represent male (D 1) and female (D2), respectively. Kusumi (1961) observed that for some living conchostracan species, females had a greater number of growth bands. He also observed that for the species of three living genera, females had an outline of the carapace "a little more expanded posteriorly than that of the male" (Kusumi, 1961, p. 7). Both of the above features are characteristics of Dimorph 2 of the new species.

39. *Cyzicus (Euestheria) rhadinis* n. sp.
(Pl. 15, Fig. 6)

Diagnosis: Eccentrically subovate valves have a straight, long dorsal margin and taper posteriorly. Anterior and posterior margins rounded; small umbo, subterminal. Growth bands in three sets (see "Discussion" for explanation); number of bands, 32 or more. Ornamentation: very fine contiguous granules on growth bands.

Measurements: (mm) Holotype: L-5.4; H-2.7; Ch-3.8; Cr-0.7; Av-0.6; Arr-1.0; a-1.3; b-1.2; c-1.3; H/L-~.500; Cr/L-.111; Av/L-.111; Arr/L-.185; Ch/L-.704; a/H-.481; b/H-.444; c/L-.241.

Material: holotype TC 9187c, paired valves, right and left interior molds, portion of original inner-valve layer preserved. A small window eroded in this inner layer affords a limited view of the right valve's impression.

Locality: Storm Peak. (See species "29" above for coordinates.) Lower Flow, Station 1, bed 4, a fish bed (holotype). On the same and/or slightly older or younger bedding planes are body scales and partially preserved fins of the actinopterygian *Oreochima ellioti* Shaeffer (1972). Several scattered conchostracan valves occur with the fish fossils. Lower Jurassic.

Discussion: The nearest Antarctic conchostracan to the above species is *Cyzicus (Lioestheria) longacardinis* n. sp., from Mauger Nunatak, which also has elongated valves. Aside from difference in ornamentation (hachure markings vs. contiguous granules, n. sp) the two species differ in several ratios: H/L (.639–.690 vs. .500 n. sp.); *"rhadinis"* has a larger Cr/L, Av/L, Arr/L and smaller Ch/L (.405–.410 vs. .704).

C. (Eu.) rhadinis occurs in same Lower Flow interbed and the same fish horizon as does *C. (Eu.) ichthystromatos* n. sp., but at a different station. It is therefore of interest to compare the two species which, though sharing some valve characteristics, differ markedly. *C. (Eu.) rhadinis* tapers posteriorly, and has a longer dorsal margin; the umbo does not arch above the dorsal margin; its H/L ratio is two to three magnitudes smaller; much smaller are all other ratios except a/H, which is close, and b/H, which overlaps. These discrepancies cannot be attributed to growth stage or size because *"ichthystromatos"* has one syntype of greater overall length.

Three sets of growth bands occur. A set of growth bands is defined by broader vs. tighter spacing of successive bands. A broadly spaced set can denote one or another stage of immature valves. A tightly spaced set can represent one or another stage of a mature valve (cf. Kusumi, 1961; Tasch, 1977b).

40. *Cyzicus (Euestheria) transantarctensis* n. sp.
(Pl. 15, Figs. 7, 8)

Diagnosis: Broadly ovate, compact valves with a short, slightly arched dorsal margin behind umbo; umbonal terminus rises slightly above dorsal margin, and the umbo is inset about one-third the length of the valve; the join of antero-dorsal margin in front of the umbo is steeper than the postero-dorsal join; anterior sector of valve has greater height than posterior; anterior and posterior margins rounded; antero-dorsal angle, 30°; valves have a posterior slant. Ornamentation of very fine granules on growth bands.

Measurements: (mm) (Holotype TC 9081C, followed by paratypes 9112B and 9112A, respectively.) L-3.5, 3.5, 4.6; H-2.2, 2.8, 3.5; Ch-1.5?, 1.4, 2.1?; Cr-1.0. 0.8, 1.0; Av-1.0, 0.8, 1.0; Arr-1.0, 1.3, 1.5; a-1.2, 1.4, 2.0; b-1.5, 1.5, 1.3; c-1.4, 1.5, 1.8; H/L-.629, .800, .761; Cr/L-.286, .229, .217; Av/L-.286, .229, .217; Arr/L-.286, .371, .326; Ch/L-.429, .400, .453; a/H-.545, .500, .571; b/H-.682, .536, .371; c/L-.400, .429, .391.

Material: Holotype TC 9081C, right valve, flattened along sector directly below dorsal margin; umbonal area carbonized. Paratype TC 9112B, left valve and portion of underlying mold. Paratype 9112A, left valve, flattened. Other specimens include TC 10099B, mold of right valve with fragments of inner layers of original valve.

Locality: Storm Peak. (See species "29" above for coordinates.) Upper Flow, bed 2 (Types), and bed 1, lower (TC 10099B). Lower Jurassic.

Discussion: The species described above differs from other Storm Peak euestheriids as follows: from *C. (Eu.) castaneus* n. sp., which is laterally subovate, has a straight dorsal margin, umbo inset one fourth vs. one-third, and with a greater height of anterior sector than posterior; from *C. (Eu.) rhadinis* n. sp., which is eccentrically ovate, has a straight dorsal margin and an H/L of .500 (vs. .629–.800 n. sp.); from *C. (Eu.) crustapatulus* n. sp., which consists of robust-ovate and subovate to subcircular valves, being dimorphic; from *C. (Eu.) ichthystromatos* n. sp., with a straight dorsal margin, small elliptical umbo that is inset but close to anterior margin, and with a larger Ch/L and smaller a/H.

The new species differs from Blizzard Heights and environs euestheriids as follows: *C. (Eu.) ellioti*, with its umbo inset less than one-third to subterminal, anterior height not greater than posterior, straight dorsal margin; whereas ratios overlap usually at the higher or lower end of the range, a/H is greater in the new species; from *C. (Eu.) juravariabalis* n. sp., which is subcircular and has a larger H/L, Av/L, c/L and a smaller a/H; from *C. (Eu.) minuta* (von Zieten) with its umbo subterminal to slightly inset from the anterior margin, a straight dorsal margin, a smaller a/H, larger c/L; from *C. (Eu.) beardmorensis* n. sp., which is subovate and tapered posteriorly, has a large dorsal margin, smaller Av/L, Cr/L, a/H, and larger Ch/L; from *C. (Eu.) for-*

mavariabalis n. sp., which is subovate to subelliptical with posterior sector of valve of greater height than anterior, umbo close to anterior margin and a larger Arr/L and c/L.

41. *Cyzicus (Euestheria) castaneus* n. sp.
(Pl. 16, Figs. 5, 6, 8)

Diagnosis: Dimorphic valves: laterally subovate (D-1) and more compactly ovate (D-2); comparatively long, straight dorsal margin; umbo inset more than one-fourth the length of the valve; anterior and posterior margins rounded; antero-dorsal angle steeper than the postero-dorsal angle. Double growth lines. Ornamentation of minute contiguous granules on growth bands.

Measurements: (mm) (Figures for syntypes 9055a, 9056a, 9056b, and 9055c are given in that order. L-6.5, 4.8, 5.5, 4.9?; H-4.5, 3.7, 4.2, 4.0?; Ch-3.2, 3.7, 4.2, 4.2; Cr-1.3, 3.7, 4.2, 4.2; Av-1.3, 1.3, 1.4?, 0.7; Arr-2.2, 1.3, 2.0, 1.6?; a-1.7, 1.9, 2.0, 1.9; b-2.1, 1.5, 2.0, --; c-2.8, 1.7, 2.0, --; H/L-.692?, .771, .764, .877?; Cr/L-.200, .271, .255, .867?; Av/L-.200, .271, .255, .167; Arr/L-.338, .271, .364, .326; Ch/L-.492, .437, .382, .867?; a/H-.378, .533, .476, .452; b/H-.467, .405, .476, --; c/L-.431, .362, .418, --.

Material: Dimorph, D-1, syntype: TC 9055a, almost complete, left valve slightly eroded. Dimorph, D-1, syntype: TC 9056a, left valve slightly covered posteriorly on dorsal margin. Dimorph, D-1, syntype: TC 9056b, crushed internal mold of left valve overlain by remnant of original shell. Dimorph D-2, Syntype: TC 9055c, right valve, part of dorsal-anterior sector eroded. Other specimens TC 9057 and TC 9122 (Station 2, bed 4).

Locality: Storm Peak. (See species "29" above for coordinates.) Lower Flow, Station 1, beds 1, 5, and 6: Station 2, beds 2 and 3; and Blizzard Heights environs, D.E site 23.57AA (= Tasch Station 0). Lower Jurassic.

Discussion: The species closest to the one described above is *Cyzicus (Euestheria) beardmorensis* n. sp. (from D. E. Site 23, Blizzard Heights; see Blizzard Heights Antarctica section, this memoir). The new species differs from *"beardmorensis"* in having an inset umbo that does not rise above the dorsal margin, a valve that does not taper, and a greater H/L, Cr/L, Av/L, and Ch/L. Some ratios overlap: a/H, b/H, and c/L (this last, while it overlaps, is greater by a whole magnitude).

Dimorph D-2, TC 9055c, differs from the D-1 dimorphs in having an H/L greater by a whole magnitude, a much greater Cr/L and Ch/L, and more tightly spaced, as well as a greater number of, growth bands.

Most of the new species' valves retain coloration—a dark brown color. This coloration is due to the haemolymph in the valve which changes color (darkens) upon demise of a given conchostracan.

42. *Cyzicus (Euestheria)* sp. 1
(Pl. 16, Fig. 7)

Diagnosis: Broadly and irregularly ovate, compact valve with a comparatively short, straight dorsal margin behind umbo that rises slightly above the dorsal margin. Margins: anterior apparently steep; posterior convex with a bulge in postero-ventral sector; ventral gently concave. Anterior sector has greater height than posterior. Growth bands, 14 or more. Ornamentation of contiguous granules on growth bands.

Measurements: (mm) L-3.8; H-3.1; Ch-1.2?; Cr-0.8; Av-0.8; Arr-1.8; a-1.3; b-1.6; c-1.5; H/L-.816; Cr/L-.210; Av/L-.210; Arr/L-.474; Ch/L-.316?; a/H-.419; b/H-.576; c/L-.395.

Material: External mold of right valve; antero-ventral sector is marginally eroded and fragment of inner layer of original valve preserved. TC 10099b.

Locality: Storm Peak, Upper Flow, bed 1, lower. Lower Jurassic.

Discussion: It is not possible to track down this irregularly ovate specimen because it is incomplete.

43. *Palaeolimnadia (Grandilimnadia)* cf. *glenleensis*
(Mitchell, 1927)
(For Mitchell's types, see below)
(Pl. 13, Fig. 8; Pl. 19)

Diagnosis: Small, subovate valves with comparatively large, subterminal umbo inset from the anterior margin approximately one-third the length of the valve. Margins: anterior steeply convex; posterior rounded; ventral gently curved. Growth bands: 9.

Measurements: (mm) (Figures for TC 9857a followed by those for TC 9033.) L-2.6, 1.8; H-1.9?, 1.2; Ch-0.9?, 0.8; Cr-0.7, 0.4; Av-0.7, 0.4; Arr-0.9, 0.6; a-0.7, 0.5; b-0.8, 0.2; c-1.1, 0.9; H/L-.731?, .667; Cr/L-.269, .222; Av/L-.269, .222; Arr/L-.346, .355; Ch/L-.346?, .444; a/H-.368?, .417; b/H-.421?, .583; c/L-.423, .500.

Material: TC 9857a, external mold of left valve, antero-ventral sector eroded; 9857c, internal mold of left valve, lower portion of valve covered; 9857d, another internal mold from a left valve but less of valve visible than in c; both 9857c and d are carbonized; TC 9033, internal mold of right valve. Another specimen is TC 10118.

Locality: Storm Peak. (See species "29" above for coordinates.) Tasch site, Upper Flow, beds 2 and 7, and Upper Flow, bed 8, lower half (for TC 9033 only). Lower Jurassic.

Discussion: The nearest palaeolimnadiid to the Storm Peak specimen described above is Mitchell's species, *Palaeolimnadia (Grandilimnadia) glenleensis* from the Permian and Triassic of Australia. (See below for Mitchell's types.) Umbonal position, size, and valve configuration are almost equivalent. Some ratios are close, such as H/L-.667 and c/L-.423; others are markedly less in the Antarctic material (Av/L, Arr/L), or greater (Ch/L). Some of these last-named differences in ratios could be due to particular sites of erosion.

The other palaeolimnadiid species suggestive of the new species is *Palaeolimnadia chauanbeiensis* Shen from the Lower Jurassic Wennan Formation, Mengyin District, Shandong, People's Republic of China (Chen, Pei-ji, 1982, p. 135 and Pl. 1,

Fig. 6). H/L is close; .666. Other parameters and ratios cannot be determined from published data. It is subovate with a comparatively large umbo inset from the anterior margin approximately one-third the valve length. Shen's (1981a, Pl. 1, Fig. 4), *Palaeolimnadia jiaodongensis* n. sp. (Lower Cretaceous Eastern Shandong) may be relevant but too poorly preserved.

44. *Palaeolimnadia (Grandilimnadia) minutula* n. sp.
(Pl. 14, Figs. 5, 6)

Diagnosis: Small, subovate valves with a prominent, elliptical, umbonal area that inclines anteriorly and is in contact with the arched dorsal margin. Growth bands, 5 to 6.

Measurements: (mm) (Holotype followed by paratype where figures are available.) L-2.0?, 0.97; H-1.5?, 0.75; Ch-0.7?, --; Cr-0.5, --; Av-0.5, --; Arr- --, --; a-0.6, --; b- --, --; c- --, --; H/L-.750?, .773; Cr/L-.250?, --; Av-.250?, --; Arr/L- --, --; Ch/L-.350?, --; a/H-.400?, --; h_u*-0.42, 0.16; l_u-0.85, 0.32; h_u/l_u-0.49, 0.50. (*h_u and l_u = height and length of umbo, respectively.)

Material: Holotype TC 10004, external mold of right valve, inner portion of valve preserved; eroded along ventral periphery and along upper posterior margin. Paratype TC 10007, internal mold of left valve, slightly crushed and eroded in the upper antero-umbonal area; zigzag medial crack near center of umbo. (The paratype was described and studied before the new suite of measurements used throughout this memoir was begun. Hence, only certain measurements and ratios are available because the paratype has been mislaid.)

Locality: Storm Peak. (See species "29" above the coordinates.) Upper Flow, bed 6 (both types). Lower Jurassic.

Discussion: The Storm Peak species differs from the Mauger Nunatak species *(Palaeolimnadia (Grandilimnadia)* sp. 1 in having a larger H/L and a smaller umbonal ratio of h_u/l_u. *P. (G.) minutula* n. sp. is unlike the Mesozoic (Jurassic and Cretaceous) *"Limnadia"* (=Palaeolimnadia) and Jurassic *"Eulimnadia"* (=Palaeolimnadia) from the USSR and other countries described and/or reassigned by Novojilov (1970). For example, among the Jurassic species, *Limnadia jaxarctica* Novojilov, 1970 (p. 96, Pl. 4, Fig. 2) has a more elongate-ovate valve, an umbo that is irregularly ovate, a longer dorsal margin that is straight, and a markedly smaller H/L (.590 vs. .773–.750 n. sp.); *Limnadia sogdianica* Novojilov (1970, p. 96, Pl. 4, Fig. 1) has an umbo that is irregularly ovate, apparently occupies most of the valve, and has a larger H/L (.780–.860 vs. .773–.750); *Limnadia dundugobica* Novojilov 1954 *in* Novojilov 1970 (p. 97, Fig. 64) has subovate valves that taper posteriorly, an ovate umbo, and a larger H/L that overlaps the upper range of the new species (.750–.800 vs. .773–.750 n. sp.)

Cretaceous species include *Limnadia pruniformis* Novojilov 1970 (p. 97, Fig. 65) with a subovate configuration, subovate umbo, steeper postero-dorsal angle, and a larger H/L (.800 vs. .773–.750 n. sp.); *Limnadia taratchiensis* (Nov., 1954, *in* Novojilov, 1970, p. 98, Fig. 66) with an irregularly subovate valve,

longer and straight dorsal margin behind the ovate umbo, and a smaller H/L (.680 vs. .773–.750 n. sp.). Of interest is the fact that all of the above Mesozoic species described by Novojilov (1970) are markedly greater in length by two to nine times.

Another specimen, TC 10015 (Storm Peak, Upper Flow, bed 11, lower) may be related to *"minutula"* but it has an umbonal ratio: h_u/l_u-.80, and is otherwise incomplete. Hence, other ratios cannot be compared.

Genus *Pseudoasmussiata* new name. Pro *Pseudoasmussia* Defretin-Le Franc, 1969. Non *Pseudoasmussia* Novojilov (1954c) who has priority. (See South America section of this study.)

45. *Pseudoasmussiata defretinae* n. sp.
(Pl. 17, Fig. 7)

Diagnosis: Valves with subquadrate-subovate aspect; short, straight dorsal margin; umbo inset, almost medial and slightly raised above dorsal margin; anterior margin steep; posterior margin rounded; valves with posterior slant.

Measurements: (mm) L-5.0; H-3.8; Ch-2.7; Cr-0.9; Av-0.9; Arr-1.6; a-1.3; b-1.5; c-2.6; H/L-.760; Cr/L-.180; Av/L-.180; Arr/L-.320; Ch/L-.540; a/H-.342; b/H-.395; c/L-.520.

Material: Holotype TC 10114a, partly carbonized inner layer of right valve and underlying mold; eroded behind umbo along dorsal margin. Erosion prevents an accurate count of growth bands.

Locality: Storm Peak. (See species "29" above for coordinates.) Upper Flow, bed 4. Lower Jurassic.

Discussion: This is a *Pseudoasmussia*-like species with the added feature of a posterior slant of the valve axis (a line through the midpoint of the umbo to the ventral margin). The straight dorsal margin with the umbo inset from the anterior margin is characteristic of the Congo basin species described by Defretin-Le Franc. These differ from *"defretinae"* as follows: *Pseudoasmussia (= Pseudoasmussiata) ndekeenensis* D.-Le Franc (Upper Wealden) is broadly ovate; anterior and posterior margins are both rounded; and it has a larger Cr/L, Arr/L, a/H, b/H, and a smaller Ch/L; *Pseudoasmussia (= Pseudoasmussiata) banduensis* D.-Le Franc (Upper Wealden) is more subovate, the anterior and posterior margins are rounded, and it has a smaller H/L, a larger Cr/L, and b/H. *Pseudoasmussia (= Pseudoasmussiata) duboisi* D.-Le F., (Upper Jurassic) is more broadly ovate with a longer dorsal margin behind the umbo, and larger Ch/L, a/H, and smaller Arr/L.

The South American *Pseudoasmussiata katooae* n. sp. (Brazil, Upper Triassic) is obliquely ovate, its Cr/L, Av/L, Arr/L, Ch/L, a/H, and b/H are larger, and H/L, Arr/L, and c/L are smaller than in *"defretinae"*.

46. *Asmussia* sp. 1
(Pl. 17, Fig. 1)

Diagnosis: Subquadrate valve with a straight, comparatively long, dorsal margin and a submedial umbo; anterior and ventral

margins are rounded; antero-dorsal angle of 45°; postero-dorsal, 30°. Ornamentation of fine, hachure markings on growth bands.
Measurements: (mm) L-5.1; H-3.8; Ch-2.2?; Cr-1.5; Av-1.5?; Arr-1.4; a-2.0; b-2.0; c-2.5?; H/L-.745; Cr/L-.294; Av/L-.294?; Arr/L-.275; Ch/L-.431?; a/H-.526; b/H-.526; c/L-.490.
Material: TC 9183A, whitened inner remnant of right valve, overlying internal mold; eroded slightly along dorsal-posterior sector and anterior margin.
Locality: Storm Peak. (See species "29" above for coordinates.) Upper Flow, bed 9. Lower Jurassic.
Discussion: The only Mesozoic species that even approaches the Storm Peak asmussid is *Asmussia dekeseensis* Defretin-Le Franc (1967) from the Lower Cretaceous Bokunga Series, Congo Basin, but it has a different overall configuration (ovate vs. subquadrate, n. sp.); umbo inset but closer to anterior margin, and a smaller H/L, a/H, b/H, Av/L, Arr/L, and a larger c/L; Cr/L overlaps.

Agate Peak (northern Victoria Land)

47. *Cyzicus (Lioestheria) disgregaris* n. sp.
(Pl. 18, Fig. 1 (see also Pls. 16, 17, 19, 20)

Diagnosis: (Repeated from the Storm Peak diagnosis of this species). Dimorphic valves: Dimorph 1 with laterally ovate valves to vertically ovate, umbonal position usually close to but inset from anterior margin and in some specimens more subterminal; steeply convex to rounded anterior margin. Growth bands variable; approximately 29. Dimorph 2, subcircular valves with umbonal position variable from subterminal to submedial. Growth bands to 33 or more. For both dimorphs, double growth lines may or may not be clearly defined. Ornamentation of hachure markings on growth bands of both dimorphs. On some growth bands, some vertical markings may be irregular; i.e., non-parallel with others, at an angle from the dominant trend. (The last type of marking cannot be attributed to shell flattening or distortion.)
Measurements: (mm). Dimorph 1: laterally ovate to vertically ovate valves. (D.E 81-1-2B, specimen "b" followed by specimen "a"). L-5.7, 5.8; H-3.4, 3.6?; Ch-2.6?, 2.0?; Cr-1.3, 1.8; Av-1.3, 1.8; Arr-1.8, 2.0; a-1.5, 1.7?; b-1.2?, 1.7; c-2.0?, --; H/L-.596, .621?; Cr/L-.228, .310; Av/L-.228, .310; Arr/L-.316, .185; Ch/L-.456?, .345?; a/H-.441, .472?; b/H-.353?, .472?; c/L-.351, --. Dimorph 2: subcircular forms. (D. E. 81-1-4, specimen "a" followed by specimen "b.") H/L-.874, .735; Cr/L-.218, .383?; Av/L-.218, .388?; Arr/L-.382, .286?; Ch/L-.400, .327; a/H- --, .555; b/H-.548, .417; c/L-.455, .306?.
Material: DE 81-1-2B (slab 2) a–c, three fairly complete specimens and other partial valves of this new species, all on the same bedding plane and close to each other. Specimen "a," left valve, eroded along dorsal sector; specimen "b," right valve, subovate, eroded, somewhat distorted marginally by inward-folded ventral growth bands; shows excellent hachure markings on the less eroded sections of anterior face of the valve; specimen "c," an

eroded left valve partly overlying a right valve. Dimorphs: D. E. 81-1-4a and b.
Locality: David Elliot Collection. D. E. 81-1-2B, Agate Peak, north Victoria Land, lat 72°57'S, long 163°48'E (Fig. 4b); and dimorphs: D. E. 81-1-4, specimen "a," slab 3; D.E 81-1-4, specimen "b," slab 1, same site. Lower Jurassic.

Carapace Nunatak (= CN)

For the present study there are collections for four Tasch stations and, in addition, the J. S. Schopf Collection (collected before 1966) from both the north and southeast spurs, generously contributed to this writer for study.

All stations and sites yielded a single, dimorphic conchostracan fossil species, *Cyzicus (Lioestheria) disgregaris* n. sp.

48. *Cyzicus (Lioestheria) disgregaris* n. sp.

Diagnosis: See Storm Peak entry for species diagnosis. (Repeated under Agate Peak, above.)
Locality: CN Tasch Station 2, Bed 1. (See Pl. 19, Figs. 4–8; Pl. 20, Figs. 1–4, for illustrations of the CN occurrences of this species.)
Measurements: (mm).

Dimorph 1
(designated a male valve)
Laterally ovate to ovate

Specimen	L	H	Ch	Cr	Av	Arr	a	b	c
80442	4.7	3.6	1.6?	1.5?	--	1.8	1.3	1.5	2.8
80504-02b	2.8	1.9	0.9?	0.9	0.9	1.0	0.8	0.9	1.4
80380	5.4	3.6	2.0?	1.7	1.7	1.8	1.6	1.6	2.7
80422b	3.2	2.1	1.0?	1.1?	1.1?	1.1	0.8	1.0	1.6

	H/L	Cr/L	Av/L	Arr/L	Ch/L	a/H	b/H	c/L
80442	.745	.306	--	.367?	.327	.301	.417	.571
80504-02b	.679	.321	.321	.357	.321	.421	.474	.500
80380	.667	.315	.315?	.333	.370?	.444	.444	.500
80422b	.656	.344	.344?	.344	.313?	.381	.476	.500

Dimorph 2
(designated a female valve)
Subcircular.

Specimen	L	H	Ch	Cr	Av	Arr	a	b	c
80430e	5.0	3.7	1.7?	1.7?	--	1.6?	1.5	1.7	2.7
80380-18a	4.6	3.5	1.1	1.5	--	1.9	2.0	1.8	2.2
80432b	4.0?	3.2	1.0?	1.2?	--	1.8?	--	1.3	1.5?
80420a	5.2	4.0	1.7	1.3?	--	2.2	2.3	2.3	2.0

	H/L	Cr/L	Av/L	Arr/L	Ch/L	a/H	b/H	c/L
80430e	.740	.340?	--	.320?	.340?	.405	.459	.540
80380-18a	.761	.326	--	.413?	.239?	.571	.514	.478
80432b	.800?	.300	--	.450?	.250?	--	.406	.375?
80420a	.769	.250?	--	.423	.327	.575	.500	.385

For range, see "Measurements" under Storm Peak entry for this species.

Locality: CN Tasch Station 2, Bed 2.
Measurements: (mm).

Dimorph
(designated a male valve)
Laterally ovate to ovate

Specimen	L	H	Ch	Cr	Av	Arr	a	b	c
80375	4.7	2.9	2.0	1.3?	--	1.4	1.2	1.3	1.8
80374b	4.6	3.1	--	--	--	--	--	1.4?	1.7

	H/L	Cr/L	Av/L	Arr/L	Ch/L	a/H	b/H	c/L
80375	.671	--	.298	.426	.277	.414	.448	.383
80374b	.674?	--	--	--	--	--	.452	.370

Range: H/L-.674 - .771; b/H-.351 - .452; c/L-.370 - .458.

Locality: CN Tasch Station 1', bed 1.
Measurements: (mm).

Dimorph 1
Laterally ovate to ovate

Specimen	L	H	Ch	Cr	Av	Arr	a	b	c
80041	6.3	4.8	2.5?	1.4?	--	2.4?	--	1.7	2.8
80065A	4.0	2.9	1.5	1.1	--	1.4	1.3	1.3	1.9
80029	4.8	2.9	2.0?	1.2?	--	1.5?	1.5?	1.5?	2.0

	H/L	Cr/L	Av/L	Arr/L	Ch/L	a/H	b/H	c/L
80041	.762	.222	--	.381	.397	--	.354	.444
80065A	.725	.272	--	.350	.375	.448	.448	.475
80029	.620?	.250?	--	.312	.417	.517	.448	.417

For range, see "Measurements" under Storm Peak entry for this species.

Dimorph 2
(designated a female valve)
Subcircular.

Specimen	L	H	Ch	Cr	Av	Arr	a	b	c
80064	3.2	2.6	1.1	0.8	0.8?	1.3	1.2	1.0	1.2?
80053a	3.1	2.2	1.1?	1.0?	--	1.0?	1.1	1.1	1.7
80030	3.3	2.6	0.9	1.1?	--	1.3?	1.2	1.1	1.3

	H/L	Cr/L	Av/L	Arr/L	Ch/L	a/H	b/H	c/L
80064	.813	.250	.250?	.406	.344	.462	.385	.375
80053a	.742	.323	--	.323?	.355?	.478	.478	.548
80030	.788	.333	--	.394?	.273?	.462	.423	.394

For range, see "Measurements" under Storm Peak entry for this species.

Locality: CN Tasch Station 1*.
Measurements: (mm).

Dimorph 1
(designated a male valve)
Laterally ovate to ovate

Specimen	L	H	Ch	Cr	Av	Arr	a	b	c
80201B	6.0	4.2	2.1?	1.3	1.3	2.5	1.7	2.3	3.0
80249b	5.5	4.0	2.0?	1.2	1.2?	2.4	1.3	1.4	2.5
80202B	4.9	3.6	2.3?	0.9	0.9	1.7	1.8	1.5	1.9
80205A	5.1	3.3?	2.2?	1.5	1.5	1.4?	1.5?	1.3	2.0

	H/L	Cr/L	Av/L	Arr/L	Ch/L	a/H	b/H	c/L
80201B	.700	.217	.217	.417	.350?	.405	.524	.500
80249b	.727	.218	.218	.436	.364?	.325	.350	.455
80202B	.735	.184	.184	.347	.469?	.500	.417	.388
80205A	.647	.294	.294	.275?	.431?	.455	.394	.392

Dimorph 2
(designated a female valve)
Subcircular

Specimen	L	H	Ch	Cr	Av	Arr	a	b	c
80221d	5.0	3.7	1.7?	1.8	---	1.5	1.5?	1.5	2.0
80204A	4.9	3.7	2.1	1.0	1.0	1.0	1.7	1.4	1.8
80243B	4.7	3.6	1.9?	1.2	---	1.7	1.7	1.2	2.4
80278b	4.7	3.5	---	1.5	1.5	1.7	1.7	1.7	2.4

	H/L	Cr/L	Av/L	Arr/L	Ch/L	a/H	b/H	c/L
80221d	.740	.360?	---	.300?	.340	.541	.541	.560
80204A	.755	.204	.204	.347	.429	.510	.486	.510
80243B	.766	.255?	---	.362?	.404?	.444	.444	.511
80278b	.745	.319	.319	.362	.319	.486	.486	.591

*TC 80205, 80204, 80201, 80249 from Station 1, bed 3; TC 80243b, 80202 from Station 1, bed 2; TC 80278 from Station 1, bed 1, TC 80221d, float.

Locality: Tasch Station 0*.
Measurements: (mm).

Dimorph 1
(designated a male valve)
Laterally ovate to ovate

Specimen	L	H	Ch	Cr	Av	Arr	a	b	c
80301b	4.9	3.1	1.3?	1.6?	1.6	2.0?	1.2	1.5	2.9
80073b	4.7	3.5	1.8?	1.6?	---	1.3?	1.6	1.8	2.3
80296-02b	4.1	2.9	1.6?	0.8?	---	1.5?	1.1	1.2	2.4

	H/L	Cr/L	Av/L	Arr/L	Ch/L	a/H	b/H	c/L
80301b	.633	.327?	.327	.408	.365?	.387	.484	.592
80073b	.745	.340?	----	.277?	.383	.457	.514	.489
80296-02b	.683	.195?	----	.366?	.390?	.393	.429	.585

Dimorph 2
(designated a female valve)
Subcircular

Specimen	L	H	Ch	Cr	Av	Arr	a	b	c
80296-03a	3.9	3.1	1.2?	1.1?	1.1?	1.5?	1.5	1.3	2.3
80011-22c	5.1	4.0	1.7?	1.4?	---	2.0	1.8	1.6	2.2

	H/L	Cr/L	Av/L	Arr/L	Ch/L	a/H	b/H	c/L
80296-03a	.795	.282?	.282	.385?	.308?	.494	.419	.590
80011-22c	.784	.275?	----	.392	.333?	.450	.400	.431

*TC 80296-03a, 80296-02b, beds 2 and 3;
TC 80011-22c and TC 80310b, bed 3;
TC 80073b, bed 4.

Locality: CN Southeast Spur* (Dr. J. S. Schopf Collection).
Measurements: (mm).

Dimorph 1
(designated a male valve)
Laterally ovate to ovate

Specimen	L	H	Ch	Cr	Av	Arr	a	b	c
80017A, b	5.0	3.2	1.3?	1.5?	---	2.2?	---	1.7	2.8
80017C, a	4.9	3.6	2.0?	1.3	---	1.6	1.3?	1.7?	2.7
80366, a	4.8	3.0	1.6	1.5?	---	1.7?	1.5	1.1	1.7
80004, b	5.8	3.9	---	1.4?	---	---	2.0?	---	2.5

	H/L	Cr/L	Av/L	Arr/L	Ch/L	a/H	b/H	c/L
80017A, b	.640	.300?	----	.440?	.260?	----	.531	.560
80017C, a	.735	.265?	----	.327	.408	.361	.472?	.551
80366, a	.625	.312	----	.354?	----	.500	.367	.354
80004, b	.672	.241?	----	----	----	.517?	----	.431

Dimorph 2
(designated a female valve)
Subcircular#

Specimen	L	H	Ch	Cr	Av	Arr	a	b	c
80011, b	4.6?	3.6	1.0?	1.5	---	1.9	---	2.0	2.4?
800365, a	4.2	3.3	1.5?	1.4?	---	1.2?	1.3	1.1	2.1

#Several others are eroded or covered in part and thus, not
measurable.

	H/L	Cr/L	Av/L	Arr/L	Ch/L	a/H	b/H	c/L
80011, b	.783?	.326	----	.413	.217?	---	.556	.522?
800365, a	.786	.333?	----	.286?	.357?	.394	.333	.520

*The S.E. spur (or ridge) of the above collection is the
same site from which Gunn and Warren (1962:111) reported
fossil conchostracans. (There are some 33+ m of sediment,
and the fossiliferous beds are in intercalated, secondarily
chertified bands).

Locality: CN North Spur* (J. S. Schopf Collection).
Measurements: (mm).

Dimorph 1
(designated a male valve)
Laterally ovate to ovate

Specimen	L	H	Ch	Cr	Av	Arr	a	b	c
80007, d	5.4	3.9	1.3?	1.7?	---	2.5?	1.6	1.7	2.3
80025, e	4.8	3.0?	---	---	---	---	1.0?	1.3	2.5
80025, f	4.2	3.0?	1.4?	0.9	0.9	1.8?	1.6?	1.4	2.2
80025, c	5.0	3.6	1.4	1.8?	---	1.8?	1.7	1.4	2.0

	H/L	Cr/L	Av/L	Arr/L	Ch/L	a/H	b/H	c/L
80007, d	.722	.315	----	.463	.241	.410	.436	.426
80025, e	.625	----	----	----	----	.333?	.433	.521
80025, f	.714	.214	.214	.429?	.333?	.533	.467	.524
80025, c	.720	.360?	----	.360	.280?	.472	.444	.400

Dimorph 2
(designated a female valve)
Subcircular

Specimen	L	H	Ch	Cr	Av	Arr	a	b	c
80025, b	4.3	3.4	1.5?	1.0?	---	1.7	1.5?	1.6	1.8

	H/L	Cr/L	Av/L	Arr/L	Ch/L	a/H	b/H	c/L
80025, b	.791	.233?	----	.395	.349?	.441?	.471	.419

*No bed data available for this collection.

AUSTRALIA (Devonian)

Buchan Caves Limestone, Victoria

Cyzicus (Euestheria) talenti n. sp.
(Pl. 4, Fig. 1)

Diagnosis: Subovate valves with the umbo inset from and near
to the anterior margin; straight dorsal margin; anterior margin
convex; posterior margin rounded. Growth-band spacing dimin-
ishes and remains so from midway on the valve to the ventral
margin. Ornamentation: many fine contiguous granules on
growth bands (seen at high power only).
Measurements: (mm).

Syntypes	L	H	Ch	Cr	Av	Arr	a	b	c
10901*	4.1	2.0	3.0	1.5	.80	.50	1.9	1.1	2.1
10903a	5.0	3.6	3.2	1.5	.90	.50	1.8	1.1	2.3

	H/L	Cr/L	Av/L	Arr/L	Ch/L	a/H	b/H	c/L
10901*	.707	.366	.195	.122	.732	.655	.378	.512
10903a	.720	.300	.180	.100	.640	.500	.306	.460

*Lectotype.

Material: Syntype: TC 10901, two slices, external mold of right valve; TC 10903, left valve, one eroded, nearly complete valve; two fragmentary valves (b, c) which show ornamentation on the growth bands.

Locality: Fairy Bed, Snowy Volcanics, Buchan Caves Limestone, Victoria. Devonian. (A locality studied by Teichert and Talent, 1958.)

Name: Named for Professor John A. Talent who discovered the conchostracan-bearing bed in the Buchan Caves Limestone.

Discussion: The new species is unlike any Devonian euestheriid described by Lutkevich (1929, Pl. 36) from the Northwest Province, USSR, or in several papers by Novojilov (1953; 1954a, Pl. 1, Fig. 2) from Kazakhstan, USSR. It differs from these on the basis of various parameters as well as those other species having either a longer dorsal margin, differently positioned umbo, more tapered posterior margin, being ovate, or other variant configuration. Maillieux's (1933) euestheriid from the Belgian Devonian has a smaller H/L and is more tapered posteriorly. The new species also differs from Rennie's Witteberg species (1934) Republic of South Africa in configuration and nonterminal umbo. An Aztec Siltstone species from the Antarctic Devonian (see Antarctica, this memoir differs from the new species in that it has a larger H/L (- .844), a subcentral umbo, shorter dorsal margin, and numerous other measured parameters.

The specimens described above were collected by this writer in company with Professor Talent.

AUSTRALIA (Carboniferous)

Western Australia

(Species described in Tasch and Jones, 1979a. Numbers that follow refer to those shown on locality map; see Fig. 7.)

15. *Leaia (Leaia) andersonae*
16. *Leaia (Hemicycloleaia) tonsa*
17. *Ellipsograpta? sp.*
18. *Leaia (Hemicycloleaia) rectangelliptica*
19. *Leaia (Hemicycloleaia) grantrangicus*
20. *Leaia (Hemicycloleaia) longacosta*
21. *Monoleaia australiata*
22. *Rostroleaia sp.*
23. *Limnadiopsileaia carbonifera*
24. *Cyzicus (Lioestheria) sp. undet. 1*
25. *Cyzicus (Lioestheria) sp. undet. 2*

(Species 15–25 are all from the Grant Range, No. 1 well, Anderson Formation, 90 km south–southeast of Derby, near Joseph Bonaparte Gulf, Northwest Territory [Visean–Namurian].)

Queensland

Drummond Basin. (Species described *in* Tasch and Jones, 1979c.)

47. *Leaia (Hemicycloleaia) drummondensis* Tasch. Narrien Range, NW of Emerald. Lower Carboniferous (Tournaisian?–Visean).

AUSTRALIA (Permian)

Newcastle Coal Measures, New South Wales

(For complete synonymy and descriptions, as well as photographs of Mitchell's types, biostratigraphy, and analysis, see Mitchell's types below.)

Cycloleaia discoidea (Mitchell) 1925
(Pl. 22, Fig. 1)

Diagnosis: Broadly ovate, bicarinate valves, with short dorsal margin.

Measurements: Preservation is too poor for meaningful measurements because all margins are eroded, but one can infer the original configuration.

Material: Tasch Collection (= TC) 10521b, external mold of left valve; 10115, external mold of right valve, overlying a larger conchostracan valve.

Locality: Station 5A, bed 2g, New South Wales, Permian (Tartarian).

Discussion: Although badly eroded, the configuration and the characteristics of the two ribs identify these valves as *C. discoidea* (cf. Mitchell's type, 1925c, Pl. 40, Fig. 6).

Leaia (Hemicycloleaia) belmontensis (Mitchell) 1925
(Pl. 22, Fig. 2)

Diagnosis: Subovate, bicarinate valve with straight dorsal margin that thickens posteriorly; anterior rib gently curved; posterior rib, straight; both cross umbo dorsad, fairly close together, and terminate on second (anterior rib), or third (posterior rib) growth band from ventral margin.

Measurements: (mm) L-4.6; H-3.1; Ch-3.5; Cr-1.3; Av-0.1; Arr-0.3; a-1.4; b-0.7; c-2.1; H/L-.674; Cr/L-.283; Av/L-.109; Arr/L-.065; Ch/L-.761; a/H-.452; b/H-.226; c/L-.457; d_1-0.2; d_2-2.9 ±.1; alpha-83° ±2°; beta-34° ±2°.

Material: TC10552a, internal mold of left valve; TC 10574, internal mold of left valve.

Locality: Substation 7a, bed 1. Quarry, 0.12 km southeast of Burton Road (TC 10552a); Darling Street, Wommara District, New South Wales (TC 10574). Permian (Tartarian).

Discussion: These specimens are quite close to the holotype in critical dimensions such as d_1 and d_2, alpha, beta, a/H, b/H, and c/L, as well as H/L, [See *L. (H.) mitchelli* to which it is now assigned.]

Leaia (Hemicycloleaia) collinsi (Mitchell) 1925
(Pl. 22, Fig. 3)

Diagnosis: Subovate, bicarinate valve with elongated dorsal margin; anterior and posterior ribs approach very close to each other as they cross the umbo dorsad; both ribs curve slightly across upper portion of valve and become straight ventrad.

Measurements: (mm) L-3.9; H-2.4; Ch-3.0; Cr-.8; Av-.4; Arr-.5; a-.9; b-?; c-1.6; H/L-.615; Cr/L-.205; Av/L-.103; Arr/L-.28; Ch/L-.769; a/H-.375; b/H-?; c/L-.410; d_1-.2; alpha-72° ±2°; beta-34° ±2°, sigma-41°.

Material: TC 10617b, internal mold of right valve flattened; ventral and lower anterior margins eroded.

Locality: Station 5, beds 5 and 6A. New South Wales, Permian (Tartarian).

Discussion: The specimen is close to the holotype in configuration and various parameters: d_1, Ch/L, H/L, a/H, and beta. The dorsal margin does not display thickening posteriorly as in the holotype, but this may be due to erosion.

Leaia (Hemicycloleaia) cf collinsi (Mitchell) 1925

A mere fragment of a valve showing the converging anterior and posterior ribs. All that can be determined is the sigma angle. (See measurements, Fig. 18: *Leaia (H.) belmontensis* -50°; *Cycloleaia discoidea* -49°; *Leaia (H.) paraleidyi* -47°; *Leaia (H.) collinsi* -40°; *Leaia (H.) sulcata* -48°.) Because the sigma angle for the fragment under discussion is 41°, it is closest to *Leaia (H.) collinsi.* (The ribbed types for which sigma angles are cited came from the chert quarry near Belmont.)

Material: TC 10502, fragment of bicarinate valve.

Locality: Belmont Quarry, behind and above the top of the large Belmont Conglomerate Quarry, New South Wales Station 3, bed 3A (upper). Permian (Tartarian).

Leaia (Hemicycloleaia) cf compta (Mitchell) 1925
(Pl. 22, Fig. 4)

Diagnosis: Several fragments; one (TC 10635b) has an apparent semicircular configuration and two ribs. The anterior rib curves more across the umbo than does the posterior rib, and the ribs do not meet. Another fragment (TC 10826) displays the two ribs and also permits some metrical comparisons with the holotype.

Measurements: Sigma (holotype)-53°; TC 10635b-53°; TC 10826-53°; and holotype, d_1-5.0 mm; TC 10826, d_1-4.5 mm.

Material: TC 10635b, external mold of left valve; posterodorsal and posteroventral sectors of valve missing; associated with *Palaeolimnadia (Palaeolimnadia) glabra* (Mitchell) on some slab. TC 10826, external mold of left valve, badly eroded except for antero-dorsal sector and the two ribs. Other poor specimens include TC 10710, 10781.

Locality: Station 8 (TC 10635b); Station 13 (TC 10826), and Station 4 (TC 10710 and TC 10781). New South Wales, Permian (Tartarian).

Discussion: Accurate determination of alpha and beta angles is difficult in poorly preserved specimens in certain forms like *compta*. In such cases the sigma angle provides a basis for comparison between the holotype and a poor specimen with little preserved except the two ribs and possibly d_1 and d_2. Even in the absence of the d_1-d_2 measurement, the sigma angle can be of comparative value for at least tentative placement of a given fragmentary specimen.

Leaia (Hemicycloleaia) elliptica (Mitchell) 1925
(Pl. 23, Fig. 4)

Diagnosis: Elliptical, bicarinate valves; anterior rib slightly curved, terminates in second (or third) growth band from the last; posterior rib straight, terminates at last growth band. Posterior margin subcircular. Median valve sector between ribs, the largest.

Measurements: (mm) L-6.6; H-3.7; Ch-4.3; Cr-2.; Av-0.7; Arr-1.3; a-1.3; b=1.7; c-3.0; d_1-0.6 ±.1; d_2-3.9 ±.2; H/L-.561; Cr/L-.313; Av/L-.106; Arr/L-.197; Ch/L-.652; a/H-.352; b/H-.459; c/L-.415; alpha-80°; beta-27° ±2.0°.

Material: TC 10564, external mold of right valve.

Locality: Station 7, bed 2. Creek, cave deposit, Mt. Hutton area. Permian (Tartarian).

Discussion: The specimen described above is close to the holotype in almost all parameters. Because only one specimen was available to Mitchell, the new specimen provides additional measurements that extend the range of the several parameters.

Leaia (Hemicycloleaia) cf etheridgei (Kobayashi) 1954
(See Mitchell's type for synonymy, below.)
(Pl. 22, Fig. 6)

Diagnosis: Although incomplete, this valve shares a number of critical characteristics with *L. (H.) etheridgei.* Chief among the shared characters are the alpha and beta angles, the larger than usual posterior sector, and the ovate, bicarinate condition; d_1 and d_2 are similar.

Measurements: (mm). L-3.15; H-2.1; Ch-?; Cr-1.0; Av-0.5; Arr-?; a-0.9; b-?; c-1.5; H/L-.677?; Cr/L-.323; Av/L-.161; Arr/L-?; Ch/L-?; a/H-.429; b/H-?; c/L-.484; d_1-0.2; d_2-1.9 ±0.1; alpha-88° ±2° (vs. 86 ±3° Mitchell); beta-41 ±2° (vs. 44° ±3° Mitchell).

Material: A single, incomplete, internal mold of a left valve (TC 10105); ventral margin and upper postero-dorsal sector missing. Permian (Tartarian).

Locality: Station 13, bed 2g. New South Wales.

Discussion: The Newcastle Coal Measures species of *Leaia* vary from 69° to 104° alpha angle, and 24° to 44° beta. The only other 88° alpha occurs in *Leaia (H.) ovata* (Mitchell), but it has a beta angle of 30°. Another species with an alpha angle of 87°, *Leaia (H.) compta* (Mitchell), has a beta angle of 34°. Thus, combining the Tasch and Mitchell specimens—86°-88° alpha, and 41°-44° beta—it is evident that the beta angle is the largest encountered among the Australian species of *Leaia* with the exception of *Leaia (Hemicycloleaia) magnumelliptica* n. sp. (TC 10623b), with alpha-95° and beta-44°.

Leaia (Hemicycloleaia) immitchelli n. sp.
(Pl. 23, Fig. 2)

Diagnosis: Broadly ovate, bicarinate valves; umbo subterminal; anterior rib thickens ventrad, slightly concave and does not straighten ventrad, fades out on the third growth band above the ventral margin; posterior rib straight along entire length, fades out on fifth or sixth growth band above ventral margin.

Measurements: (mm).

Specimen	L	H	Ch	Cr	Av	Arr	a	b	c	d_1	d_2
Holotype 10547	5.5	4.3	4.0	1.0	0.4	1.0	1.8	1.8	2.0	0.3±0.0	3.8±0.2
Paratype 10116a	7.0	5.0	4.5	1.5	1.0	1.2	2.0	2.0	3.2	0.4±0.0	3.8±0.1
10587	5.8	4.1	4.9	1.0	0.5	0.5	1.7	1.4	3.1	0.3±0.0	3.9±0.0

Specimen	H/L	Cr/L	Av/L	Arr/L	Ch/L	a/H	b/H	c/L	a*	b
10547	.782	.182	.073	.182	.727	.419	.419	.364	81°±0.0	30°±2.0
10116a	.714	.114	.143	.171	.643	.400	.400	.457	77°±0.0	32°±2.0
10587	.707	.172	.086	.086	.890	.415	.317	.534	78°±0.0	33°±2.0

*a = alpha angle; b = beta angle.

Material: TC 10547 (holotype), and TC 10116a, 10587, and 10589 (paratypes). Other specimens: TC 10112, 10554d, and 10540.

Locality: Station 7a, bed 1 (TC 10547); Station 5, bed 6f (TC 10116a); Station 4, bed 2c (TC 10587). Upper Permian (Tartarian).

Discussion: When metrically compared to *"mitchelli", "im-mitchelli"* has a greater H/L (.707–.782 vs. .615–.702), smaller Cr/L (.114–.182 vs. .245–.354), and smaller a/H (.400–.419 vs. .424–.450). Where overlap occurs, in several instances it is in the upper part of the range: Av/L (.073–.143 vs. .140–.169) and c/L (.364–.534 vs. .526–.551); this type of overlap also applies to *"mitchelli"* for Ch/L (.714–.737 vs. .643–.890 *"immitchelli"*). Alpha and beta angles overlap, but the ribs terminate higher above the ventral margin and the anterior rib does not straighten ventrad, in which direction it thickens—both characteristics lacking in *"mitchelli"*. The difference of overall shape (broadly ovate vs. subovate *"mitchelli"*) accounts for the greater H/L in *"immit-chelli."* The indicated differences—morphologically and metrically—including the nature of overlap, in several instances may denote evolutionary divergence from a common species population, during which only certain valve aspects vary while alpha and beta angles and other features are unchanged.

Leaia (Hemicycloleaia) kahibahensis n. sp.
(Pl. 24, Fig. 3)

Diagnosis: Subovate, bicarinate valves with gently curved dorsal margin; both ribs curve across umbo and approach each other but are not in contact; both ribs terminate at about the fifth from last growth band; anterior rib the more arcuate; medial sector of valve largest, anterior sector, smallest.

Measurements: (mm) L-7.0; H-5.0; Ch-5.0; Cr-1.9; Av-0.6; Arr-1.6; a-2.0; b-1.8; c-3.5?; H/L-.714; Cr/L-.271; Av/L-.086; Arr/L-.143; Ch/L-.714; a/H-.400; b/H-.360; c/L-.500; d_1-0.3 ±0.1; d_2-4.0 ±0.0; a-78° ±2.0°; b-31° ±2.0°.

Material: Holotype TC 10101, internal mold of left valve.

Locality: Station 13, bed 2g. New South Wales, Permian (Tartarian).

Discussion: Among subovate leaiids, one of Mitchell's types, *Leaia (Hemicycloleaia) pincombei* M. 1925, can be compared to the new species. *L. (H.) pincombei* M. has a dorsal margin that is slightly sinuate posteriorly; this differs from the arcuate character of *"kahibahensis,"* but both d_2 and beta angles are close. Larger ratios of *"pincombei"* include H/L, Cr/L, Av/L, Ch/L, b/H, and d_1; Arr/L, a/H, and c/L are smaller.

Kusumi (1961, p. 7) observed that in the females of some living species of conchostracans in Japan "the dorsal margin is more arcuate" than in the male. Possibly the valve of the new species might represent a female.

Two specimens of Novojilov's (1956a, p. 34, Pl. 5, Figs. 6 and 8) species *Hemicycloleaia pruniformis* N, 1956 (= *Leaia (Hemicycloleaia)* have a slightly arched dorsal margin which, as remarked above, may denote a female; their d_1 and d_2 separation is larger than in the new species, and the alpha angle is 108° (vs. 78° n. sp.), and the beta angle is 43°–44° (vs. 31° n. sp.).

Cyzicus (Euestheria) lata (Mitchell)
(Pl. 23, Fig. 3)

Diagnosis: Obliquely subovate valves with subterminal umbo; greatest height posterior to midline of valve; margins: dorsal slightly curved; anterior convex; posterior slightly rounded; ventral gently curved.

Measurements: (mm).

Specimen	L	H	Ch	Cr	Av	Arr	a	b	c
10595	5.5	4.4	4.0	1.5	0.5	0.4	1.5	1.7	2.9
10591-01	4.5	3.0	3.1	1.5	0.6	0.4?	1.0	---	2.5

Specimen	H/L	Ch/L	Cr/L	Av/L	Arr/L	a/H	b/H	c/L
10595	.800	.727	.273	.091	.073	.341	.386	.527
10591-01	.667	.689	.333	.133	.089?	.333	----	.555

Material: TC 10595, slice number 3, right valve overlain by fragment of left valve at postero-dorsal sector. TC 10591-01, internal mold of right valve.

Locality: Station 14, both specimens from same bed, 4.75 m below Dudley seam. (Other specimens 5.27 m below seam.) Permian (Tartarian).

Discussion: These fairly well preserved valves are close to Mitchell's type in most parameters. (See below for these types.)

Leaia (Hemicycloleaia) magnumelliptica n. sp.
(Pl. 22, Fig. 5)

Diagnosis: Broadly elliptical, bicarinate valves with subterminal umbo; anterior rib gently curved, terminates on second from last growth band. Posterior rib straight (wide apart from anterior rib on umbo), and thicker that anterior rib; it terminates on third from last growth band. Margins: anterior rounded; posterior convex; ventral gently curved.

Measurements: (mm).
Material: Holotype TC 10618a, external mold of left valve; paratype TC 10623b, internal mold of left valve.

Specimen	L	H	Ch	Cr	Av	Arr	a	b	c	d_1	d_2
Holotype 10618a	9.2	5.1	6.8	2.4	1.7	0.5	2.1	2.0	4.0	1.8±0.2	6.1±0.2
Paratype 10623b	9.2	5.1	6.8	3.5	2.7	0.5	2.5	2.0	4.0	2.4±0.1	6.1±0.2

	H/L	Ch/L	Cr/L	Av/L	Arr/L	a/H	b/H	c/L	a*	b
10618a	.554	.739	.261	.185	.054	.412	.392	.435	104°+0.0	37°+1.0
10623b	.554	.729	.380	.238	.065	.490	.451	.489	95°+0.0	44°+2.0

*a = alpha angle; b = beta angle.

Locality: Station 1, bed 4, float, Darling Street outcrop near entrance to colliery. New South Wales, Permian (Tartarian).
Discussion: Three outstanding features of *"magnumelliptica"* are its broadly elliptical configuration compared to *Leaia (H.) elliptica* M., 1925, the wider separation of the two ribs on the umbo (d_1-2.4 ±0.1 vs. 1.8 ±0.2 *"elliptica"*), and the larger alpha angle (95°–104° vs. 83° ± *"elliptica"*) and beta angle (37°–44° vs. 24°).

Leaia (Hemicycloleaia) mitchelli (Etheridge, Jr.) 1892
(Pl. 23, Fig. 1)

Diagnosis: Subovate, bicarinate valves, with subterminal umbo; anterior rib concave, straightens ventrad and fades at third from last growth band; it may or may not thicken; posterior rib mostly straight but slightly curved dorsad; terminates on second or third growth band above ventral margin.
Measurements: (mm).

Specimen	L	H	Ch	Cr	Av	Arr	a	b	c	d_1	d_2
10623a	5.2	3.6	3.6	4.1	0.8	0.3	1.3	1.1	2.5	0.4±0.1	3.3±0.1
10619	6.0	3.8	4.7	1.0	0.5	0.5	1.2	1.3	2.6	0.5±0.1	3.6±0.1

	H/L	Ch/L	Cr/L	Av/L	Arr/L	a/H	b/H	c/L	a*	b
10623a	.692	.788	.288	.184	.058	.391	.306	.491	80°+0.0	30°+1.0
10619	.633	.783	.167	.083	.083	.316	.342	.333	83°+1.5	37°+1.5

*a = alpha angle; b = beta angle.

Material: TC 10623a, external mold of left valve; TC 10619, external mold of right valve; TC 10102, internal mold of right valve, and dozens of other specimens. Possible equivalents of the broadly ovate form* of this species, TC 10519a and TC 10529; occurs at Station 5, bed 6. (*See discussion below.)
Locality: Station 1, bed 3, Darling Street outcrop, TC 10623a; Station 4, bed 5 (Kimble Hill), TC 10619, and several other stations (Fig. 10, lower panel). Permian (Tartarian).
Discussion: Minor variations in rib termination may be attributed to erosion of the last one or two sectors. However, rib

termination is not subjective determination because identation of the missing pieces and/or some coloration corresponding to the lost pieces on growth bands are generally present. *Leaia (M.) mitchelli* (Etheridge, Jr.) is the most abundant species in both the Mitchell and Tasch collections and was the dominant species in collections from many stations in this study. The above collections (exclusive of the type for *"mitchelli"*) had no specimens with "incipient ribs" (cf. Mitchell's type, below) indicative of their nonpersistence in the *"mitchelli"* population.

Mitchell (1925, Pl. 40, Fig. 1) figured a broadly ovate form (F 24426, Australian Museum) which is an incomplete specimen. Originally this writer thought it might be a dimorph of *"mitchelli,"* because the alpha angle, 87°, exceeded that of both *"mitchelli"* (81°) and *"immitchelli* n. sp." (77°–81°). The beta angle overlapped in both of these species, and the posterior rib of F 24426 differed from that of *"immitchelli* n. sp.," which terminates on a different growth band (fifth instead of third). However, finding only two broadly ovate specimens among the *"mitchelli"* population available in the Tasch Collection would argue against the dimorph assignment. In addition, the Mitchell specimen is incomplete and does not allow for a complete set of measurements. Accordingly this specimen is tentatively left in *"mitchelli"*.

Cyzicus (Euestheria) novocastrensis (Mitchell) 1925
(Pl. 23, Fig. 5)

Diagnosis: Broadly subovate valves with umbo inset from the anterior margin; straight dorsal margin; rounded anterior margin; steep dorsal-posterior join; closely spaced growth bands.
Measurements: (mm).

Specimen	L	H	Cr	c	H/L	Cr/L	c/L
10593	4.6+	3.5+	2.3+	2.0	.711+	.236+	.434
10597	4.6+	3.3+	1.8+	1.9	.717+	.391+	.413

Note: + means approximately.

Material: One crushed but complete right valve (10594), two incomplete valves: right valve, TC 10597; left valve, TC 10593, both in ironstones; upper dorsal quarter of one valve missing and both valves crushed.
Locality: Station 14, Merewether Beach, New South Wales, 4.75 m below Dudley seam. Newcastle Coal Measures. Permian (Tartarian).
Discussion: Despite limitation on metrical determinations due to poor preservation, there is sufficient closeness in general morphology and the measurable parameters to place the above specimens under Mitchell's species. (Examples: H/L and c/L.)

Leaia (Leaia) oblonga (Mitchell) 1925
(Pl. 23, Fig. 6)

Diagnosis: Suboblong, bicarinate valves; posterior sector largest; anterior sector smallest; length markedly greater than height; straight dorsal margin; posterior rib curved anteriorly, straight thereafter; terminates on ventral margin; anterior rib slightly concave; terminates on ventral margin; both ribs converge and contact on the umbo.

Measurements: (mm) L-3.6; H-2.0; Ch-3.1; Cr-0.9; Av-0.6; Arr-0.1; a-0.8; b-0.6; c-1.6; H/L-.556; Ch/L-.861; Cr/L-.250; Av/L-.167; Arr/L-.028; a/H-.400; b/H-.300; c/L-.444; d_1-1.0; d_2-1.8 ±0.1; a-79° ±2.0; b-37° ±2.0.

Material: TC 10625a, external mold of right valve, as well as four other specimens at same or other stations.

Locality: Station 5, bed 18, James Street Quarry. Also Stations 4 and 5, beds 17d, 17e, and 17f. Newcastle Coal Measures, New South Wales, Permian (Tartarian).

Discussion: Compaction of these specimens has created some distortion of the metrical value of some parameters; for example, angle beta differs from the holotype by 8°, and angle alpha differs by 1°. The configuration (length about two times greater than height) relates these specimens to Mitchell's species, and the suboblong shape compared to the elongate rectangular shape and lack of a curved dorsal margin distinguishes it from *L. (L.) oblongoidea.*

Leaia (Leaia) oblongoidea n. sp.
(Pl. 24, Fig. 1)

Leaia oblonga Mitchell 1925 (p. 441, Pl. 43, Fig. 18; non Pl. 41, Fig. 5 = *L. (L.) oblonga* Mitchell).

Diagnosis: Elongate, rectangular bicarinate valves with slightly curved dorsal margin. Anterior rib gently arcuate, terminates on fourth from last growth band; posterior rib straight, same termination ventrad as anterior rib, but of greater overall length. Greatest height, anterior third.

Measurements: (mm) First listed number is the holotype, the second is the paratype: L-6.6, 5.8; H-3.9, 3.3; Ch-5.3, 4.6; Cr-1.5, 1.3; Av-0.7, 0.7; Arr-0.1, 0.5; a-1.5, 1.5; b-1.5, 1.3; c-3.2, 2.8; H/L-.591, .569; Cr/L-.227, .224; Av/L-.106, .017; Ch/L-.803, .793; a/H-.385, .455; b/H-.385, .394; c/L-.485, .483; d_1-0.3, 0.3; d_2-4.1 ±0.2, 3.8 ±0.2; a-76°, 76° ±2.0°; b-31° ±0.0, 32° ±2.0.

Material: Holotype, right valve (Mitchell, 1927, Pl. 43, Fig. 18*, Australian Museum, F 25420); paratype, right valve, TC 10608. (*See Pl. 40, Fig. 7, herein.)

Locality: Holotype, chert quarry near Belmont. Paratype, Tasch Station 4, bed 3. Permian (Tartarian).

Discussion: The new species is close to but distinguishable from Mitchell's *Leaia (Leaia) oblonga* by its elongate rectangular configuration, slightly concave dorsal margin, and the larger posterior sector of the valve.

Leaia (Hemicycloleaia) ovata (Mitchell) 1925
(Pl. 24, Fig. 2)

Diagnosis: Subovate, bicarinate valves; umbo inset from anterior margin; anterior rib gently concave from umbo to termination on second from last growth band; posterior rib apparently curved as

it crossed umbo; straight subsequently, and terminates on second from last growth band. The two ribs do not meet on the umbo.

Measurements: (mm).

Specimen	L	H	Ch	Cr	Av	Arr	a	b	c	d_1	d_2
10622	6.0	4.0	4.1±	1.5	0.8	--	1.3	--	3.0	0.5±	3.7+0.1
10583	3.8	2.6	3.0	1.1	0.2	0.2	0.9	0.8	2.0	0.4±	2.6+0.1

	H/L	Ch/L	Cr/L	Av/L	Arr/L	a/H	b/H	c/L	a	b
10622	.667	.683	.250	.133	---	.325	---	.500	82°+0.0	37°+1.0
10583	.684	.789	.289	.158	---	.346	.308	.526	81°+2.0	36°+2.0

Material: TC 10583, TC 10622 and several other specimens (10535, Station 1).

Locality: Station 4 (10583), Station 1 (10622), and Stations 2, 9A, and 11. Newcastle Coal Measures, New South Wales, Permian (Tartarian).

Discussion: In addition to the configuration and rib characteristics, the following metrical characteristics are similar, close, or overlap: d_1 (0.4 type vs. 0.4–0.5); d_2 (4.0 type vs. 2.6 ±0.1-3.7 ±0.1); Cr/L (.257 type vs. .250–.289); H/L (.600 type vs. .667–.684). An alpha angle difference of some 6°–7° (approx. 88° type vs. 81°–82°), and beta angle difference (30° type vs. 36° ±1°) may have been influenced by flattening of the valves on the bedding plane. This is suggested by d_1 and d_2, which represent the distance between ribs extended to dorsal margin and ventral margin, respectively, and are either identical with Mitchell's type, or effectively so.

Leaia (Hemicycloleaia) cf paraleidyi (Mitchell) 1925
(Pl. 24, Fig. 4)

Diagnosis: Subovate, bicarinate valves with straight dorsal margin. The two ribs cross the subterminal umbo and approach fairly close; both thicken ventrad; the posterior rib is straight, the anterior slightly arcuate, and both terminate on the last or next to last growth band.

Measurements: (mm) L-4.6; H-2.9; Ch-?; Cr-1.1; Av-0.7; Arr-?; a-0.9; b-1.1; c-2.0; H/L-.630; Cr/L-.239; Av/L-.152; Arr/L-?; Ch/L-?; a/H-.310; b/H-.379; c/L-.435; d_1-0.2; d_2-3.6 ±0.1; a-74° ±2.0; b-33° ±2.0°; sigma-43°.

Material: TC 10533, external mold of right valve; posterior margin and adjacent area eroded; anterior and ventral margins slightly eroded.

Locality: Station 6a, bed 5. Also found at Stations 5, 8, and 9. Newcastle Coal Measures, New South Wales, Permian (Tartarian).

Discussion: The measured specimen is close to the holotype in alpha and beta angles, H/L ratio, and in termination and direction of ribs on the umbo. It does not correspond to any other of Mitchell's leaiids in these features. The discrepancy in the d_1 measurement (larger d_1 than in holotype) may be due to the fact that the holotype is slightly eroded above the umbonal area where the ribs curve anteriorly, and as a consequence the extension of these ribs had to be inferred.

Palaeolimnadia (Palaeolimnadia) glabra (Mitchell) 1927
(Pl. 24, Fig. 5)

Diagnosis: The large, ovate, eccentric umbonal area of the valve, and its few growth bands are the chief distinctive features of this species, along with the metrical data.

Measurements: (mm).

Specimen	L	H	Ch	Cr	Av	Arr	a	b	c	l_u	h_u
10110	3.0	2.2	1.7	1.0	0.6	0.7	0.9	1.0	1.3	2.1	1.6
10577	2.5	1.8	1.3	1.0	0.4	0.6	0.7	0.7	1.3	1.5	1.3

Specimen	H/L	Ch/L	Cr/L	Av/L	Arr/L	a/H	b/H	c/L	h_u^*/l_u
10110	.733	.566	.333	.200	.233	.410	.450	.430	.761
10577	.720	.520	.400	.160	.240	.390	.390	.520	.866

*Height of umbo (h_u) over length of umbo (l_u).

Material: TC 10110, right valve; TC 10577, left valve.

Locality: Station 1, bed 4, lower conchostracan bed (TC 10110); Station 12B, bed 3 (TC 10577). Other specimens occurred at Stations 4, 5, 7a, and 8. (See Fig. 9 for station data.) Newcastle Coal Measures, New South Wales, Permian (Tartarian).

Discussion: There is a range of variation in the umbonal ratio of about 0.2 relative to the holotype. It is to be expected that as more specimens are found, the ranges will vary and be extended in the upper and lower limits.

Palaeolimnadia (Palaeolimnadia) wianamattensis
(Mitchell)
(Pl. 24, Fig. 6)

Diagnosis: Ovate valves with large umbo and only three growth bands.

Measurements: (mm).

Specimen	L	H	Ch	Cr	Av	Arr	a	b	c	h_u	l_u
10603A	3.5	1.7	2.5?	1.6	0.5	0.4	1.2	1.0	1.4	1.8	2.1
10609	2.2	1.5	1.4?	1.0	0.4	0.4	0.6	0.7	1.1	0.8	1.0

Specimen	H/L	Ch/L	Cr/L	Av/L	Arr/L	a/H	b/H	c/L	h_u^*/l_u
10603A	.771	.714	.457	.143	.114	.444	.370	.400	.857
10609	.682	.636	.455	.182	.182	.400	.467	.500	.800

Material: TC 10603A, right valve; TC 10609, right valve.

Locality: Station 4, bed 1 (10603A) and bed 5 (10609); also found at Station 8, unit B, and Station 1, bed 2, lower conchostracan horizon. Newcastle Coal Measures, New South Wales, Permian (Tartarian).

Discussion: The ratio of umbonal height (h_u) to umbonal length (l_u) is variable in this species. The Kimble Hill specimens (Station 4) are associated with fragments of bicarinate leaiids on the same bedding plane. The specimens are flattened, which can account for some deviation from the holotype in a few parameters.

Novojilov assigned large-umbo types from the Russian Mesozoic (Triassic and Jurassic) to both living genus *Limnadia*

(Brongniart, 1820; Novojilov, 1970, Figs. 27, 58, and 99, for example) and to the living genus (*Eulimnadia* Packard, 1874; Novojilov, Figs. 97 and 99) from the Russian Permian and Triassic. Neither of these assignments are acceptable because of the misfit that results. Several of Novojilov's limnadiids have straight dorsal margins (1970, Figs. 24, 25, 27, 30, 31, 53, 57, 62, 63, for example), whereas modern *Limnadia* species (Daday de Deés, 1925) and Bishop (1968, Fig. 42) are never straight, but are prominently arched. The umbo in modern species is large and usually subterminal, or in a few species occupies most of the valve (*L. rivolensis* Brady- Sayce; e.g., Daday de Deés, 1925).

A variety of conchostracan valves with umbo inset from the anterior margin (Novojilov, 1970, Figs. 15, 21, 34–36, 58, 61, 66 and 67) were assigned to *Limnadia*, yet that is a condition generally absent in the modern genus. Similarly Novojilov's assignments to *Eulimnadia* consist of a variety of fossil valves markedly variable in umbonal size and position as well as valve configuration (Novojilov, 1970, Figs. 87–98, 99–104, and 105–110) none of which fit *Eulimnadia* Packard, 1874.

The kind of taxonomic tangle that arises using Novojilov's interpretation of *Eulimnadia* can be illustrated by a Paraguayan living species, *Eulimnadia chacoensis* Gurney (1931, p. 269, Figs. 8–9). The large female valves of this species would be assigned to *Limnadia* and the smaller male valves to *Eulimnadia*, yet they are dimorphs of the same species. (For other data on living *Eulimnadia* see Tasch and Shaffer, [1964].)

Cyzicus (Euestheria) sp. undet.
(Pl. 23, Fig. 7)

Diagnosis: Subovate valves with straight dorsal margin; posterior margin obliquely joins dorsal margin; anterior margin rounded; ventral margin gently curved; umbo subterminal; ornamentation, polygonal granules on growth bands.

Measurements: (mm) L-3.1; H-2.0; Ch-1.4; Cr-1.4; Av-0.7; Arr-1.0; a-1.1; b-0.9; c-1.6; H/L-.645; Ch/L-.452; Cr/L-.452; Av/L-.226; Arr/L-.323; a/H-.550; b/H-.450; c/L-.516. Growth-band number cannot be determined due to erosion.

Material: TC 10113, a single badly eroded and partially crushed specimen of a left valve. Nevertheless, essential measurements were possible because the configuration was intact and umbonal portion determinable.

Locality: Station 13, bed 2g.

Discussion: This specimen seems to be unlike any other euestheriid in the Newcastle Coal Measures or in the Triassic of the Bonaparte Basin (Tasch *in* Tasch and Jones, 1979b). In addition, it does not match any euestheriids described by Zaspelova (1966), Novojilov, and colleagues.

Leaia (Hemicycloleaia) cf. *sulcata* (Mitchell)
(Pl. 24, Fig. 7)

Diagnosis: The specimen is incomplete, but what is preserved indicates an ovate, bicarinate right valve with a straight dorsal margin. Assignment made is based on the d_1-d_2 measurements given below.

Material: TC 10829, external mold of part of right valve, incompletely preserved.

Locality: New South Wales Station 13, bed 2e. Permian (Tartarian).

Measurements: (mm) The only reliable measurements are those of d_1 and d_2. They are given first for the specimen described above, and second for Mitchell's holotype: d_1-0.8, 0.6 ±0.1; d_2-4.8, 4.5 ±0.4. (H/L measured by estimating extent of missing valve portion - .685 and .616.)

Discussion: Because the d_1-d_2 measurements are reliable indicators of placement of ribs on the valve and are sufficiently close to those of the holotype, affinity is suggested with Mitchell's type.

AUSTRALIA (Other Permian Conchostracan Species)

(Location Map No. in Fig. 7)

76.* Cyzicus *(Lioestheria) bellambiensis.* Upper Permian. South Bulli (Bellambi). (For description, see Mitchell's types below.)

Numbering System: Numbers 1–14 used on map (Fig. 8) are the same as Tasch Stations number 1 to 14 in Fig. 10. Numbers 15–92 on the location map listings below and subsequently do not correlate with any Tasch Station numbers. For details of the Lake Macquarie area, see Figures 8–11, in that order.

Bowen Basin, Queensland

50. *Palaeolimnadia (Grandilimnadia)* cf. *glenleensis* Mitchell, 1927. Upper Permian, Blackwater Group. Bureau of Mineral Resources locality 314/6, 4.2 km south-southwest of Winchester Homestead.

50. *Leaia (Hemicycloleaia) deflectomarginis* Tasch 1979. Upper Permian. Blackwater Group. Same locality as *"glenleensis"* above. Upper Bowen Coal Measures. (Tasch, *in* Tasch and Jones, 1979c, Pl. 6, Figs. 1, 2).

50. *Leaia (Hemicycloleaia)* sp. Same locality and age as *"deflectomarginis"* above. (Tasch, *in* Tasch and Jones, 1979c, Pl. 6, Fig. 3.)

Newcastle, New South Wales

73. *Cornia* sp., south of Blackwater at Weoroona, headwater of Sirius Creek. Upper Permian (Australian Museum No. 51987).

70. *Cyzicus* sp. Probably Merewether Beach near Newcastle (Australian Museum No. 28085).

70. *Cyzicus lata?* Probably Merewether Beach near Newcastle (Australian Museum No. 28099).

Benelong, New South Wales

78. *Cyzicus* cf. *mangliensis?* Benelong near Dubbo, New South Wales (Australian Museum No. 28833).

Bowning, New South Wales

69. *Palaeolimnadiopsis?* sp. Bowning near Yass. Upper Permian. (Australian Museum No. 29396.)

AUSTRALIA (Triassic)

Queensland

Palaeolimnadia (Grandilimnadia) glenleensis (Mitchell) 1927
(Pl. 4, Fig. 2)

Diagnosis: Small, ovate valves with a comparatively large subterminal umbo and short straight dorsal margin. Twelve growth bands.

Measurements: (mm).

	L	H	Ch	Cr	Av	Arr	a	b	c
Lectotype	3.0	2.0	2.1	1.1	0.4	0.3	0.7	0.7	1.3
10591	2.2	1.6	1.5	0.7	0.4	0.3	0.5	0.6	1.1
MU 3682	2.2	1.5	1.2	1.0	0.4	0.3	0.6	0.6	1.3

	H/L	Ch/L	Cr/L	Av/L	Arr/L	a/H	b/H	c/L
Lectotype	.666	.700	.367	.133	.100	.350	.350	.467
10591	.727	.602	.318	.182	.136	.313	.375	.500
MU 3682	.682	.545	.455	.182	.136	.353	.353	.591

Material: Tasch Collection No. 10591. (Collected by Anne Warren, La Trobe University, Victoria.) Macquarie University No. 3682 (collected by P. B. Mitchell, Macquarie University, New South Wales).

Locality: Duckworth Creek near Bluff, Queensland; Arcadia Formation (formerly Upper Rewan Formation). Lower Triassic. Also Hornsby diatreme (Hawkesbury Sandstone fragments), Hornsby, New South Wales, Middle Triassic. (See Mitchell's types below for the lectotype of this species.)

Discussion: Morphologically and metrically, the Arcadia Formation species is close to the lectotype. Another form, transversely ovate, that was placed in *Palaeolimnadia (Grandilimnadia)* cf. *glenleensis* (Mitchell) came from the Upper Permian Blackwater Group (Tasch *in* Tasch and Jones, 1979b, p. 42, Pl. 7, Fig. 1). The Blackwater species has a higher H/L ratio (.820 vs. .666–.727) and a smaller umbo than either the lectotype or specimen 10591. Morphologically similar to *"glenleensis"*, the Blackwater species' smaller umbo leaves that assignment indecisive. The Hawkesbury Sandstone specimen is close to the lectotype in form and measured parameters.

Palaeolimnadia (Grandilimnadia) arcadiensis n. sp.
(Pl. 25, Fig. 1)

Diagnosis: Ovate valves with comparatively large umbonal area; anterior third of valve has greatest height; umbo broadly ovate and closer to anterior margin; growth bands variable, 7–10.

Measurements: (mm).

Specimen	L	H	Ch	Cr	Av	Arr	a	b	c	h_u*	l_u
Holotype											
10592A	2.3	1.7	1.1	0.8	0.4	0.4	0.7	1.0	1.0	1.1	1.5
Paratypes											
10592B	2.5	1.8	1.3	1.0	0.5	0.2	0.8	0.8	1.0	0.9	1.5
10592E	2.6	1.8	1.8	0.7	0.3	0.4	0.6	0.8	1.0	1.0	1.6
10592F	2.8	2.0	1.5	1.0	0.5	0.3	0.8	0.9	1.3	0.9	1.3

*h_u = height of umbo; l_u = length of umbo.

	H/L	Ch/L	Cr/L	Av/L	Arr/L	a/H	b/H	c/L	h_u/l_u*
10592A	.739	.478	.348	.174	.175	.412	.588	.435	.733
10592B	.720	.520	.400	.200	.080	.444	.444	.400	.600
10592E	.692	.692	.269	.115	.154	.333	.444	.385	.625
10592F	.714	.536	.357	.179	.107	.400	.450	.464	.692

*h_u - umbonal height over l_u - umbonal length.

Material: Arcadia Formation, coquina of shells collected by Anne Warren. Holotype TC 10592A, crushed left valve; paratypes 10529B, and 10529E, crushed right valve; 10529F, crushed left valve. Other specimens in Narrabeen Group: TC 10650E, 10587B and F, 10651E, 10651F.

Locality: Duckworth Creek near Bluff, Queensland; Arcadia Formation, Rewan Group. Comparable large umbo types also occur at Garie Beach (Lower Triassic, Narrabeen Group).

Discussion: Valves with a comparatively large umbo (range h_u/l_u.600–.733) surrounded by multiple growth bands that are confined to a narrow, peripheral concentric belt, are found in the Russian Permian (Novojilov, 1970, Pl. 1, Fig. 415. [*Limnadia nekhoroashevi* Nov. and *Limnadia karaungirica* Nov. are both reassigned here to *Palaeolimnadia (Grandilimnadia)*.]). The umbo in each of these two species is larger than that of the Australian species described above. None of the grandilimnadiids described by Tasch (*in* Tasch and Jones, 1979a, 1979b, 1979c) compare in umbonal size to the new species.

Cyzicus (Euestheria) ipsviciensis (Mitchell), Denmark Hill, Ipsvich, Upper Triassic. (See Mitchell's types below.)

Sydney Basin, New South Wales (Triassic).

***Palaeolimnadiopsis bassi* Webb, 1978**
(Pl. 25, Fig. 8)

Palaeolimnadiopsis bassi Webb 1978, p. 265–266, Figs. 1 and 2).

Diagnosis (slightly modified): Moderately large, robustly ovate valve with small umbo (larval valve) situated, relative to the anterior end, one-third the length of the slightly concave dorsal margin. Small concavity on the postero-dorsal margin (recurved

growth bands). Posterior height greater than anterior height. Paired growth lines. Ornamentation not observed.

Measurements: (mm) These have been extended beyond the original (indicated by askerisks) to bring them in concordance with the other taxa described in this memoir. L-17.5*; H-12.0*; Ch-11.1; Cr-4.90; Av-1.70; Arr-2.76; a-3.40; b-3.80; c-8.30; H/L-.686*; Ch/L-.630; Cr/L-.280; Av/L-.097; Arr/L-.160; a/H-.283; b/H-.318; c/L-.474.

Locality: Hawkesbury Sandstone, Middle Triassic, Brookvale, Sydney, New South Wales.

Material: Holotype F36275, mold of left valve. Australian Museum Collection.

Discussion: Webb (1978, p. 266) thought the "paired growth lines" should be regarded "as a genetically determined character, sufficiently definitive to warrant specific designation." He further noted that there was only one other palaeolimnadiopsid with paired growth lines from the Permian of Korea, *Koreolimnadiopsis tokioi* Osaki, (1970, p. 76-77, Pl. 12, Figs. 1, 2) but these growth lines were "undulating and tuberculate", and had a distinctive dorsal-posterior angle. *P. bassi* Webb differed from other species in larger size, slightly concave dorsal margin, nonterminal umbo, and small larval valve (umbo), as well as the H/L ratio.

Paired (or double) growth lines occur not only in the cited palaeolimnadiopsid species, but in species of *Cyzicus, Glyptoasmussia,* and numerous other genera, for example *Cyzicus (Euestheria) lefranci* n. sp. (Pl. 3, herein) from the Algerian Sahara (Cretaceous), and *Glyptoasmussia meridionalis* n. sp. from the Antarctic (Jurassic). In none of the species with double growth lines described herein was this valve character the sole basis for species designation—rather it was but one of a suite of characteristics. However, the Korean double growth lines with their distinctive other features (undulant, tuberculate) merit specific designation.

Other Australian palaeolimnadiopsids are now known. *Palaeolimnadiopsis tasmani* Tasch *(nom. correct.)* from the Knocklofty Formation (Lower Triassic of Tasmania), a correlate of the Blina Shale, Australia (Tasch, 1975). There is a probable *Palaeolimnadiopsis* sp. undet. (Australian Museum Collection No. F 29296, from Merewether Beach, Bowning, New South Wales). This specimen, labeled "Estheria sp., Permian," has the last few growth bands recurved. Unfortunately, the upper quarter of the valve is missing, so useful measurements could not be made.

***Cyzicus (Euestheria) triassibrevis* n. sp.**
(Pl. 25, Figs. 3, 4)

Diagnosis: Slightly oblique, ovate valves with a short, straight, dorsal margin; posterior sector of greater height than the anterior sector; umbo subterminal. Sharp decline from umbonal area to anterior margin; the anterior junction with the dorsal margin is less steep. Growth bands variable: 21±. Dimorphs A (ovate valves), and B, subcircular valves with shorter dorsal margin than ovate forms. Both types and dimorph specimens have ornamentation of contiguous granules.

Measurements: (mm).

Specimen	L	H	Ch	Cr	Av	Arr	a	b	c
Holotype 10581b	3.5	2.4	1.7	1.2	0.6	0.7	1.0	1.3	1.8
Paratype 10582d	3.3	2.5	1.8	1.3	0.7	0.6	1.2	1.0	1.9
Paratype 10646a	2.8	2.0	---	1.0	0.6	0.6	1.2	1.0	1.5
Dimorph A 10581r	3.2	2.8	1.5	1.3	0.6	0.6	1.2	1.0	1.5
Dimorph B 10582s	3.2	2.7	1.5	1.1	0.4	0.6	0.7	1.1	1.8

	H/L	Ch/L	Cr/L	Av/L	Arr/L	a/H	b/H	c/L
Holotype 10581b	.686	.486	.343	.171	.200	.417	.542	.514
Paratype 10582d	.758	.545	.394	.212	.182	.485	.400	.576
Paratype 10646a	.714	----	.357	.179	.250	.500	.500	.500
Dimorph A 10581r	.875	.530	.403	.187	.187	.429	.357	.468
Dimorph B 10582s	.853	.555	.343	.125	.188	.395	.423	.562

Material: Holotype TC 10581b, external mold of right valve; paratypes TC 10582d, external mold of right valve; TC 10646a, external mold of right valve, unflattened, eroded at uppermost portion of umbo, and many crushed or fragmented specimens. Fossils collected by Anne Warren.

Locality: Garie Beach, 10 km southwest of Port Hacking. Lat 34°8.2′S, long 151°7.3′E, Narrabeen Group, Lower Triassic, southeast Australia (Sydney Basin).

Discussion: Lower Triassic euestheriids from the Bonaparte Gulf Basin, Australia, differ from the new species in having a longer dorsal margin, greater valve length, and posterior sector not of greater height than anterior (Tasch *in* Tasch and Jones, 1979b, Pl. 5, Figs. 1–4); or in having less robust ovateness of the valve (Tasch *in* Tasch and Jones, 1979b, Pl. 5, Fig. 6). An Upper Triassic specimen described by Mitchell (1927, and see Mitchell's types below), *Cyzicus (Euestheria) ipsviciensis* (Mitchell), also has a longer dorsal margin (3.2 mm vs. 1.5–1.8 mm), and is laterally subovate, with a smaller H/L.

Other Triassic *Cyzicus* species are under subgenus *Lioestheria*. Among Mitchell's Permian euestheriids, most have longer dorsal margins by a factor of 2 or 3; all differ in configuration; for example, *C. (Eu.) trigonellaris* (Mitchell) is eccentrically ovate; *C. (Eu.) obliqua* (Mitchell) is transversely ovate; *C. (Eu.) novocastrensis* (Mitchell) has a robust, broadly subovate valve; and *C. (Eu.) belmontensis* (Mitchell) has its greatest height in the anterior sector behind the umbo.

The main feature of *"triassibrevis"* is its slightly oblique subovate to subcircular configuration, inset umbo with a short dorsal margin, and a posterior sector of greater height than the anterior, in addition to its metrical data.

AUSTRALIA (Cretaceous)

Victoria, Australia

58. *Cyzicus (Lioestheria) branchocarus* Talent 1965
Pl. 25, Figs. 6, 7

Cyzicus? branchocarus Talent 1965, p. 197–203, Pl. 26.

Diagnosis (emended): Subovate valves with straight dorsal margin; umbo inset slightly from rounded anterior margin and terminus of umbo rises slightly above it; posterior margin convex, and joins dorsal margin at steep angle; valves expand posteriorly; anterior and ventral margins gently curved; anterior sector of valve has lesser height than posterior; ornamentation: punctate. Growth bands, 15 or more.

Measurements: (mm).

Specimen	L	H	Ch	Cr	Av	Arr	a	b	c
NMV-34291	5.4	2.3	3.1	1.8	1.1	0.95	1.24	1.62	3.00
NMV-34290	3.7	2.2	2.2	0.8	0.6	0.92	0.83	1.00	1.67
TC-10652b	4.0	2.7	2.3	1.3	0.7	0.7	1.20	1.60	1.60
TC-10652c	6.2	4.1	3.5	1.7	0.8	1.7	1.30	1.80	3.00
TC-10652d	5.3	3.6	2.7	1.7	1.0	1.5	1.30	1.60	2.80

	H/L	Ch/L	Cr/L	Av/L	Arr/L	a/H	b/H	c/L
NMV-31291	.635	.596	.346	.211	.183	.376	.491	.577
NMV-34290	.594	.594	.216	.157	.249	.377	.454	.451
TC-10652b	.675	.575	.325	.200	.175	.444	.370	.400
TC-10652c	.661	.565	.274	.129	.274	.317	.439	.414
TC-10652d	.679	.509	.321	.189	.283	.361	.414	.528

Material: Holotype, National Museum Victoria NMV 34291, right valve; paratype NMV 34290, paired valves flattened on bedding plane. Tasch Collection (TC): topotypes, TC 10625b and c; TC 10652d (contributed by Peter Duncan, Melbourne), all left valves.

Locality: North side of South Gippland Highway; road cutting 2.4 km west of Tarwin and 4 km east of Koonwarra; 148 km by road northeast of Melbourne, Victoria. Lower Cretaceous (Valangian–Aptian).

Discussion: Two lioestheriids from the Brazilian Mesozoic (Cardoso, 1962) share some characteristics of the Australian species. *Lioestheria florianensis* Cardoso (1962, p. 32–35, Pl. 1, Fig. 5) from the Upper Triassic Motuca Formation also has a straight dorsal margin, umbo inset and rising above that margin, a subovate configuration, and rounded anterior margin. It differs in that the valve tapers posteriorly and H/L is larger (.710–.730 vs. .594–.679). The above morphological characters are also found in *Lioestheria codoensis* Cardoso (1962, p. 35–37, Pl. 1, Fig. 4) from the Cretaceous Codo Formation; however, the posterior taper of the valves is even more pronounced than in the Upper Triassic form described above, it has hachure markings instead of being punctate, and H/L overlaps (.570–.710 vs. .594–.679).

Fossil Conchostracan Fauna
(Paleozoic-Mesozoic) Australia

	Group	Formation	Species
Lower Cretaceous	Korumburra Group		Valanginian-Aptian: <u>Cyzicus</u> (<u>Lio.</u>) <u>branchocarus</u>
Middle-Upper Triassic	Wianamatta		Glenlee Homestead: <u>Cornia coghlani</u>, <u>Paleolimnadia</u> (<u>G.</u>) <u>mitchelli</u>, <u>Paleolimnadia</u> (<u>P.</u>) <u>wianamattensis</u> Denmark Hill, Ipsvich, Queensland: <u>Cyzicus</u> (<u>Lio.</u>) <u>ipsvichensis</u>, <u>Paleolimnadia</u> (<u>G.</u>) <u>sp.</u>, <u>Cyzicus</u> (<u>Lio.</u>) <u>australensis</u>
Middle Triassic	Hawksbury Sandstone		Brookvale: <u>Cyzicus</u> (<u>Lio.</u>) <u>wianamattensis</u> Beacon Hill Quarry: <u>Cyzicus sp.</u>, <u>Paleolimnadiopsis bassi</u> Hornsby diatreme: <u>Paleolimnadia</u> (<u>G.</u>) <u>glenleensis</u> Sydney Basin: <u>Paleolimnadia</u> (<u>G.</u>) <u>mitchelli</u>
Lower Triassic	Narrabeen Group		Garie Beach: <u>Cyzicus</u> (<u>Lio.</u>) <u>triassibrevis</u>, <u>Paleolimnadia</u> (<u>G.</u>) <u>arcadensis</u>
		Arcadia Fmtn.	<u>Paleolimnadia</u> (<u>G.</u>) <u>glenleensis</u>, Bowen Basin: (Rewan Group): <u>Cyzicus</u> (<u>Lio.</u>) <u>sp. undet.</u> I to 3 , <u>Paleolimnadia</u> (<u>P.</u>) <u>wianamattensis</u>
		Blina Shale	Canning Basin: <u>Estheriina</u> (<u>N.</u>) <u>rewanensis</u>, <u>Paleolimnadia</u> (<u>G.</u>) <u>sp. undet.</u> <u>Paleolimnadia</u> (<u>P.</u>) <u>mediensis</u>, <u>Estheriina</u> (<u>N.</u>) <u>blina</u>, <u>Cyzicus</u> (<u>Lio.</u>) <u>sp. undet.</u>, <u>Cyzicus</u> <u>sp.</u>, <u>Cyzicus</u> (<u>Lio.</u>) <u>erskinehillensis</u>, <u>Cyzicus</u> (<u>Lio.</u>) <u>australensis</u>, <u>Cyzicus</u> (<u>Lio.</u>) <u>sp. undet.</u> 3 and 4, <u>Paleolimnadia</u> (<u>P.</u>) <u>wianamattensis</u> Sydney Basin: <u>Cornia coghlani</u> Narrabeen Bore: <u>Cyzicus</u> (<u>Lio.</u>) <u>fictacoghlani</u> <u>Cyzicus</u> (<u>Lio.</u>) <u>wianamattensis</u> Bonaparte Basin: <u>Cyzicus</u> (<u>Eu.</u>) <u>dickinsoni</u>, <u>Cyzicus</u> (<u>Eu.</u>) <u>sp. undet.</u> I to 3, <u>Paleolimnadia</u> (<u>G.</u>) <u>arcoensis</u>, <u>Paleolimnadia</u> (<u>G.</u>) <u>profunda</u>, <u>Cyzicus</u> (<u>Lio.</u>) <u>sp. undet.</u> I to 3, <u>Paleolimnadia</u> (<u>G.</u>) <u>sp. undet.</u>, <u>Estheriina</u> (<u>N.</u>) <u>circula</u>, <u>Estheriina</u> (<u>N.</u>) <u>rewanensis</u>

	Group	Formation	Species
Lower Triassic	Narrabeen Group	Blina Shale	Carnarvon Basin: (Kockatea sh.): <u>Cyzicus</u> (<u>Eu.</u>) cf. <u>minuta</u> (?) Tasmania (Knocklofty Fmtn.): <u>Paleolimnadiopsis tasmanii</u>, <u>Paleolimnadia</u> (<u>P.</u>) <u>sp.</u>, <u>Paleolimnadia</u> (<u>G.</u>) <u>sp.</u>, <u>Cyzicus</u> (<u>Lio.</u>) <u>sp.</u> I to 3 (Ross Sandstone): <u>Paleolimnadia</u> (<u>P.</u>) <u>poatinis</u>

Location/ Station		Seam below	
M15	Eleebana Fmtn	Fassifern Seam	(Mitchell): <u>Leaia</u> (<u>H.</u>) <u>elliptica</u>, <u>Leaia</u> (<u>H.</u>) <u>latissima</u>, <u>Leaia</u> (<u>H.</u>) <u>ovata</u>, <u>Leaia</u> (<u>H.</u>) <u>paraleidyi</u>, <u>Leaia</u> (<u>H.</u>) <u>pincombei</u> Bowen Basin: (Kazanian-Tartarian) (Blackwater Group): <u>Leaia</u> (<u>H.</u>) <u>deflectomarginis</u>, <u>Leaia</u> (<u>H.</u>) <u>sp. undet.</u>, <u>Paleolimnadia</u> (<u>G.</u>) cf. <u>glenleensis</u>
6,10	Eleebana Fmtn	Fassifern Seam	<u>Leaia</u> (<u>H.</u>) <u>mitchelli</u>, <u>Leaia</u> (<u>H.</u>) <u>immitchelli</u>, <u>Leaia</u> (<u>H.</u>) <u>paraleidyi</u>
1,2,3,4,5,6 7,7A,8,9, 12A,B, 13	Croudace Bay Fmtn	Upper Pilot Seam	(Mitchell): <u>Cycloleaia discoidea</u>, <u>Cyzicus</u> (<u>Eu.</u>) <u>linguaformis</u>, <u>Cyzicus</u> (<u>Eu.</u>) <u>bellambiensis</u>, <u>Leaia</u> (<u>H.</u>) <u>mitchelli</u>, <u>Leaia</u> (<u>H.</u>) <u>etheridgei</u>, <u>Leaia</u> (<u>H.</u>) <u>intermedia</u> (=<u>mitchelli</u>), <u>Leaia</u> (<u>H.</u>) <u>oblonga</u>, <u>Leaia</u> (<u>H.</u>) <u>collinsi</u>, <u>Leaia</u> (<u>L.</u>) <u>quadrata</u> (Tasch): <u>Leaia</u> (<u>H.</u>) <u>belmontensis</u>, <u>Leaia</u> (<u>H.</u>) <u>paraleidyi</u>, <u>Leaia</u> (<u>L.</u>) <u>oblonga</u>, <u>Leaia</u> (<u>L.</u>) <u>oblongoidea</u>, <u>Leaia</u> (<u>H.</u>) <u>mitchelli</u>, <u>Leaia</u> (<u>H.</u>) <u>immitchelli</u>, <u>Leaia</u> (<u>H.</u>) <u>elliptica</u>, <u>Paleolimnadia</u> (<u>P.</u>) <u>glabra</u>, <u>Cycloleaia discoidea</u>, <u>Leaia</u> (<u>H.</u>) <u>compta</u>, <u>Leaia</u> (<u>H.</u>) <u>kahibahensis</u>, <u>Leaia</u> (<u>H.</u>) <u>ovata</u>, <u>Leaia</u> (<u>H.</u>) <u>magnumelliptica</u>
14	Bar Beach Fmtn (above seam)	Dudley Seam	<u>Leaia</u> (<u>H.</u>) <u>compta</u>, <u>Cyzicus</u> (<u>Lio.</u>) <u>lenticularis</u>
	Bogey Hole Fmtn (below seam)	Dudley Seam	<u>Cyzicus</u> (<u>Eu.</u>) <u>lata</u>, <u>Cyzicus</u> (<u>Lio.</u>) <u>trigonellaris</u>, <u>Cyzicus</u> (<u>Lio.</u>) <u>obliqua</u>, <u>Cyzicus</u> (<u>Eu.</u>) <u>novacastrensis</u>

Canning Basin	Anderson Fmtn	Lower Visean- Lower Namurian	<u>Leaia</u> (<u>L.</u>) <u>andersonae</u>, <u>Leaia</u> (<u>H.</u>) <u>tonsa</u>, <u>Leaia</u> (<u>H.</u>) <u>rectangulata</u>, <u>Leaia</u> (<u>L.</u>) <u>grantrangicus</u>, <u>Monoleaia australiata</u>, <u>Rostroleaia</u> sp., <u>Limnadiopsileaia carboniferae</u>, <u>Cyzicus</u> (<u>Lio.</u>) sp. undet. I and 2	(P) Paleolimnadia (G) Grandilimnadia (Lio) Lioestheria (N) Nudusia (Eu) Euestheria (H) Hemicycloleaia (L) Leaia
Drummond Basin	Raymond Fmtn	Upper Tournasian Lower Visean	<u>Leaia</u> (<u>H.</u>) <u>drummondensis</u>	

Left margin labels: Cretaceous — Triassic — Triassic — Newcastle Coal Measures (Sydney Basin) Permian (Tartarian) — Carboniferous

Figure 26. Australia. Carboniferous-to-Cretaceous fossil conchostracan species by group and/or formation. (Note absence of Jurassic species.) *Palaeolimnadia (Palaeolimnadia) wianamattensis* (Mitchell)
was found at three Tasch stations (Stations 1, 4, 8), in the Croudace Bay Formation, Permian (Tartarian). It was originally described from the Triassic. During drafting of this figure, it was inadvertently
omitted. Corrections: *Cyzicus (Euestheria) triassibrevis* n. sp. *(non Lioestheria)*, Triassic, Garie Beach;
Cyzicus (Euestheria) obliqua Mitchell *(non Lioestheria),* below Dudley seam. [*Cyclestherioides (Cy-
cloitherioides) lenticularis non Cyzicus* (Lio.), station 14.)]

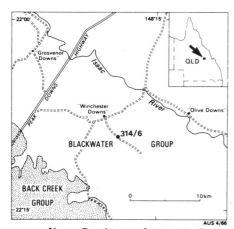

Upper Permian conchostracans—Bowen Basin: *Leaia (Hemicycloleaia) deflectomarginus* **sp. nov.,** *L. (Hemicycloleaia)* **sp. undet. 1,** and *Palaeolimnadia (Grandilimnadia)* **cf.** *glenleensis* (Mitchell).

A

Lower Triassic conchostracan localities— Bowen Basin.

B

LOCALITY	CONCHOSTRACAN FAUNA
A 27	*Cyzicus (Lioestheria)* sp. undet. 1 *Palaeolimnadia (Palaeolimnadia) wianamattensis* (Mitchell)
B 63	*Estheriina (Nudusia) rewanensis* sp. nov. *Estheriina (Nudusia) circula* sp. nov.
BHP Blackwater No. 2 Bore 338 ft	*Palaeolimnadia (Palaeolimnadia) wianamattensis* (Mitchell) *Palaeolimnadia (Grandilimnadia)* sp. undet. 1 *Cyzicus (Lioestheria)* sp. undet. 3
350 ft	(Carbonised plants)
647⁺ ft	*Cyzicus (Lioestheria)* sp. undet. 2

AUS 4/67

Lower Triassic conchostracans at three Bowen Basin localities.

C

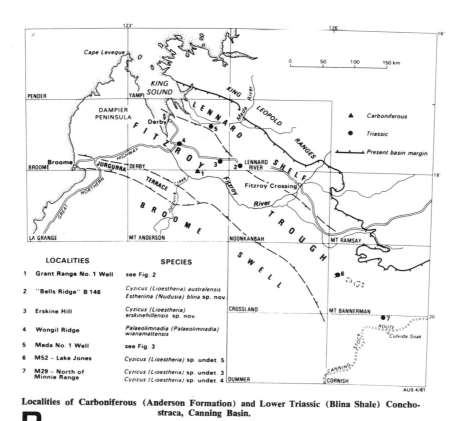

Localities of Carboniferous (Anderson Formation) and Lower Triassic (Blina Shale) Conchostraca, Canning Basin.

D

Figure 27. Australia. A: Bowen Basin (eastern Australia) conchostracan fauna and associated biota. Upper Permian. B: Bowen Basin, Lower Triassic conchostracan fauna and sites. C: Lower Triassic conchostracan species at three Bowen Basin sites. D: Carboniferous and Triassic species and localities in Western Australia.

The above measurements extend the original H/L given at the time the species was described. Sparsity of reports on Cretaceous conchostracans from Australia is notable especially in light of the complete absence of reports on Jurassic occurrences in the Australian nonmarine deposits. (See earlier text discussion on this theme.)

INDIA (Lower Carboniferous)

Himalaya Mountains (Himachal Pradesh)

Spiti District

6. *Estheria* sp. (cf. *striata* Muenst). Diener, 1915 (= *?Cyzicus (Euestheria)* sp.). Slab K51/290, Geological Survey of India (GSI). Coquina of shells. H/L-.666 (Pl. 30, Fig. 4 this memoir). Described but not figured in Diener, *Palaeontologia Indica*, n.s., 1915, Memoir 5, part 2; section on p. 113, description, p. 115–116. Lipak River section, Horizon G. shale. (Paleozoic conchostrans reported in Po Series, Lipak Section of Kashmir; Ghosh, 1973, p. 5.)

INDIA (Upper Permian)

Kawadsi

Estheria sp. (Hughes, 1877). (= *?Cyzicus* sp.) Kawadsi Village on the bank of the Wardhi River; bedrock 0.5 m below base of soil bank. *Glossopteris* flora and estheriids. Originally reported by Hughes, 1877. Data on zonation in Tasch and others, 1975, p. 445–446. Upper Permian.

Pumwat

Estheria sp. (Blanford, 1860). (= *?Cyzicus* sp.). Pumwat, upstream from Hughes (1877) locality (see Kawadsi entry above); estheriids reported by Blanford, 1860. Restudied by Tasch and others (1975). Upper Permian.

INDIA (Lower Triassic)

Raniganj Coal Field (for sites and sections, see Appendix).

Panchet Formation (Tasch Collection from surface outcrops).

2. *Cyzicus (Euestheria) basbatiliensis* n. sp. Station RB-2, bed 5.
2. *Cyclestherioides (Sphaerestheria)* sp. Station RN-1-A, bed 4.
2. *Pseudoasmussiata bengaliensis* n. sp. Station RN-1-B, bed 2.
2. *Cyzicus (Lioestheria) miculis* n. sp. Station RN-1-B, bed 4.
2. *Cyzicus (Euestheria) raniganjis* n. sp. Station RN-2, bed 4.
3. *Cyzicus (Euestheria) mangaliensis* (Jones) 1862. Redescribed herein based on types and new collections. Station M-1, beds 1 to 5 and float.

3. *Cornia panchetella* n. sp. Station M-1, beds 1, 2, and float.
2. *Cyclestherioides (Cyclestherioides) machkandaensis* n. sp. Station RM-1, bed 3b.
2. *Cyzicus (Euestheria) dualis* n. sp. Station RN-2(?), Nama River.
3. *Estheria* sp. (Hislop, S.) (= *Cyzicus (Euestheria) mangaliensis* (Jones) 1862) *in* Hislop and Hunter, 1854, p. 472; 1855, p. 371; Hislop, 1861, p. 347, 353. Mangli Village; Jones, 1862, p. 78–81, Pl. 2, Fig. 16.

2. *Cyzicus (Euestheria) basbatiliensis* n. sp.
(Pl. 29, Fig. 5)

Diagnosis: Compact, ovate valves with subterminal umbo and short dorsal margin. Height of posterior third of valve greater than anterior third. Ornamentation of contiguous granules on growth bands.

Material: Syntype: TC 40011b, internal mold of right valve slightly eroded along posterior margin; TC 40011a, mold of right valve, clear display of ornamentation, and two additional valve fragments. Another specimen, TC 40195, assigned to this species.

Locality: Raniganj Coal Field along Basbatili nala (= creek). Tasch site RB-2, bed 5, Panchet Formation (Lower Triassic); (syntypes only). RN-1-B bed 2 (TC 40195 only). The difference in elevation between RB-2 and RN-1-B above the *Glossopteris-Schizoneura* datum is 30.7 m (see Fig. 14). (This datum marks the boundary between the Raniganj Formation [Upper Permian] and the Panchet Formation [Lower Triassic] in the surface deposits of the Raniganj Basin and the subsurface of the east Bokaro Basin.)

Measurements: (mm) Syntype TC 40011b followed by TC 40011a. L-3.9?, 3.4; H-2.8?, 2.8?; Ch-2.1?, 1.4; Cr-0.8, 0.7; Av-0.8, 0.7; Arr-1.0, 1.2; a-1.4, 1.0; b- --, 1.1; c-2.1?, 1.4?; H/L-.718?, .824?; Cr/L-.205?, 206; Av/L-.205?, .206; Arr/L-.256?, .353; Ch/L-.538?, .412; a/H=.500?, .357?; b/H- --, .393?; c/L-.538, .412.

Discussion: The only euestheriid in the Panchet Formation with which to compare the above species is *Cyzicus (Euestheria) mangaliensis* (Jones), 1862 from Mangli Village (Station M-1). The new species differs in configuration in that it is more compact, not elongate, and has a short, not a long, dorsal margin. Difference in configuration is reflected best in Cr/L and Av/L ratios, both of which are greater in "*mangaliensis.*" Otherwise, parameters are close and/or overlap; TC 40235 ("*mangaliensis*") has a length closest to the new species. It is probable that the new species and "*mangaliensis*" derived from a common ancestor.

2. *Pseudoasmussiata bengaliensis* n. sp.
(Pl. 27, Fig. 7; Pl. 30)

Diagnosis: Dimorph valves: one broadly ovate with a short, straight dorsal margin behind the umbo, and a curved join with the anterior margin, presumably a female valve; posterior margin of valve gently arched, whereas anterior margin is convex. The presumably male valve is smaller, subovate but not broadly so, and hence has different parameters, except H/L which overlaps the ratio for the female valve. It also has a straight dorsal margin, umbo inset from anterior margin; posterior and anterior margins rounded. Numerous growth bands. In both dimorphs the axis through umbo plunges posteriorly.

Measurements: (mm) Syntype parameters for TC 40024 (female), followed by TC 40074b, and TC 40113 (males). L-4.0, 2.7, 2.6; H-3.0, 2.1, 1.9; Ch-1.9?, 1.0, 1.2?; Cr-0.9, 1.0, 0.7; Av-0.9, 1.0, 0.7; Arr-1.2, 0.6, 0.6; a-1.2, 1.0, 1.0; b-1.5, 1.0, 0.8; c-2.3, 1.2, 1.2; H/L-.750, .778, .731; Cr/L-.225, .370, .269; Av/L-.225, .370, .269; Arr/L-.300, .222, .231; Ch/L-.475, .370, .462; a/H-.400, .476, .526; b/H-.500, .476, .421; c/L-.575, .444, .462.

Material: Syntype TC 40024, internal mold of right valve (female); TC 40113 and TC 40074b, carbonized right valve (both male).

Locality: All components of the syntype come from the Panchet Formation, Raniganj Basin, Station RN-1-B, bed 2. Lower Triassic.

Discussion: The new species described above differs from *Pseudoasmussiata indicyclestheria* n. sp. (described in this section) in configuration (broadly subovate and ovate vs. subcircular *"indicyclestheria"*); umbonal size and placement (small umbo inset from the anterior margin vs. large umbo, submedial in position), and curvature of the posterior margin (gently arched to rounded vs steeply arched). Moreover, for *"bengaliensis"*, H/L and Arr/L are greater and b/H and c/L are smaller than in *"indicyclestheria,"* whereas other ratios overlap at the upper part of the size range; for example, Ch/L (.370–.475 n. sp. vs. .435), or overlap at the lower part of the size range for a/H (.444 n. sp. vs. .400–.526). The two species were vertically separated by some 32 m in the Panchet Formation.

2. *Cyzicus (Lioestheria) miculis* n. sp.
(Pl. 29, Fig. 8; Pl. 30)

Diagnosis: Dimorphic valves. One valve (male) is irregularly ovate (i.e., there is an oblique join from the umbonal terminus to the posterior margin), with a short dorsal margin behind umbo; the umbo is almost submedial and rises above the dorsal margin; greatest height in the anterior third of the valve; margins rounded. The other valves (female) are subelliptical to laterally ovate in configuration. Both dimorphs display punctate ornamentation on growth bands.

Measurements: (mm) Parameters for TC 40079, male, followed by TC 40999 and TC 40078, females. L-2.5, 3.3, 2.5; H-2.0, 3.3, 2.5; Ch-1.1?, 1.7?, 1.1?; Cr-0.8, 0.7, 0.7?; Av-0.8, 0.7, 0.7?; Arr-0.6, 0.9, 0.7?; a-0.8, 1.0, 1.0?; b-0.9, 0.8, 0.6?; c-1.1, 1.3, 1.0; H/L-.800, .636, .640; Cr/L-.320, .212, .280; Av/L-.320, .212, .280; Arr/L-.240, .273, .280; Ch/L-.440?, .575?, .440?; a/H-.400, .476, .625?; b/H-.450, .381, .375?; c/L-.440, .394, .400.

Material: Syntype TC 40999, internal mold of right valve, female, and TC 40079, male, internal mold of left valve; TC 40078,

female. All of these and other specimens from the Raniganj Coal Field are embedded in a highly micaceous matrix; this explains their comparatively poor preservation.

Locality: Station RN-1-B, bed 4, down-dip from Station RN-1-A; i.e., younger beds. Numa River, Raniganj Coal Field, Panchet Formation. (See Figs. 13 and 14.) Lower Triassic.

Discussion: The new species differs from *Cyzicus (Lioestheria) bokaroensis* n. sp., Bokaro Basin, described in this section, in its almost submedial umbo that is raised above the dorsal margin; steep descent from umbonal terminus to posterior margin, and a smaller Arr/L, b/H, and c/L; Ch/L is larger (.440–.575 n. sp. vs. .430). Although H/L overlaps at the lower end of the range, the new species differs by one-plus magnitude at the upper end (.640–.800 n. sp. vs. .677 *"bokaroensis"*).

2. *Cyzicus (Euestheria) raniganjis* n. sp.
(Pl. 29, Fig. 2)

Diagnosis: Ovate valve with short dorsal margin; umbo inset from anterior margin approximately one-third the length of the valve; posterior margin rounded; anterior margin uncertain (partly covered). Ornamentation of contiguous granules on growth bands. Growth bands: 21, last five more tightly spaced.

Measurements: (mm) L-3.3; H-2.7; Ch-2.1?; Cr-0.6; Arr-0.6; a-1.0; b-1.1; c-1.7; H/L-.818; Cr/L-.182; Av/L-.182; Arr/L-.182; Ch/L-.636?; a/H-.370; b/H-.407; c/L-.515.

Material: Holotype TC 40094, internal mold of left valve.

Locality: Station RN-2, bed 4. Raniganj Coal Field, Numa River. Distance between RN-1 and RN-2, approximately 400 m upstream. Panchet Formation, Lower Triassic.

Discussion: Parameters of the above species are closest to *Cyzicus (Euestheria) basbatiliensis* from site RB-2, the base of which section is 35 m above the *Glossopteris-Schizoneura* datum, compared to RN-2, the base of which section is 97 m above the datum. However, *"basbatiliensis"* has a subterminal umbo and differs in several parameters: Cr/L, Av/L, and Arr/L are larger than in the new species above, and Ch/L is smaller.

2. *Cyzicus (Euestheria) dualis* n. sp.
(Pl. 28, Fig. 1)

Diagnosis: Dimorphic valves. The smaller valve is presumed male; it is irregularly ovate (i.e., tapers posteriorly), with short, straight dorsal margin and subterminal umbo; the larger valve (female) is laterally ovate with umbo inset from anterior margin; dorsal margin as in the male valve. Both dimorphs have double growth lines (i.e., the edge of two successive growth bands are distinct and appeared as grooves in the original valve). Ornamentation of contiguous granules on growth bands.

Measurements: (mm) TC 40092a, male, followed by TC 40092b, female. L-3.3, 3.4; H-2.3, 2.6; Ch-1.3, 1.5; Cr-0.7, 0.8; Av-0.7, 0.8; Arr-1.1, 1.1; a-1.1, 1.2; b-0.9, 1.4; c-1.5, 1.5; H/L-.697, .765; Cr/L-.212, .235; Av/L-.212, .232; Arr/L-.333, .324; Ch/L-.394, .441; a/H-.482, .462; b/H-.391, .538; c/L-.455,

.441. (Female valves are based on a reconstruction of the crushed valve.)

Material: Syntype TC 40092a (male), a complete external mold of right valve fairly well preserved; syntype TC 40092b (female), an external mold of right valve, slightly eroded on portion of dorsal margin, more eroded on ventral margin, and crushed.

Locality: Station RN-2(?), Numa River, Raniganj Coal Field, northwest of Asansol, west Bengal. Panchet Formation. Lower Triassic.

Discussion: Compared to *Cyzicus (Euestheria) raniganjis,* the new species has a smaller H/L, Cr/L, Ch/L, a/H, and c/L, and a greater Av/L and Arr/L.

Family CYCLESTHERIDAE Sars, 1899

Genus *Cyclestherioides* Raymond, 1946
Subgenus *Cyclestherioides* Raymond, 1946

Diagnosis: Ovate valves with nonterminal umbo and distinctly unequal height and length.

Type: *Estheria lenticularis* Mitchell, 1927.

Subgenus *Sphaerestheria* Novojilov, 1954c

Diagnosis: Trigonal to subcircular valves with height and length actually or nearly coequal.

Type: *Estheria koreana* Ozawa and Watanabe, 1923.

Cyclestherioides (Cyclestherioides) machkandaensis n. sp.
(Pl. 28, Fig. 4)

Diagnosis: Small, ovate to subcircular valves; umbo is close to, but inset from, the anterior margin; dorsal margin short behind umbo; growth-band number variable from approximately 13 to 20.

Measurements: (mm) Holotype (TC 41002), followed by paratypes (TC 40019a and 40019b). L-2.0, 2.3, 1.9; H-1.4, 1.7?, 1.3; Ch-0.9?, 1.2?, 1.0?; Cr-0.9, 0.5, 0.5?; Av-0.6?, 0.5, 0.5; Arr-0.5, 0.6, 0.4; a-0.6?, 0.6, 0.6; b-0.8, 0.8?, 0.5; c-0.8, --, 0.6?; H/L-.700, .740?, .684; Cr/L-.450, .217, .263; Av/L-.300?, .217, .263; Arr/L-.250, .261, .211; Ch/L-.300?, .522?, .526; a/H-.429?, .353, .462; b/H-.571, .471?, .385; c/L-.400, --, .316.

Material: TC 41002, holotype, internal mold of left valve. TC 40019a, paratype, internal mold of left valve, portion of antero-ventral sector missing. TC 40019b, paratype, external mold of right valve. Additional paratypes: TC 41001, 41003, 41004, 41005.

Locality: Station RM-1, bed 3b. Raniganj Coal Field on Machkanda Nala (= creek), Upper Panchet Formation. (See Figs. 13 and 14.) Lower Triassic.

Discussion: This is the only species of the genus *Cyclestherioides (Cyclestherioides)* in the Raniganj Coal Field conchostracan collection. One can compare the species *Cyclestherioides (Cyclestherioides) pintoi* n. sp. from the Estrada Nova Formation (= ENF) (Permian) of Brazil. (See South America section, this memoir.)

The Raniganj species described above has a smaller ratio H/L (.700–.740 vs. .845–.810), Av/L, Arr/L, and c/L; other ratios are close or overlap. Because Av/L and Arr/L ratios relate to the configuration, these are sufficiently distinct to distinguish the new species: Av/L-.217–.300? n. sp. vs. .350–.362; Arr/L-.211–.261 n. sp. vs. .296–.345.

2. *Cyclestherioides (Sphaerestheria)* sp.
(Pl. 29, Fig. 6)

Diagnosis: Subcircular valve with length and height almost equal; short, straight dorsal margin and subterminal umbo; numerous growth bands (19 below eroded area).

Measurements: (mm) TC 40077a. L-2.5; H-2.2; Ch-1.2; Cr-0.5; Av-0.5; Arr-0.8; a-1.1; b-1.1; c-1.4; H/L-.880; Cr/L-.200; Av/L-.200; Arr/L-.320; Ch/L-.480; a/H-.409; b/H-.500; c/L-.560. TC 40091: H/L-.840; Cr/L-.240; Ch/L-.480; b/H-.474.

Material: TC 40077a, external mold of left valve; eroded around and below the umbonal area and along the antero-ventral margin; also TC 40091 (97 m above datum).

Locality: Station RN-1-A, bed 4; Station RN-2, bed 7d; Station RN-1-B, bed 10. Raniganj Coal Field. (See Fig. 14.)

Discussion: The khaki-green micaceous, silty shale originally was, in most instances, a very poor substrate for preservation of conchostracan valves and their molds. The incompleteness of the specimen described above precludes species identification. Of interest is the recurrence of this species over a vertical gap of approximately 32 m. Parameters of the younger occurrence of this species (RN 2, bed 7d, 97 m above datum) are close to those of the geologically older specimen (i.e., RN-1-A, bed 4, 65 m above the *Glossopteris-Schizoneura* datum).

Panchet Formation. (Subsurface sections and sites; see Fig. 15.) Borehole data, Geological Survey of India (GSI), examined by Tasch at the S.E.M. Laboratory, GSI (Calcutta). Figures that follow are in metres and denote depth below surface.

Andal area, west Bengal: (RNM 2, 3, and 4 and CMPDI are boreholes.)

4.	RNM-2	*Cyzicus* sp. (47.5–50.0)
4.	RNM-3	*Palaeolimnadia* (161.0)
		Cyzicus sp. (47.0)
		Palaeolimnadia sp. (47.0)
		Estheriella sp. (9.0)
		Cyzicus sp. (9.0)
4.	RNM-4	*Cornia panchetella* n. sp. (250). (Described herein = dh.)
		Cyzicus sp. A (250.0)
		Gabonestheria sp. (106.0)] *A corniid–]gabones-]theriid]assemblage.
		Cornia sp. (106.0)
		Vertexia? sp. (106.0)
		Cyzicus sp. (40.0)
		Palaeolimnadia sp. (40.0)

4.   CMPDI   *Cyzicus* sp. A (249.0)
            Cyzicus sp. (200.0)
            Cornia panchetella n. sp. (185.0)
            Cyzicus sp. (145.0)

Raniganj Coal Field

            Cyzicus (Euestheria) minuta (v. Zieten),
            Ghosh, 1973, p. 4, 5. (Also found in
            North Karanpura Coal Field. Lower
            Triassic.)

   ### *East Bokaro Coal Field.* (EBP = boreholes.)
1.   EBP-4   *Cornutestheriella taschi* (Ghosh and
            Shah), Tasch n. g. (250.0)
            Palaeolimnadia sp. (250.0)
            Cyzicus sp. (250.0)
            Cyzicus (Lioestheria) bokaroensis n. sp.
            (dh) Hazaribagh, Bihar.
1.   EBP-5   (geologically younger than EBP-4)
            Cornia sp. (cf. *Cornia panchetella*
            n. sp.)

Mangli Village (100 km south of Nagpur)

Panchet Formation Equivalent

3. *Estheria mangaliensis* Jones [= *Cyzicus (Euestheria) mangaliensis* (Jones)]. Mangli, 92 m above *Glossopteris–Schizoneura* Zone. Jones, 1862, p. 78–81, Pl. 2, Figs. 16–23. (See Fig. 14 for Tasch section of M-1.) Topotype (dh).

3. *Pseudoasmussiata indicyclestheria* n. sp. Mangli, Tasch Station M-1.

3. *Cornia panchetella* n. sp. Mangli, Tasch Station M-1, beds 2, 4, 8.

Mahadeva Formation

Estheriid (Ghosh, 1973, p. 4, 5, text fig. and cover photo.). (= *Cyzicus (Euestheria)* sp. (See Pl. 30, Fig. 6 herein.) Mahadeva Formation, supra Panchet, approximately 25 m above the Mahadeva/Panchet contact. Upper Triassic. (The indicated species occurs in a shale bed on top of Lugu Hill.)

Andal Area, West Bengal

4. *Cornia panchetella* n. sp.
(Pl. 26, Figs. 3–8; also Pl. 30)

Diagnosis: Small, dimorphic, subovate to subcircular valves, with a node (female? valve) or a vertical small spine (male? valve), dorsal on the umbo, that rises from the subcircular to subovate umbo, which is inset from the anterior margin; growth bands: 11 or more.

Material: Syntype (2 core pieces: A-slab with five specimens; B-slab with three). Raniganj borehole specimens: TC 40197A and TC 40197B, internal molds of dimorphic left valves, re-spectively, with umbonal spine (male?) and umbonal node (female?). Other specimens are incomplete fragments with umbonal spine. (Additional specimens of the same species occur in the Mangli Collection [TC 40204B, TC 40210B, TC 41088-22, TC 40228, and TC 40217]; see Fig. 14.)

Locality: Syntype, Geological Survey of India (GSI). RNM-4, borehole 250 m above Panchet Formation/Raniganj Formation contact. Other specimens from Raniganj Coal Field, west Bengal; lat 23°31′30″N, long 87°13′02″, and also northeast of the G. T. Road from Asansol to Calcutta (see Fig. 15).

Measurements: (mm) B-slab 41097B followed by 40197A. L-1.9, 1.6?; H-1.4, 1.3; Ch-0.8?, 1.0?; Cr-0.51, 0.5; Arr-0.5?, 0.22?; a-0.6, 0.5; b-0.5, 0.5; c-0.9, 0.7; H/L-.737, .812; Ch/L-.526, .500; Cr/L-.293, .312; Av/L-.263, .312; Arr/L-.263, .125; a/H.429, .385; b/H-.357, .385; c/L-.368, .563.

Range for all syntypes (including the Mangli specimens listed above): H/L-.720–.812; Ch/L-.400–.526; Cr/L-.250–.320; Av/L-.250–.300; Arr/L-.125–.280; a/H-.385–.600; b/H-.357–.467; c/L-.368–.563. (All of the Raniganj Coal Field type ratios, except Arr/L, *overlap* those of the Mangli Village specimens. These are accordingly assigned to the new corniid species described above.)

Discussion: Field evidence at the Mangli Village site and in the Raniganj Coal Field in the past indicated that the Mangli beds were Triassic and correlated with the Triassic of the Raniganj Coal Field (Tasch and others, 1975). Now, however, borehole samples (RNM-4) at 250 m above the Panchet Formation/Raniganj Formation (Triassic/Permian) contact indicate occurrence of the same corniid species at both localities, even though they are separated in time (92 m Mangli vs. 25 m RNM, above the contact). A considerable distance separates Mangli Village from the Raniganj Basin (near Asansol; see Fig. 12).

Among the nine *Cornia* species from the Triassic and one from the Jurassic, described by Novojilov and colleagues from the USSR (*in* Novojilov, 1970, p. 135–149, Table 136), five of the most comparable differ from the new species as follows.

Cornia portenta Nov. (1970, p. 149, text Fig. 138, Pl. 6, Fig. 1) from the Russian Jurassic differs in having robust, irregularly subovate valves, a less prominent umbo that rises above the dorsal margin, and a thicker umbonal spine, placed more in an anterior direction on the umbo (vs. dorsad in *"panchetella"*). Among the Triassic species, *Cornia vosini* Molin, 1965 (Molin and Novojilov, 1965, p. 71, Pl. 9, Fig. 6; Novojilov, 1970, Fig. 134) differs in having a subterminal umbo bearing an oblique and comparatively very large and thick node; *Cornia jugensis* Nov. (Novojilov, 1970, Pl. 4, Fig. 8, text Fig. 135) differs in that it has a very small umbo and minute node; the umbo is situated close to the dorso-anterior valve sector; *Cornia samarica* Nov. (Novojilov, 1970, text Fig. 136) has a subterminal umbo that is smaller than in the new species, as well as a thicker node situated anteriorly on the umbo; *Cornia buzulukensis* Nov. (Novojilov, 1970, text Fig. 137) has a subterminal umbo that is very large, elliptical, and obliquely placed on the valve, and bears a small, thick node closest to the upper posterior sector of the umbo.

Cornia haughtoni Tasch, 1984 (Pl. 3, Fig. 1) from the Triassic Cave Sandstone of Lesotho (Thabaning) differs from the new species in having a less prominent and smaller umbo and bearing a smaller, tapered to blunt node, placed more dorsad. *Cornia angolata* n. sp., from the Triassic of northern Angola, differs in its prominent submedial umbo and much smaller node or spine that is placed higher up on the umbo; it also has a greater Ch/L and smaller a/H and b/H; H/L overlaps.

The fact that the dimorphic spine or node characters occur together on separate valves and on the same bedding plane is indicative of the coexistence of the two forms. (Spine= male, node= female, as interpreted here; see Pl. 26, Fig. 7; see also Tasch, 1969, p. R144–145, section on dimorphism.)

Regarding living conchostracans, Pennak (1953; and see discussion in Tasch, 1969, p. R147) observed that in the same small water situation, more than one species of the same genus rarely occurred. This further sustains the above inference. This writer's collecting experience of living forms in Kansas confirms Pennak's observation, although it may be qualified for somewhat larger water situations such as ponds, lake margins, and extensive river flats.

This writer is unaware that dimorphism has been previously noted in *Cornia*. This no doubt is due to the isolated recovery of either spine or node types, i.e., insufficient materials.

Under S.E.M. magnification (Pl. 26, Fig. 4), the node topography reflects an internal muscle attachment site; the interior, hollow area of the external node represents the actual locus of internal muscle insertion. Umbonal spines could have been used for breaking out of dried mud substrates; the irregular undulations of the external node may have been adaptive for the same purpose.

Under high S.E.M. magnification, the corniid spine appears to consist of a series of concentric rings, and a honeycomb pattern has been observed on the surface of the spines where better preserved; similarly, nodose forms (Pl. 26, Fig. 4) have the same horizontal pattern that appears on the umbo. Further, at 5-10K (SEM) magnification, the polygonal ornamentation of the valves of the spined forms, and, though less clearly defined, a probable central pore is seen in each polygon.

Family ESTHERIELLIDAE Kobayashi, 1954

Carapace bearing variable number of radial costae which, in general, become obsolete near the umbo.

Subfamily ESTHERIELLINAE Kobayashi, 1954

Carapace bearing five or more costae.

Cornutestheriella Tasch, n.g.
(Pl. 26, Figs. 1, 2)

Estheriellid valves with partial nodose ribs and a small curved umbonal spine.

Type: *Estheriella taschi* Ghosh and Shah, 1977. Geological Survey of India, Calcutta, India. GSI Borehole No. EBP-4. (See Fig. 14.)

1. *Cornestheriella taschi* (Ghosh and Shah)

Estheriella? Ghosh and Shah, 1973, p. 4–5, Fig. 3, and cover page.
Estheriella taschi Ghosh and Shah, 1977, p. 14–18.

Diagnosis: Subelliptical valves; initial valve (umbo) is small and ovate and inset from the anterior margin about one-fourth the length of the adult valve. Originating near the center of the umbo, posteriorly, a curved, thin, elongate type of umbonal spine reaches the second growth band in the flattened condition of the valve. Under low power (binocular microscope) the spine appears to be irregularly segmented. Under S.E.M. magnification, the spine is seen to be composed of a series of interlocking, basally expanded bundles of fibers. There are 25 or more nodose, partial costae of the estheriellid type; these costae margin the adult valve with greater relief where they cross the tighter spaced growth bands posteriorly. Anteriorly costae continue as unconnected nodes on each growth band. The first seven growth bands have no nodes.

Measurements: (mm) L-3.2; H-2.07; Ch-1.27?; Cr-1.07; Av-1.07?; Arr-0.96?; a-1.42?; b-0.71; c-1.27; H/L-.647; Ch/L-.397; Cr/L-.334; Av/L-.334; Arr/L-.300; a/H-.686; b/H-.343; c/L-.397.

Material: Holotype: *Estheriella taschi* Ghosh and Shah is on deposit with the GSI, Calcutta, India. GSI No. EBP-4. (This specimen was studied at the GSI S.E.M. laboratory during June, 1979.) The Tasch Collection (TC 46000) contains an internal mold of a right valve and several fragments showing partial estheriellid nodose costae.

Locality: Upper reach of Ohardharwa Nala (lat 25°45′N, long 88°46′E) in the eastern part of the East Bokaro Coal Field district, Hazaribagh, Bihar. Upper part of the Panchet Formation. Lower Triassic.

Discussion: The new species bears on the origin of four genera: *Cornia* Lutkevich, 1937 (with node or spine on umbo); *Curvacornutus* Tasch, 1961 (with curved umbonal spine); *Estheriella* Weiss, 1875 (with multiple nodose, radiating costae) and *Leaia* Jones, 1862 (bearing two or three radial ribs).

Before explaining these relationships, the distinction and affinity of "costae" and "ribs" needs clarification. They both have a radial placement on the valve; continuous costae are generally nodose and nodes may occur even where the continuous costae do not (see Pl. 26, Figs. 1 and 2). The critical term is "node(s)" that underlay both costae and ribs. This last characteristic leads back to *Cornia*, which bears a single node on the umbo. From one species to another, such nodes are variable in position on the umbo. Moreover, they vary in thickness from barely perceptible to quite prominent.

This writer (1963a, p. 145–157, Fig. 66) considered the above observation and others that follow below relevant to the origin of the leaiid rib. A nodose substructure of the leaiid rib has been revealed in many instances, for example, *Leaia ingens* (Novojilov) 1956a (formerly *Igorvarentsovia* Nov.; reproduced in Tasch, 1958, text Fig. 30A and B) from the Russian Paleozoic, and is not uncommon in Mississippian and Pennsylvanian leaiids. Accordingly, the origin of the leaiid rib was postulated (Tasch, 1963a, Fig. 66) as follows: (1) The *Cornia*-type nipple-like node was displaced to the periphery of the larval valve by a mutation. (2) Subsequently, this displaced node was repeated on successive valve growth bands by the genetic mechanism of redundance. Later, these nodes in contact with one another formed the leaiid rib. (3) Continuous redundance (with an effect equivalent to polyploidy) led to two or three leaiid ribs.

Equivalents of the segmented, curved umbonal spine of the new species occur in several of Novojilov's *Curvacornutus* species (Novojilov, 1970, Pl. 7), as well as in the segmented (uncurved) partial rib of *Protomonocarina* Tasch, 1962 (*in* Tasch 1969, p. R161, Fig. 54; 4).

The new species, which has two valve features not previously recorded in the same valve (a curved segmented umbonal spine and estheriellid nodose costae), immediately links *Cornia, Curvacornutus,* and *Estheriella,* which respectively have an umbonal node, a curved spine, and partial nodose costae. The link to *Leaia* was outlined above.

How are the thicker and more continuous *Leaia* ribs and the thinner, more numerous partial nodose costae of *Estheriella* related? Obviously, multiple costae are due to redundance as are repeated leaiid ribs. Because both are based on a repetition of the individual node, and because the new species has a segmented umbonal spine (a derivative of the corniid node), as well as nodose costae, it appears that the *Estheriella* and *Leaia* lines diverged from the same *Cornia*-like progenitor.

Beyond the first seven growth bands barren of nodes are node-bearing bands. Ventrad, these are followed by partial costae which are fused nodes. *Partial* absence of nodes plainly denote a preestheriellid stage of evolution. The estheriellid stage commences with the appearance of fusion of isolated nodes. This earliest event in development is a relict of a more primitive state before development of complete costae.

An attempt by Shen Yan-bin (1978, Fig. 1) to decipher leaiid and estheriellid evolution indicated the line of descent for the Estheriellidae by a question mark. Data from the new species can resolve the puzzle of estheriellid origin discussed above.

1. *Cyzicus (Lioestheria) bokaroensis* n. sp.
(Pl. 27, Fig. 3)

Diagnosis: Broadly subovate valve, ventrad, both anterior and posterior margins are expanded beyond the length of the dorsal margin with comparatively short, straight dorsal margin behind the umbo, which is inset from the anterior margin almost one-third the length of the valve; margins: anterior rounded up to and including join to dorsal margin; posterior rounded ventrad and oblique along postero-dorsal section. Growth bands: 16 or more. Ornamentation: hachure-type markings on growth bands.

Measurements: (mm) L-3.0; H-2.0; Ch-1.3; Cr-0.8; Av-0.8; Arr-0.9; a-0.9; b-1.0; c-1.4; H/L-.667; Ch/L-.433; Cr/L-.267; Av/L-.267; Arr/L-.300; a/H-.450; b/H-.500; c/L-.467.

Material: Holotype TC 40117, external mold of left valve.

Locality: Site No. EBP-4 (TC 40117), Geological Survey of India, East Bokaro Coal Field. Upper reach of Ohardharwa Nala, lat 25°45′N, long 88°46′E, Hazaribagh, Bihar. Upper part of Panchet Formation. Lower Triassic.

Discussion: The only species in the Lower Triassic Panchet Formation to which *"bokaroensis"* can be compared is *Cyzicus (Lioestheria) miculis* n. sp. (Pl. 29, Fig. 8). The latter species differs from *"bokaroensis"* in that it is irregularly ovate (male), subelliptical to laterally ovate (female), has a unique oblique posterior-dorsal join, and a submedial umbo (vs. inset one-third the length of the valve) that is raised above the dorsal margin. It also has a smaller b/H and c/L. Some *"bokaroensis"* ratios are greater than those of the female valve and smaller than the male valve of *"miculis"*; H/L, Cr/L, and Av/L. This writer's notes on the borehole slices from this site (EBP-4) record the presence of contemporaneous biota—fish scales and two conchostracan genera, *Cornutestheriella* and *Palaeolimnadia.*

3. *Pseudoasmussiata indicyclestheria* n. sp.
(Pl. 27, Fig. 5)

Diagnosis: Subcircular valves with axis through umbo plunging posteriorly; straight dorsal margin and comparatively large umbo covering the medial upper sector of the valve and raised above dorsal margin; ventrad of the umbo 15 growth bands; margins: anterior rounded; posterior steeply arched.

Measurements: (mm) L-2.3; H-1.8; Ch-1.0; Cr-0.7; Av-0.7; Arr-0.5; a-0.8; b-1.0?; c-1.5; H/L-.713; Ch/L-.435; Cr/L-.304; Av/L-.304; Arr/L-.217; a/H-.444; b/H-.555; c/L-.652.

Material: Holotype TC 40208 (M-1 float, slab M 25), internal mold of right valve in red shale with limonite covering bedding planes. *Cornia* sp. is on same bedding plane as the above new species. (See description of Mangli corniids herein.)

Locality: Tasch Station M-1, Mangli Formation (a correlate of the Panchet Formation, Raniganj and other coal basins); Lower Gondwana, Damuda Series, Mangli, 100 km south of Nágpur. Lower Triassic.

Discussion: This species is closest to the Greenland Triassic *Pseudoasmussia grasmücki* Defretin-Le Franc (*"Pseudoasmussia"* is a preoccupied name and renamed in this memoir to *Pseudoasmussiata*). It shares the chief characteristic of this genus in having the axis through the umbo plunge in a posterior-ventral direction, and in having a nonterminal umbo and subcircular form. It is close to the new species. However, parameters differ; for example, Cr/L, Arr/L, and Ch/L are smaller in the new species, whereas Av/L and c/L are larger.

3. *Cyzicus (Euestheria) mangaliensis* (Jones)
(Pl. 29, Fig. 1; Pl. 27, Fig. 2)

Estheria sp. Hislop in Hislop and Hunter, 1854, p. 472; 1855, p. 371; Hislop, 1861, p. 347, 353.
Estheria mangaliensis Jones, 1862, p. 78–81, Pl. 2, Figs. 16-23.

Diagnosis: (Jones, 1862, original description abbreviated.) "Carapace valves, usually broadly subovate but varying from subtriangular to suboblong according to age, sex(?), and state of preservation . . . Normally the valve somewhat narrower in front than behind" (i.e., posterior). "Hingeline . . . equal to rather more than half the length of the valve . . . Obscurely hexagonal reticulations of small meshes," on growth bands.

(Emendations herein.) Ovate valves, more or less elongate, umbo subterminal; dorsal margin straight; posterior margin rounded; anterior margin more steeply curved. (Other valves of variable shape as reported by Jones are either juveniles, distorted, or eroded, and/or partially covered along margin or hingeline.)

Measurements: (mm) Topotype TC 40215 followed by Jones' syntype, BM In-28223. L-7.6, 6.8; H-5.7, 4.7; Ch-3.6, 3.8?; Cr-1.2, 1.1; Av-1.2, 1.0; Arr-2.8, 2.7; a-1.3, 1.8; b-2.4, 1.8; c-3.0, 2.9; H/L-.750, .691; Ch/L-.474, .559?; Cr/L-.158, .162; Av/L-.158, .162; Arr/L-.368, .294; a/H-.404, .383; b/H-.421, .383; c/L-.395, .426; Antero-dorsal angle-45°, 45°.

Juvenile specimen (TC 40235B, topotype). L-3.3; H-2.6; Ch-1.6; Cr-0.9; Av-0.9?; Arr-1.0; a-0.9; b-1.2; c-1.6; H/L-.788; Ch/L-.485; Cr/L-.273; Av/L-.273; Arr/L-.303; a/H-.346; b/H-.461; c/L-.485. Antero-dorsal angle-25°.

Material: Tasch Collection, eight slabs of hard red shale, almost all bearing whole valves as molds or eroded fragments of this species. Fossils and slabs coated with limonite on bedding planes. Topotypes include TC 40214 (m-1, float), TC 40215 (M-1, bed 5, lower), and TC 40235 (M-1, bed 4, lower), which may be a juvenile represented by an internal mold with submedial umbo. Figured specimens herein include Jones' syntype (BM In-28223) and one of the topotypes, TC 40215.

Locality: Station M-1, Mangli Village, 100 km south of Nágpur, central India. Lower Triassic. (See Appendix for measured section.)

Discussion: Jones (1862) incorporated a variety of odd-shaped fossils on a single bedding plane as different configurations attributable to age and/or sex or preservation of this species. However, it is not clear whether these small valves are all juveniles of *"mangaliensis"* because there are also present small ovate to subcircular valves with a node or spinous umbo (often eroded) that occur on the same bedding plane; these valves are attributable to corniids.

If we interpret TC 40235B as a juvenile valve of *"mangaliensis"* it will be seen that H/L, Cr/L, Av/L, and c/L ratios are somewhat larger than in the adult, Arr/L and Ch/L overlap, and a/H and b/H are smaller, as is the antero-dorsal angle. It has long been established that during growth, the shape of the individual changes. Where numerous ratios overlap, it is usually due to heterochroneity of the species' variants.

Kotá Formation

In lieu of a number, D2 is used as a map coordinate for all Kotá Formation species. (See Fig. 12 for India locality map.)

D2. *Estheria kotahensis* Jones [= *Cyzicus (Lioestheria) kotahensis* (Jones)]. Kotá on the Pranhítá River. Jones, 1862, p. 81–83, Pls. 24, 25. (Emended description and newly figured syntypes given below.) Hislop's (1861) site is equivalent, in part, to Tasch Station K-7, which is southwest of Kotá Village. (See Appendix for measured Kotá Formation sections.)
D2. *Estheria* sp. Hislop. [= *Cyzicus (Lioestheria)* sp.] Kotá Formation, Lower Jurassic. Hislop, 1861, p. 348; 1862, p. 201. Site: Kotá Village on the Pranhítá River. (Placed in synonymy of *Estheria kotahensis* Jones [1862, p. 81].) Other specimens from Katanapali, 24 km north of Kotá Village. (See Fig. 12, D2, inset.)

D2. *?Cyzicus (Lioestheria?) kotahensis* (Jones), 1862
(Pl. 29, Fig. 3)

Estheria kotahensis Jones 1862, p. 81–83, Pl. 2, Figs. 24, 25.

Diagnosis (emended): Subovate to subcircular valves; growth bands numerous and closely spaced. Umbo inferred to be subterminal by following trend of partially eroded growth bands. Ornamentation poorly defined. (According to Jones, "interspaces usually smooth . . . but occasionally ornamented towards the ventral border by a pattern consisting of slight, vertical, anastomosing wrinkles with accompanying rows of minute pits" (Jones, 1862, p. 82).

Measurements: (mm) (Measurements that follow are all approximate because of the poor preservation and/or incompleteness.) In-28268 followed by the nongraphite-coated specimen nearest to it on the slab. L-3.5?, 4.0; H-2.3?, 3.2?; Ch-1.5, --; Cr-0.9, --; Av-0.9, --; Arr-1.1, 1.5; a-1.2, 1.3; b-1.0?, 1.4; c-1.4, 1.7; H/L-.675, .800; Ch/L-.429, --; Cr/L-.257, --; Av/L-.257, --; Arr/L-.314, --; a/H-.522, .406; b/H-.435, .437; c/L-.400, .425.

Material: Syntype (British Museum-BM In-28268) consists of the internal mold of a broadly subovate right valve so badly eroded that almost all characteristics have been removed; details are therefore impossible to determine. Two other valve fragments (I-6837) which preserve only three growth bands are also lacking in definitive details. One of these fragments is graphite coated. The syntype figured, being the only nearly complete specimen, is illustrated here (Pl. 29, Fig. 3). This is included in order to elucidate the kind of evidence available for this species, even though it cannot be of any use for comparative purposes.

Locality: Kotá Formation on the Pranhítá River. Jones' site is equivalent, in part, to Tasch Station K-7. (See Appendix for measured sections of this formation.) Lower Jurassic.

Discussion: The material on which Jones based this species is exceptionally poor; included are BM specimens of syntype In-28268 and two undecipherable incomplete valves, I-6837. The specimen In-28268 was not figured by Jones, nor is the incompleteness of the valve fragments (I-6837) indicated in his whole valve reconstruction. His published illustrations are difficult to reconcile with the fossils. Nor could the ornamentation he described be seen by examination of the very poor syntypes. Accordingly, Jones' species, reviewed and restudied herein, is questionable on all levels: generic, subgeneric, and specific.

D2. *Pseudoasmussiata andhrapradeshia* n. sp.
(Pl. 27, Fig. 8)

Diagnosis: Subovate valves with a short, dorsal margin, the anterior and posterior portions of which are equal; elliptical, smooth, submedial umbo that rises above the margin. (Eight growth bands terminate on the dorsal margin on either side of the umbo.) Anterior margin steeply arched; posterior margin bulging, convex midway. The axis of the valve through the umbo has a bias posteriorly. Ornamentation of small, contiguous granules.

Measurements: (mm) L-2.1; H-1.5; Ch-0.9; Cr-0.6; Av-0.6; Arr-0.5; a-0.5; b-0.6; c-0.9; H/L-.714; Ch/L-.429; Cr/L-.286; Av/L-.286; Arr/L-.231; a/H-.333; b/H-.400; c/L-.429; antero-dorsal angle approximately 30°; postero-dorsal angle approximately 40° (both angles measured on photograph).

Material: Holotype TC 40002, internal mold of right valve, covered slightly along antero-ventral margin; upper sector of posterior margin eroded. A beetle elytron occurs on the upper left of the valve.

Locality: Tasch Station K-1, bed 6a, upper, Kotá Formation, 2.5 km east-southeast of Kadamba Village (Sirpur Taluka), Adilabad District, Andhra Pradesh State. Lower Jurassic.

Discussion: This writer has been unable to locate any Mesozoic species that match this new species. The prominent elliptical submedial umbo raised above the short, straight dorsal margin that flanks it on either side, combined with the posterior bias of the valve and the posterior margin bulge, are distinctive features.

D2. *Estheriina (Nudusia) adilabadensis* n. sp.
(Pl. 27, Fig. 1; Pl. 28, Fig. 3)

Diagnosis: Laterally ovate valves with a comparatively prominent subterminal, subovate umbo that is smooth, swollen, slightly tapered dorsad, and raised above the dorsal margin. The latter is short and straight; the antero-dorsal join is steep; posterior margin convex. Growth bands: 11.

Measurements: (mm) TC 40001 followed by TC 40198. L-1.7, 2.3; H-1.0?, 1.5; Ch-0.8?, 1.0; Cr-0.4, 0.4; Av-0.4, 0.4; Arr-0.5, 0.8; a-0.5?, 0.5; b-0.6, 0.6; c-0.5, 1.0; H/L-.588?, .652; Ch/L-.471, .435; Cr/L-.235, .174; Av/L-.235, .174; Arr-.294, .348; a/H-.500?, .333; b/H-.600?, .400; c/L-.294, .435; h_u-0.2, 0.3; l_u-0.5, 0.7; h_u/l_u-.400, .429 (h_u, l_u = height and length of umbo).

Material: Holotype TC 40001, internal mold of left valve, last growth band as well as uppermost portion of postero-dorsal margin covered. Paratype TC 40198, external mold of left valve. Paratype TC 40061, external mold of left valve, well preserved except for eroded ventral-anterior sector.

Locality: Tasch Station K-1, bed 8A (TC 40198, TC 40001) and bed 8B (TC 40061). Kotá Formation. Same locality as *Pseudoasmussiata andhrapradeshia* (see above). Lower Jurassic.

Discussion: This estheriinid species is unlike any other in the Kotá Formation. *Estheriina (Nudusia) indijurassica* n. sp. (Site K-2, bed 2, lower) differs from the above species in configuration (broadly ovate vs. laterally ovate n. sp.) and several parameters and ratios; larger H/L, Cr/L, Av/L, and smaller Arr/L, Ch/L, and so on. The new species is distinct from estheriinids (subgenus *"Nudusia"*) described from the Australian Lower Triassic (Tasch *in* Tasch and Jones, 1979c) as follows: from *Estheriina (Nudusia) circula* Tasch in configuration (laterally ovate vs. circular), umbonal shape (subovate vs. elliptical), and a smaller H/L (.588–.652 vs. .780–.810), and smaller h_u/l_u; from *Estheriina (Nudusia) rewanensis* Tasch in configuration (more elongate), in umbonal position (subterminal vs. submedial), smaller h_u/l_u and H/L, which overlaps only in the upper part of the range (.586–.652 vs. .650–.800); from *Estheriina (Nudusia) blina* Tasch (in Tasch and Jones, 1979a) in having a smaller umbo h_u/l_u (.588–.652 vs. .800–.900), smaller H/L (.588–.652 vs. 0.76–0.80), and a steeper dorsal-anterior join.

Closest in horizontal elongation of the valve is *Estheria kawasakii* Ozawa and Watanabe, 1923 (Pl. 5, Fig. 4b) from the Korean Triassic, reassigned to *Estheriina* Jones by Novojilov (1954c and cf. Novojilov and Kapel'ka, 1960, text Fig. 17). This species apparently has a naked umbo that is raised above the dorsal margin, but the new species differs from it in configuration; not as elongate, and valve not anteriorly tapered; an umbo that is large and tapered vs. a comparatively small, nontapered umbo; and a larger H/L (.588–.652 vs. .409).

D2. *Palaeolimnadia (Grandilimnadia?)* sp.
(Pl. 29, Fig. 4)

Diagnosis: Subovate valve with a comparatively large, ovate umbo inset from but close to anterior margin; slightly arched dorsal margin; anterior and posterior margin rounded; five or more widely spaced growth bands below umbo.

Measurements: (mm) Due to erosion and cover, measurements made are suggestive rather than definitive. L-2.5?; H-1.6?; a-0.8?; c-0.7?; H/L-.640?; a/H-.320?; c/L-.280?

Material: TC 48001, internal mold of a left valve, badly eroded along posterior margin; antero-ventral margin partly covered.

Locality: Tasch Station K-2, bed 2, about 500 m upstream from Station K-1. Same locality as *Pseudoasmussiata andhrapradeshia* (see above). Lower Jurassic.

Discussion: The presence of this genus is recorded solely to provide a precise stratigraphic context for its occurrence. The large smooth umbo of 5 or more growth bands signifies *Palaeolimnadia (Grandilimnadia)*.

D2. *Estheriina (Nudusia) indijurassica* n. sp.
(Pl. 27, Figs. 4, 6; Pl. 28, Fig. 5)

Diagnosis: Broadly ovate valves with smooth umbo, swollen below and tapered above that rises above the short, arched, dorsal margin and is inset slightly from the dorsal-anterior margin to submedial in position. Six to eleven growth-band terminations on the dorsal margin occur behind the umbo. Posterior margin convex; anterior margin rounded.

Measurements: (mm) Holotype: L-2.1; H-1.4; Ch-0.7; Cr-0.8; Av-0.7; Arr-0.6; a-0.5; b-0.4; c-0.8; H/L-.667; Ch/L-.333; Cr/L-.381; Av/L-.333; Arr/L-.286; a/H-.357; b/H-.286; c/L-.381.

Material: Holotype TC 40055, internal mold of left valve and remnant of inner layer of right valve; covered in the upper anterodorsal sector slightly behind umbo; last two growth bands missing in part along postero-ventral margin. Paratype TC 48002, internal mold of left valve (naupliid), and TC 40052a, b.

Locality: Tasch Station K-2, bed 2, lower; about 500 m upstream from Station K-1. Same locality as *Pseudoasmussiata andhrapradeshia* n. sp. (see above). Lower Jurassic.

Discussion: This species above differs from Australian Lower Triassic estheriinids (subgenus *Nudusia*) (Tasch *in* Tasch and Jones, 1979c) as follows: *Estheriina (N.) circula* Tasch has a larger H/L, a circular configuration, and its umbo is elliptical; *Estheriina (N.) blina* Tasch (1979a) has a larger H/L (0.76–0.80 vs. .677 n. sp.) and an umbo ranging from subterminal to inset; *Estheriina (N.) rewanensis* Tasch generally has a larger H/L that overlaps in the lower part of the range (.650–.800 vs. .667 n. sp.) as well as a subterminal umbo (vs. inset to submedial).

The new species differs from Storm Peak TC 10121, *Estheriina (Nudusia) stormpeakensis* n. sp., in configuration (ovate vs. elliptical), smaller size, and less tapered umbonal terminus. H/L, a/H, b/H, and c/L are fairly close in the India and Antarctic species, although parameters and other ratios differ by several magnitudes (e.g., Ch/L-.765 [Storm Peak], .333 [Kotá]). Estheriinids from the Kotá Formation and Storm Peak Upper Flow interbed are both Lower Jurassic in age; the respective formations have been shown to share several faunal characteristics (Tasch and others, 1975; Jain, 1980).

Compared to *"adilabadensis"* n. sp., the Kotá Formation species *Estheriina (Nudusia) adilabadensis* (Station K-1, bed 8A) *"indijurassica"*, differs in having a slightly arched dorsal margin, and umbo inset from the anterior margin, as well as having a greater H/L, Cr/L, and Av/L. In addition, *"adilabadensis"* has a greater Arr/L, Ch/L, and b/H. Only two ratios, a/H and c/L, overlap.

D2. *Cyzicus (Euestheria) crustabundis* n. sp.
(Pl. 28, Figs. 6–8)

Diagnosis: Laterally subovate valves with a comparatively long, straight, dorsal margin behind the umbo; umbo inset from ante-

rior margin almost one-fourth the length of the valve; anterodorsal join angle is steeper than postero-dorsal; anterior margin steeply arched; posterior margin bulging and convex. Growth bands, 20 or more. Ornamentation of contiguous granules on growth bands.

Measurements: (mm) Syntype TC 40189a followed by TC 40189b, and then TC 40189c. L-4.3, 4.6, 4.3; H-3.0, 3.3, 2.9; Ch-1.8, 2.4?, 2.1; Cr-0.9, 1.0, 1.0; Av-0.9, 1.0, 1.0; Arr-1.6, 1.2, 1.2; a-1.2, 1.2, 1.1; b-1.4, 1.3, 1.1?; c-1.9, 2.0, 2.0; H/L-.689, .717, .674; Ch/L-.419, .522, .488; Cr/L-.209, .217, .233; Av/L-.209, .217, .213; Arr/L-.372, .266, .279; a/H-.400, .364, .379; b/H-.467, .394, .379; c/L-.443, .435, .465.

Material: Syntype TC 40189a, internal mold of right valve, growth bands of lower portion of valve close to anterior margin (crushed). Syntype TC 40189c, the best preserved of the valves, internal mold of right valve. Other syntype material includes: TC 40185c, TC 40200a and b, TC 40129, 40189b, 40177, and 40207a, in addition, TC 40666c, TC 40667a, b, TC 40106c (from K-4, upper conch bed), and TC 48005 (K-7, bed 6). There are abundant representatives in varying states of preservation throughout the several layers; for the most part, they are in a condition of low relief.

Locality: From the stream section of the eastern side of a tank, 1–2 km southeast of Metapalli, Kotá formation, Station K-4, bed 11. Coquinas of this species also occur at Stations K-5 and K-6 in layer after layer of flattened valves, and also at Station K-7. Lower Jurassic.

Discussion: This species, which is abundant in coquinas in the Station K-4 bed, is distinct in configuration, parameters, and ratios from other euestheriids; for example, it differs from another coquina euestheriid from the Lower Triassic Bonaparte Gulf Basin, Australia, *Cyzicus (Eu.) bonapartensis* Tasch, 1979, which has a subterminal umbo (vs. inset one-fourth the length of the valve); a longer dorsal margin; and in configuration (robust ovate vs. laterally subovate). It also differs from another coquina euestheriid from the Port Keats area (same basin as above), *Cyzicus (Eu.) dickinsi* Tasch, 1979, which has a larger H/L, longer dorsal margin, umbo subterminal to submedial (vs. inset almost one-fourth), and a less steep anterior-dorsal join than the new species.

TC 40666 and TC 40667 are two subcircular valves (K-1, float) that also occur throughout the K-5 beds; their H/L is .868 and .822, respectively. They are poorly preserved and are here questionably assigned as probable juveniles of the above new species. Exclusive of the probable juveniles (also found in the matrix noted below), TC 40106c (K-4, upper conchostracan bed) and TC 48005 (K-7, bed 6) overlap the new species' ratios that could be determined, but are more poorly preserved, as they are recrystallized in a ferruginous siltstone and shale matrix.

Because the new species appears in abundance in coquinas at K-4, K-5, K-6, and K-7, it appears to indicate a correlation of the K-4 Station (above the 19° parallel) with Stations K-5, K-6, and K-7 (below the 19° parallel). (See inset, Figs. 12 and 16.)

D2. *Estheriina (Estheriina) pranhitaensis* n. sp.
(Pl. 28, Fig. 2)

Diagnosis: Broadly ovate valves with long, straight dorsal margin behind subterminal umbo. Below the terminal umbonal region, there are six growth bands which are coarser than the much closer spaced ones that follow to the ventral margin and comprise a broad, apron-like continuation of the valve. The naupliid valve (umbo) also appears superimposed on the rest of the valve.

Measurements: (mm) Holotype: L-3.5; H-2.8; Ch-1.9; Cr-0.5; Av-0.5; Arr-1.1; a-1.1; b-1.1; c-2.0; H/L-.771; Ch/L-.543; Cr/L-.143; Av/L-.143; Arr/L-.314; a/H-.407; b/H-.407; c/L-.571.

Material: Holotype TC 48004b, external mold of left valve.

Locality: Tasch Station K-7, bed 6. Station K-7 is the type section of the Kotá Formation, southwest of Kotá Village, District Chandrapur, Maharastra State. Lower Jurassic.

Discussion: Compared to Jones' (1897a) original type of the genus, i.e., *Estheriina (Estheriina) bresiliensis* from the Brazilian Lower Cretaceous, the species described above has smaller ratios for Cr/L, Av/L, Arr/L, and c/L. Two ratios are close, Ch/L and a/H; H/L is greater. The new species also differs from *Estheriina (Estheriina)* sp. 1 from the Antarctic (Blizzard Heights, Station 1) in configuration (broadly ovate vs. eccentrically ovate) and umbo (subterminal n. sp. vs. inset one-third), a smaller H/L, Cr/L, Av/L, while Arr/L is the same, and greater Ch/L (.543 n. sp. vs. .371?).

D2. *Estheriina (Nudusia) bullata* n. sp.
(Pl. 29, Fig. 7)

Diagnosis: Small valves with a near triangular aspect due to an elevated submedial prominent umbo that is small, smooth, bubble-shaped, and distinct from rest of the valve. Nine growth bands below umbo.

Measurements: (mm) L-2.4; H-1.7; Ch-1.0?; Cr-0.5; Av-0.5; Arr-0.8; a-0.8; b-0.8; c- --?; H/L-.708; Ch/L-.417?; Cr/L-.208; Av/L-.208; Arr/L-.333; a/H-.333; b/H-.333; c/L-?.

Material: Holotype TC 40155a, external mold of left valve.

Locality: Kotá Formation, Station K-2, bed 3, upper "c", about 500 m upstream from Station K-1. (See *Pseudoasmussiata andhrapradeshia* for locality data and Fig. 12 and 16 for map and section data.) Lower Jurassic.

Discussion: The major characteristics of this species which make it distinct from all Kotá and other Jurassic estheriinids are the prominent submedial, bubble-shaped, smooth umbo, overall triangularoid configuration, and measurements.

INDIA (Cretaceous)*

Himalaya Mountains (Uttar Pradesh-U.P.)

Srivastava, 1973, p. 193.

---

*Personal communication, December, 1979, from Sri S. C. Ghosh, senior geologist, Geological Survey India; he indicated that the Tal Formation has "Cretaceous estheriids."

"The fossil collection described from the dark grey micaceous shale/slate succession of the Lr. Tal Formation, exposed approximately 3.2 Km N.E. of Banali Village (30°20'; 78°8') in Topo. Sheet No. 53J/SW . . . closely resembles the forms described by Bock (1953) from Triassic beds of Venezuela and Colombia."

SOUTH AMERICA

(Number preceding species name is a South American locality map number [Fig. 17].)

Argentina

Chubut

3. *Cyzicus (Euestheria) volkheimeri* Tasch (Tasch *in* Tasch and Volkheimer [= T & V], 1970, p. 5, Pl. 1, Figs. 1–3). Syntype, locality VAs₂ (Tasch in Tasch and Volkeimer, 1970; see map, Pl. 13, Figs. 4 and 5), Cañadón Asfalto Formation (Jurassic–Callovian), southern facies. Types and other specimens on deposit at Museo Argentino de Ciencias Naturales, Buenos Aires.

3. *Cyzicus (Lioestheria) patagoniensis* Tasch (locality, age, reference as in *"volkheimeri";* Tasch *in* Tasch and Volkheimer, 1970, p. 8, Fig. 2, holotype. In VAs₃).

3. *Cyzicus (Lioestheria)* sp. A. Same data as *"volkheimeri"* (Tasch *in* Tasch and Volkheimer, 1970, Pl. 3, Fig. 1, VAs₉).

3. *Cyzicus (Lioestheria)* sp. B. Same reference as *"volkheimeri"* (Tasch *in* Tasch and Volkheimer, p. 8, Pl. 3, Fig. 1). Locality: Colan Conhue, Museo de la Plata.

3. *Cyzicus (Lioestheria)* sp. C. Same reference as *"volkheimeri"* (Tasch *in* Tasch and Volkheimer, 1970, p. 8). Locality, VCh 15, Cañadón La Chacra, (northern facies of the Cañadón Asfalto Formation).

3. *Cyzicus (Euestheria)* sp. A. Same reference as *"volkheimeri"* (Tasch *in* Tasch and Volkheimer, 1970, p. 5 and 8). Locality, VCh₆. Same facies as *C. (Lio.)* sp. C.

Mendoza

62. *Cyzicus (Euestheria) forbesi* (Jones), 1862
(Pl. 31, Figs. 1, 2)

Cypridina Burmeister 1861, p. 77, 1876, p. 262.
Estheria forbesi Jones 1862, p. 109, Pl. 4, Figs. 8–11; 1897b, p. 263–264, Pl. 11, Figs. 1a, b, 2.
Estheria? mendocina Philippi 1887, p. 233, Pl. L, Fig. 12.
Estheria megalensis (sic.) (*?mangaliensis*) (*partim*) Geinitz, 1876, p. 3, Pl. 1, Fig. 5.
Euestheria forbesi (Jones) Raymond 1946, p. 242–243.
Euestheria forbesi (Jones), Kobayashi 1954, Pl. 9, Pt. 1, p. 157.
Cyzicus (Euestheria) forbesi (Jones), Morris 1980, p. 35, Pl. 2, Figs. 1, 2.

Diagnosis (emended from Jones, 1862): (Adult) - Comparatively large valves, subovate, with a slight obliquity along a line from center of umbo to postero-ventral sector, resulting in a convex bulge of posterior margin. (Juvenile) - Orbicular valves with a more rounded posterior margin. The umbo is slightly inset from the anterior margin; dorsal margin is straight and the join with the posterior margin is less steep than the anterior. Adult valves have more than 28 growth bands, whereas juvenile valves have 10–16. Ornamentation of contiguous granules on growth bands (= Jones "irregularly hexagonal reticulations").

Measurements: (mm) BM 50521, no. 8, followed by the orbicular form, same slab, no. 9. L-11.6, 8.2; H-7.8, 6.3; Ch-6.3, 4.0; Cr-1.8, 2.7; Av-1.5, 2.1; Arr-3.8, 2.1; a-2.2, 2.0; b-4.2, 2.5; c-6.8, 3.6; H/L-.672, 768; Ch/L- --, .543; Cr/L-.155, .256; Av/L-.129, .256; Arr/L-.328, .256; a/H-.282, .317; b/H-.538, .397; c/L-.586, .439.

Material: Original syntype, British Museum (BM). Three pieces, two of which contain numerous *"forbesi"* valves, broken, eroded, partially covered, and incomplete. BM 50521-one piece with this number bears both an orbicular juvenile valve that is an internal mold of a right valve (no. 9), and an external mold of left valve (no. 8 here designated lectotype); a second piece, BM 50521, bears an internal mold of a right valve and a smaller valve showing the ornamentation; BM 2170 (in two separate parts) bears an internal mold of a left valve with dorsal margin eroded, and bears a few incomplete valves.

Locality: Cacheuta is approximately 32 km southwest of Mendoza (lat 33°), Argentina, at the eastern slope of the Andes; specimens in brown-grey, finely laminated shale. Upper Triassic (Rhaetic).

Discussion: The existence of different configurations in the same species on the same bedding plane is not unusual. This may have several explanations: structural distortion, dimorphism, compaction, selective erosion, or coexistence of juvenile and adult individuals of the same population. The last appears to be the case with the type-suite of *"forbesi."*

10. *"Estheria"* sp. Hedberg, 1964, p. 1796. Upper part of Unit B of the Mendoza Oil Field, Triassic.

10. *Estheriopsis bayensis* Rusconi 1948a. (= *?Estheriopsis bayensis* Rusconi 1948a, questionable, a probable pelecypod.) Rusconi, 1948a, p. 200–201, Fig. 1. Twenty km from city of Mendoza. Museo de Historia Natural de Mendoza, No. 4086. Middle Triassic.

10. *Euestheria striolatissima* Rusconi 1948a. [*Cyzicus (Lioestheria) striolatissima* (Rusconi)]. Quebrada de los Leones, el Challao. Six km from Mendoza. Museo de Historia Natural de Mendoza, No. 3136. Middle Triassic.

10. *Pseudestheria* (sic.) *contorta* Rusconi 1948a. (*?Pseudestheria contorta* Rusconi). 1948a, 1948b; same locality as *"striolatissima."* Middle Triassic.

10. *Pseudestheria* (sic.) *leonense* Rusconi 1948a. (*?Pseudestheria leonense* Rusconi.) Rusconi, 1948a, 1948b; locality same as *"striolatissima."* Middle Triassic.

63. *Estheria minoprioi* Rusconi 1946a. (= *Pseudestheria* [sic.] *minoprioi* Rusconi 1948b.) *?Estheria minoprioi* Rusconi 1946a of questionable generic assignment. Unlike *Cyzicus (Euestheria) mangaliensis* to which Rusconi compared it. Rusconi, 1946a, p. 755–758, Figs. 1, 2. Museo de Historia Natural de Mendoza, No. 2926, 2927. Locality: Aqua de la Zorra, Paramillos de Uspallata, San Lorenzo District. Lower Triassic.

50, 51, 62. *Estheria mangaliensis* Jones, Geinitz 1876. [= *Cyzicus (Euestheria) forbesi?* (Jones), Rusconi, 1948a, p. 202, Fig. 4.] (= *Cyzicus* sp.) Geinitz (1876, p. 3, Figs. 1–6) provided line drawings only. The six specimens assigned to *"mangaliensis"* are as drawn, quite different in configuration, placement of the umbo, and measured parameters (using Geinitz' scale). It is not possible to unravel this mixed collection of valves without more detailed data. It is possible that the specimens figured by Geinitz represented bituminous shale collections from several Mendoza localities; R. Challao, Agua salada, Cacheuta, and Agua de la Zorra, all in San Lorenzo District. Middle Triassic.

Cordoba

65. *Leaia (Hemicycloleaia) leanzai* Leguizamón (Pl. 31, Figs. 4, 6)

Leaia leanzai Leguizamón 1975, p. 358–359, Pl. 1, Figs. 1–2, Tucuman.

Diagnosis (emended): Subovate, bicarinate valves with subterminal umbo and straight dorsal margin: two ribs originate approximately midway on the umbo; the anterior rib curves slightly where it crosses the umbo, then straightens and fades out before reaching the ventral margin; the straight posterior rib does likewise. On the paralectotype where this can be most clearly seen, the anterior rib fades at about the sixth growth band above the ventral margin and the posterior rib at about the seventh.

Measurements: (mm) These measurements are new and more extensive than the original.

Specimen	L	H	Ch	Cr	Av	Arr	a	b	c
PZ 1071a	8.3	6.0	4.8?	2.5	2.5	1.5	2.2	3.5	4.5
1071b	8.0	4.3?	4.5?	2.5	2.5	1.2	2.0	3.0?	---
1080	10.0?	2.3	3.0	3.0	3.0	4.7	2.3	2.7	---
1077	13.3	8.0	4.0	3.0	3.0	5.0?	3.7	4.0	6.3?

	H/L	Ch/L	Cr/L	Av/L	Arr/L	a/H	b/H	c/L
1071a	.723	.578	.301	.301	.181	.367	.583	.542
1071b	.538?	.563?	.313	.313	.150	.465	.698	----
1080	.770?	.230?	.300	.300	.470?	.299	.381	----
1077	.602	.301	.226	.226	.376	.463	.500	.474

	a*	b*	s*	d₁	d₂
1071a	100°	45°	55°	0.5	6.5
1071b	93°	45°	48°	1.0	8.0
1080	105°	50°	55°	1.0	6.8
1077	105°	45°	60°	0.5	9.0

*alpha, beta, sigma angles.

Material: Repository, Museo de la Catedra de Paleontologie de la Universidad Nacional de Cordoba (Cord-PZ) Argentina. Lequizamón's (hereafter R.R.L.) holotype (Cord-PZ-1071) is a deformed external mold of the right valve, partially covered ventrally. As R.R.L. noted, the medial sector between ribs is depressed creating the illusion of a third rib. Because of this and the eroded and partly covered condition of the available fossils, a syntype is created here to include the best specimens of the lot, namely Cord-PZ-1071 (Lequizamon, 1974, Pl. 1, Figs. 1, 1a, 2). A lectotype is designated here, Cord-PZ-1071 (cf. R.R.L.'s Figs. 1, 1a, and 2 and Pl. 31, Fig. 4 herein). A paralectotype designated here includes one of two valves of the same individual, Cord-PZ-1080 (R.R.L.'s Pl. 1, Figs. 6 and 7; cf. Pl. 31, Fig. 6 herein). Another syntype, Cord-PZ-1077, is also an external mold. These fossils are intercalated with mudstones and argillites bearing a *Glossopteris* taphoflora and leaiids.

Locality: Tasa Cuna Formation, northwest of the Province of Cordoba, Argentina, lat 65°18′W, long 30°46′S. Lower Permian (probably Sakmarian–Artinskian).

Discussion: The range of variation in the syntypes can be attributed to greater distortion, erosion, and cover of some valves relative to others. R. R. Lequizamón provided photographs and data as well as his papers on the biota of the Tasa Cuna Formation. As noted earlier, his holotype has been designated the lectotype. He compared his species to *Leaia pruvosti* Reed and *Leaia gondwanella* Tasch and found overlap with both, although his species is generally larger. Comparative measurements for the Argentine (listed first below) and Antarctic (Ohio Range) Permian specimens (listed second) show overlap (except for d_2) for the two species in the following angles and measurements: alpha angle (93°–105° vs. 75°–103°); beta angle (45°–50° vs. 38°–60°); sigma angle (45°–60° vs. 38°–48°); H/L (.602–.770 vs. .590–.790); d_1 (0.5–1.0 vs. 0.0–1.0). In the Argentine species, d_2 is larger (6.5–9.0 vs. 3.2–5.6). It will be noted that alpha and sigma angles overlap in the upper part of the range. (See new measurements for *Leaia (Hemicycloleaia) gondwanella* Tasch, under Permian Antarctica, Ohio Range, herein.)

Santa Cruz

Cyzicus (Lioestheria) malacaraensis n. sp.
(Pl. 31, Figs. 3, 5)

Diagnosis: Subovate valve with umbo inset from anterior margin about one-third the valve length; dorsal margin straight, with steeper descent to anterior margin than to the posterior margin. Greatest height anterior third of valve; multiple growth bands. Ornamentation: coarse hachure markings on growth bands.

Measurements: (mm) Holotype followed by paratype. L-5.8, 5.0; H-3.9, 3.6; Ch-2.5, 2.6; Cr- --, 2.2?; Av-1.7, 1.7; Arr- .05, .07; a-1.4, 1.3; b-1.9, 1.5; c-3.2, 2.8; H/L-.672, .720; Ch/L-.431, .520; Cr/L- --, .379?; Av/L-.293, .340; Arr/L-.293, .340; a/H-.359, .361; b/H-.487, .418; c/L-.552, .560.

Material: Slab TC 70509, holotype, with multiple equivalent valves, generally incomplete. Slab TC 70513, paratype, better display of coarse lioestheriid ornamentation. All valves flattened. Several other slabs with eroded, covered and incomplete valves of the same species. Both types are internal molds of left valves, with the relict of original valve as an overlay in upper two-thirds of the anterior margin. (Fossils collected by Wolfgang Volkheimer.)

Locality: Malacara Limestone, Malacara, Santa Cruz Province, Argentina. Jurassic.

Discussion: This lioestheriid species differs from those described in Tasch and Volkheimer (1970) from the Jurassic of Patagonia as follows: H/L is greater (.672–.720 vs. .570, *Cyzicus (Lio.) patagoniensis* Tasch); umbonal placement (inset about one-third vs. subcentral to subterminal); ornamentation (coarse hachure markings vs. very fine hachure markings, *"patagoniensis"*). The new species differs from *Cyzicus (Lio.) sp. A* in configuration (subovate vs. variable from ovate to subelliptical); dorsal margin straight vs. arched; ornamentation (hachure markings vs. pustules); from *C. (Lio.) sp. B,* in umbonal position (inset about one-third vs. subterminal); from *Cyzicus (Lio.) sp. C* in a smaller H/L (.672–.720 vs. .830).

Roca Blanca Formation

70. *"Estheria"* sp. (= *Cyzicus*? sp.) Roca Blanca Formation. Estáncia El Tranquilo, approximately lat 47°50′ and 48°10′S, long 68°30′ and 68°42′W. Profile C and Section C, bed 20. Lower Jurassic (Middle to Upper Liassic). *"Estheria"* is briefly referred to in Herbst, 1965, p. 21, 54–56; H/L = .600. (Estheriids not figured).

Bolivia

Ipaguazu Formation

48. *Cyzicus (Lioestheria)* sp. Ipaguazu Formation, Gulf Oil Corporation, Y.P.F.B. Carandaiti, No. 2, at 2592 m depth. Permian–Tartarian? Specimens provided by Gulf Oil Corporation.

Vitiacua Limestone

48. *Palaeolimnadia* sp. Vitiacua Limestone, Aguaraque Range near Villa Montes; near intersection of lat 21°10′S and long 63°40′W. Permian (Middle or Upper) or Permian-Triassic. Specimens provided by Gulf Oil Corporation.

48. ?*Palaeolimnadiopsis* (cf. *P. eichwaldi* [Netshajev, 1894*]). (Vitiacua Limestone [Novojilov, 1958d, p. 101, Pl. 1, Fig. 8, and 8a]. Same locality and age as *Palaeolimnadia* sp. above.)

Palermo Canyon

"Estherids" (= ?*Cyzicus* sp.) Palermo Canyon 177 km north of Bogota (Rhaetic to Jurassic), in Langenheim (1961).

Brazil

Bahía

56. *Estheriina (Estheriina) expansa* Jones, 1897
(Pl. 34, Fig. 7)

Estheriina expansa Jones 1897. Jones, 1897, p. 201, Pl. 8, Figs. 6a, b.

Diagnosis (emended): Elongate ovate valves with comparatively long, straight dorsal margin, umbo inset slightly from anterior margin, and valve subdivided into two distinct sectors: the umbonal sector with a very small initial valve, convex in life, surrounded by closely spaced growth bands (approximately eleven), and a larger sector of the valve less convex in life, with wider spaced growth bands, except for the last few which are more closely spaced. Both posterior and anterior margins are rounded.
Measurements: (mm) New measurements based on restudy of holotype. L-7.2; H-4.5; Ch-2.7?; Cr-1.6; Av-1.5; Arr-3.0; a-2.0; b-2.6; c-3.8; H/L-.625; Ch/L-.375; Cr/L-.222; Av/L-.208; Arr/L-.417; a/H-.444; b/H-.577; c/L-.528.
Material: Holotype, British Museum, BM-I-3475, internal mold of right valve.
Locality: at km 4 to 5, Pedra Furada, along cutting on railroad, 83 km from Salvador, near Pojuca, Bahía, Brazil, Ilhas Formation. Lower Cretaceous.
Discussion: Two subgenera of *Estheriina* were established (Tasch *in* Tasch, and Jones, 1979a, p. 19a) to accommodate different types of convex (swollen) umbonal regions: with growth bands (subgenus *Estheriina*), and without growth bands (subgenus *Nudusia*). Jones' species belongs to *Estheriina*. *Estheriina (Estheriina) pranhitaensis* n. sp., from the Lower Jurassic Kotá Formation of India differs from the new species in almost all valve characteristics, broadly ovate vs. elongate ovate n. sp.; with dorsal margin not so long and a subterminal umbo (vs. umbo inset from anterior margin), a greater H/L (.777 vs. .625 n. sp.), Ch/L (.543 vs. .375 n. sp.), and c/L, and also a smaller Cr/L, Av/L, Arr/L, and b/H; *Estheriina (Estheriina) bresiliensis* Jones differs from the new species in having a greater a/H (.469–.474 vs. .444 n. sp); Ch/L (.456–.522 vs. .375 n. sp.); H/L (.667–.696 vs. .625, n.sp.); smaller Av/L, b/H (.406–.477 vs. .577), c/L (.351–.391 vs. .538 n. sp.). Umbo rises above dorsal margin (vs. no rise of umbo). (See below for *"bresiliensis."*)

31-33, 54. *Cyzicus (Lioestheria) mawsoni* (Jones)
(Pl. 35, Figs. 3, 4, 5)

Estheria mawsoni Jones, 1897, p. 290–292; Pl. XI, Figs. 3a–g, 4, 6.
Bairdestheria mawsoni (Jones), Raymond 1946, p. 230.
Euestheria mawsoni (Jones) Kobayashi, 1954, p. 161.
Bairdestheria mawsoni (Jones), Cardoso 1966, p. 63–64, Fig. 9.

Cyzicus (Euestheria) mawsoni (Jones), Morris 1980, p. 35, Pl. 2, Fig. 13.

Diagnosis (emended): Subovate valves with subterminal umbo, terminus of the umbo rises above a short dorsal margin; height of anterior sector greater than that of the posterior; anterior margin rounded, posterior margin more steeply rounded; multiple thickened growth bands bearing closely spaced hachure-type markings. (A pseudocancellate appearance of the last several growth bands occurs due to erosion and tighter spacing.)
Measurements: (mm) Lectotype I 6804 followed by I 6805. L-5.3, 5.5; H-3.6, 3.6?; Ch-2.8?, 3.2?; Cr-1.5, 2.1; Av-0.7, 0.8; Arr-1.7, 1.6; a-1.5, 1.2; b-1.8, 1.4; c-2.4, 2.3; H/L-.692, .673?; Ch/L-.538?, .582?; Cr/L-.288, .382; Av/L-.135, .145; Arr/L-.327, .291; a/H-.417, .324?; b/H-.500, .378?; c/L-.461, .418.
Material: Syntype, British Museum. BM I 6803-6805, 6807, consisting of three pairs of valves with some original brownish coloration retained, and the interior infill of sediment. Valves eroded variously but all retain characteristics of the species. I 6804 is here designated lectotype. Ornamentation can be seen on the lectotype and I 6805. A crushed left valve (I 6807), with a fragment of a different valve, also shows excellent lioestheriid ornamentation (i.e., hachure markings on growth bands).
Locality: Syntype BM I 6803-6805. Shale in a cutting on the Bahía and São Francisco Railroad, 12–13 km from Bahía on the seaside between Periperi and Olaria. Another specimen found 73 to 74 km from Bahía on the railroad (Pitanga cutting). Lower Cretaceous.
Discussion: *Bairdestheria* Raymond 1946 is a synonym of *Cyzicus (Lioestheria)*; the Kobayashi (1954) and Morris (*1980) attribution to *Euestheria* is excluded because of the lioestheriid ornamentation (i.e., hachure markings). The brown coloration in the Cretaceous valves probably represents alteration of the original valve haemolymph. Slight variations in configuration of the three paired valves are due to differential compaction.

On I 6807, the eroded ends of the hachure-like markings create an illustory, wavy aspect in lower bands of the valve.

34, 35. *Estheriina (Estheriina) bresiliensis* Jones
(Pl. 35, Figs. 1, 2; Pl. 34, 35)

Estheriina bresiliensis Jones 1897a. p. 198–199, Pl. 8, Figs. 1–5.
Estheriina bresiliensis Jones, Raymond, 1946, p. 267–268, Pl. 4, Figs. 5, 6.
Estheriina (Estheriina) bresiliensis Jones, Tasch *in* Tasch and Jones, 1979a, p. 19.
Estheriina bresiliensis Jones, Morris, 1980, p. 38.

Diagnosis (emended): Subovate to elongate ovate valves with umbonal sector much more convex than subsequent portions of the valve; umbo inset from, yet near to the anterior margin, and with subround to tapered terminus directed anteriorly and slightly elevated above dorsal margin. Dorsal margin straight and, depending on growth stage (in more mature specimens), may extend slightly beyond the most anterior part of the umbo. Ornamentation: punctate on growth bands.

Measurements: (mm) Lectotype followed by paralectotype (I 6822B). The following are all new measurements. L-4.6, 5.7; H-3.2, 3.8; Ch-2.4, 2.6; Cr-1.0?, 1.6; Av-1.0?, 1.6; Arr-1.2, 1.5; a-1.5, 1.8; b-1.3, 1.6; c-1.8, 2.0; H/L-.696, .667; Ch/L-.522, .456; Cr/L-.217, .281; Av/L-.217, .281; Arr/L-.261, .263; a/H-.469, .474; b/H-.406, .447; c/L-.391, .351.

Material: British Museum. BM I 6828, designated here as the lectotype. (This replaces a previously designated lectotype by Tasch *in* Tasch and Jones, 1979a, p. 19, based on BM I 6822A [Jones, 1897a, Pl. 8, Fig. 1] intended to be a correction of Raymond's [p. 46] selection of Jones, 1897a, Pl. 8, Fig. 5, as the "genotype.") The new lectotype (I 6828) is actually the better specimen to display the chief features of this species and genus; it is a right valve. BM-I-6822B thus becomes a paralectotype. I-6827 consists of paired valves. All valves are crushed and partially eroded. There are several other incomplete specimens of the species on slabs which constitute part of the original collection.

Locality: 3.85, 4.0 and 5.0 km from Bahía, Brazil on the São Francisco Railroad. The new lectotype apparently occurred at the 4.0 km mark. Lower Cretaceous.

Discussion: Jones (1897a) made much of the variable spacing of growth bands, i.e., of the umbonal sector compared to the rest of the valve. However, this is a common occurrence and cannot have either generic or specific taxonomic value. It is apparently related to seasonal shrinking of the water level or decrease in oxygen and/or food supply resulting therefrom (tighter spacing of last growth bands compared to the wider spacing higher on the valve when presumably contrary conditions prevailed). The paralectotype (I 6822 B), an external mold, slightly crushed, shows quite well the characteristics of the umbonal sector.

The ornamentation of the closely spaced, terminal set of growth bands is difficult to decipher, but in several valves of the syntypes a lateral array of minute beadlike pustules is visible on the external mold. These no doubt correspond to setae or spine attachment sites. Jones' (1897a, Pl. 8, Fig. 1) reconstruction inadvertently shows a posterior recurvature of the early valve which does not occur in I 6822 from which the illustration was made.

36. *Estheriina (Estheriina) astartoides* Jones, 1897a
(Pl. 38, Figs. 5, 6, 7)

Estheriina astartoides Jones, 1897a, p. 201–202, Pl. 8, Figs. 7 and 8.
Estheriina astartoides Jones. Morris, 1980, p. 38, Pl. 2, Fig. 16.

Diagnosis (emended): Dimorphic valves (subovate-male?, subcircular-female?): umbonal sector with widely spaced growth bands appears distinct from rest of the valve (flattened or not) in which more closely spaced growth bands occur; dorsal margin brief behind umbo and short in front of it; umbo closer to anterior margin.

Measurements: (mm) Holotype I-6816 (male?) followed by I-6817 (female?): L-2.5, 2.6; H-2.0, 2.3; Ch-0.8?, 0.7; Cr-1.0, 1.0; Av- --, 0.6; Arr-0.7, 1.1; a-0.9, 1.1; b-0.9, 1.0; c-1.1, 1.3; H/L-.769, .920; Cr/L-.385, .400; Av/L- --, .240; Arr/L-.269, .440; Ch/L-.308?, .280; a/H-.450, .478; b/H-.450, .435; c/L-.423, .520.

Material: British Museum. BM I-6816 (Jones holotype, 1897a, Pl. 8, Fig. 7; male?), mold of right valve; and paratype (= allotype, i.e., opposite sex to holotype) BM I-6817 (Jones, 1897a, Pl. 8, Fig. 8; female?), mold of right valve; other paratype BM I-6820 (Morris, 1980, Pl. 2, Fig. 16). Lower Cretaceous.

Locality: Cutting on railroad 83 km from Bahía, between Pojuca and San Thiago, Brazil.

Discussion: Jones' species has two configurations which are interpreted here to represent probable sexual dimorphs. This species is distinct from the other estheriinids Jones described in configuration (dimorphs), size, and/or placement of umbo, as well as a shorter dorsal margin.

13, 14, 15. *Estheriella brasiliensis* Oliveira, 1953. [=*Graptoestheriella brasiliensis* (Oliveira).] Cardoso, 1965, p. 18–21, Pl. 3, Figs. 1–3; city of Serrinha; road between Queimada Grande (lat 11°58′S, long 38°49′W; Cacimbo do Carvao). Lower Cretaceous.

37,40. *Aculestheria novojilovi* Cardoso, 1962 (Holotype, Universidad de São Paulo.) (Only the paratype of *Aculestheria novojilovi* clearly displays both anterior and posterior recurvature ending in spinous projections, i.e., the original paratype [DGP 7-934, Universidad São Paulo] [Cardoso, 1962, Pl. 1, Fig. 2].) The original holotype (DGP 7-985) displays a prominent dorso-posterior spinous projection but only dorso-anterior recurvature. (The writer has photographs and specimens of this species generously contributed by Cardoso.) Other specimens contributed by Cardoso clearly have posterior recurvature distinct from the species above, while lacking anterior recurvature. Spinous projections are also absent. These species are tentatively assigned to *Palaeolimnadiopsis* sp. (TC 90025a and b and TC 90022). (Cardoso, 1962, p. 23–26, map no. 40, 5 km from Santa Amaro, Candeias Formation. Map No. 37, km 77–78 on railroad from Santa Amaro to Candeias [Cardoso, 1966].) Upper Jurassic–Lower Cretacous. (See P. 35, Fig. 6)

29. *Pseudestheria iphygenioi* Cardoso, 1966 [= *Cyzicus (Lioestheria) iphygenioi* (Cardoso)]. Holotype, Divisao de Geologia e Mineralogia Departamento Nacional de Producão Mineral. Rio de Janeiro. (DGM 4875; paratypes include DGM 4873, 4874, 4877, 4881. Cardoso, 1966, p. 70–72, text-Fig. 15, Pl. 3, Fig. 2.) Candeias Formation, Riacho, São Paulo, near Candeias, Bahía. Lower Cretaceous.

29. *Pseudestheria pricei* Cardoso, 1966 [= *Cyzicus (Lioestheria) pricei* (Cardoso)]. Holotype, DGM 4865. Cardoso, 1966, text-Fig. 16, Pl. 3, Fig. 1. Same locality, formation, and age as *"iphygenioi."*

30. *Pseudograpta erichseni* Cardoso, 1966 [= *Cyzicus (Lioestheria) erichseni* (Cardoso)]. Holotype DMG 4405-B. Paratypes include DMG Series 4000 and 4400. Cardoso, 1966, text-Fig. 12, Pl. 2, Fig. 3. Ilhas Formation, Fazenda Joazeiro, City of Cicero Dantas, Bahía. Lower Cretaceous (Middle Wealden).

31, 32, 33. *Pseudograpta* sp. Cardoso, 1966 [= *Cyzicus (Lioestheria)* cf *mawsoni*? (Jones).] Cardoso figured drawings of Jones (1897a, Pl. 11, Figs. 6a–7). Ilhas Formation, at any one of three places along the São Francisco railroad, at km 12–13 from Bahía on seaside between Periperi and Olaria, shale in railroad cutting. Lower Cretaceous (Middle Wealden).

37. *Palaeolimnadiopsis* sp. Cardoso, 1962 (Departamento Geologia e Paleontologia; [DGP], no. 7-987, Universidad de São Paulo.) Cardoso, 1963, p. 26–28, text-Fig. 7, Pl. 1, Fig. 3. Occurs with *Aculestheria novojilovi* Cardoso at km 77–78 on railroad from Santo Amaro to Candeias. Upper Jurassic–Lower Cretaceous.

37. *Palaeolimnadiopsis linoi* Cardoso, 1966 (Holotype, DMG, Escola Federal de Minas de Ouro Preto [EFMOP] 3998-A.) Cardoso, 1966, p. 57–58, text-Fig. 2, Pl. 1, Fig. 3. Same formation, locality, and age as no. 30 above.

Maranhão

43. *Lioestheria codoensis* Cardoso, 1962 [= *Estheriina (Nudusia?) codoensis* (Cardoso)]. Syntypes Departamento Geologia e Paleontologia DGP 7-992 to 996 (Universidad de São Paulo; U.S.P.). Lectotype designated here, DGP 7-996. U.S.P.; Cardoso, 1962, p. 35–37, text-Fig. 6, Pl. 1, Fig. 5. Codo Formation, approximately 5 km from Imperatriz. Cretaceous.

Mato Grosso

9. *Palaeolimnadiopsis* Raymond, 1946 *in* Rocha-Campos and Farjallat, 1966, p. 102–103, Figs. 6, 9 (figures not of fossils). Botucatú Formation. Locality 8, road from Jardim to Bela Vista, km. 51.5. Upper Triassic.

9. *Eustheria* Depéret and Mazeran, 1912 [=?*Cyzicus** (Euestheria) sp. (*cf. Tasch, 1969, p. R151)]. (Undescribed, unfigured *in* Rocha-Campos and Farjallat, 1966, p. 102–103.)

9. *Orthotemos (sic.)* Raymond, 1946 [= ?*Cyzicus** (Euestheria) sp.] (Cf. Tasch, 1969, p. R151.) (Undescribed, unfigured, *in* Rocha-Campos and Farjallat, 1966, p. 102–103.)

9. *Palaeolimnadia* sp. (Identified by this writer during study of the Palaeontological Collection at the Universidad Federal Minas Gerais.) Unnumbered. A prominent, submedial, smooth umbonal sector with a brief, straight dorsal margin on either side. L = 2.8 mm. Botucatú Formation, same locality as above, km 51.5, from Jardim. Upper Triassic.

9. ?*Echinestheria* sp. (Examined by this writer. A possible *Echinestheria* sp.) Right valve incomplete. Same collection as *Palaeolomnadia* sp. above. Same locality and formation at km 51.5, from Jardim (Fig. 29, lower, left.)

Minas Gerais

44. *Palaeolimnadiopsis freybergi* Cardoso, 1971 (Holotype, Universidad Federal Minas Gerais [UFMG], No. .001.) On road MG-51 from Patos de Minas to Pirapora, km 40. Cardoso, 1971, text-Fig. 10, p. 28–30, Pl. 2, Figs. 1, 2. Areado Formation. Lower Cretaceous (Fig. 29, lower, right.)

44. *Pteriograpta* cf. *reali* Teixeira, 1960, Cardoso, 1971. (= *Palaeolimnadiopsis* cf. *reali* Teixeira; *non Pteriograpta*; Cardoso, 1971, text-Fig. 9, p. 30–32, Pl. 1, Figs. 1-3. DMG No. 4627. São Jose near Varjão on same road as "*freybergi*" above (Areado Formation. Lower Cretaceous.

46. *Pseudestheria abaetensis* Cardoso, 1971 [*Cyzicus (Lioestheria) abaetensis* (Cardoso). Cardoso, 1971, p. 32–35. Syntypes, one figured, Pl. 1, Fig. 4 (= lectotype designated here). Museo de Historia Natural de la Universidad Federal Minas Gerais (UFMG), No. .005. Near São Goncalo do Abaete, about 1 km from MG-51 (Patos de Minas-Pirapora). Areado Formation. Lower Cretaceous.

46. *Pseudograpta* cf. *barbosai* (Almeida), Cardoso, 1971 [*Cyzicus (Lioestheria)* cf. *barbosai* (Almeida)]. Museo de Historia Natural, UFMG, No. .003. Cardoso, 1971, text-Fig. 11, p. 36–37, Pl. 2, Figs. 3, 4. Areado Formation. Lower Cretaceous; 20 km from Quintinos to Carmo do Paranaíba.

Paraíba

Pterograpta cf. *reali* Teixeira (Tinoco and Katoo, 1975). [= *Palaeolimnadiopsis* cf. *reali* (Teixeira, 1960)]. See Pl. 32, Fig. 5 herein, based on rubber replica prepared by Yoco Katoo.) UFMG No. 2522 and 2523. Tinoco and Katoo, 1975, Pl. 1, Fig. 1; Fig. 2, p. 141. Last 8 to 9 growth bands recurved. H/L-.430. Souza Formation (Upper Jurassic–Lower Cretaceous). Pedreguiho.

Palaeolimnadiopsis freybergi Cardoso, 1971 (Tinoco and Katoo, 1975). (See Pl. 32, Fig. 4 herein, based on rubber replica prepared by Yoco Katoo.) UFMG No. 2522 and 2523. Km 40, Road MG-51 (Patos to Pirapora). Last several growth bands recurved. (Cf. Minas Gerais map entry No. 44.) Tinoco and Katoo, 1975, Pl. 1, Fig. 6. Lower Cretaceous.

?*Pseudograptus barbosai* (Almeida), 1950, Tinoco and Katoo, 1975. [= *Cyzicus (Lioestheria)* cf. barbosai (Almeida)]. Broadly ovate; right valve; subterminal umbo; hachure-type markings. 1 km north of Brejo dos Freiras, along road to Uiraúna. Lower Cretaceous.

"Estheriellids" [Tinoco and Katoo, 1975 (= ?*Graptoestheriella* sp.). Unnumbered; two cores, depth 775–780′. One km north of Brejos dos Freiras at Rio Passagen, on either side of the Brejo dos Freiras-Uiraúna road. Carbonized specimens; strong ridges for costellae. (See Pl. 37, Fig. 3, herein.) Lower Cretaceous.

Graptoestheriella fernandoi Cardoso, (Tinoco and Katoo, 1975). UFMG No. 2524 and 2525; cf. holotype Departamento Geologia e Mineralogie No. 4617. Along length of Souza-Aparecida Road at 12 km from Souza. Lower Cretaceous. Cf Pl. 36, Fig. 3.

Measured Section of Conchostracan-Bearing Beds
(Topographic Profile of the São Pascoal Area, Brazil by Prof. Henrique Papp)

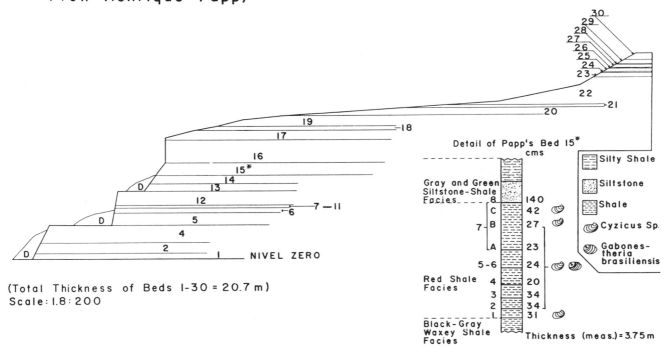

(Total Thickness of Beds 1-30 = 20.7 m)
Scale: 1.8 : 200

Detail of Papp's Bed 15*

NIVEL ZERO

CONCHOSTRACAN—BEARING BEDS ALONG HIGHWAY BR 116, SIERRA ESPIGÃO SANTA CATARINA, BRAZIL
(RIO DO RASTO FORMATION, UPPER PERMIAN)

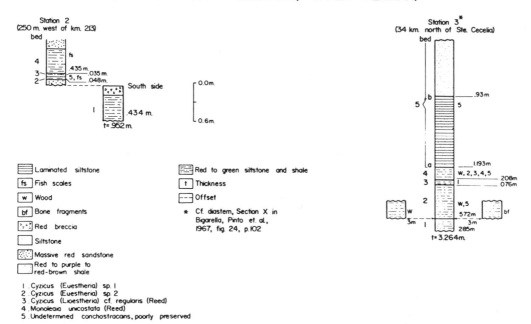

1. Cyzicus (Euestheria) sp. 1
2. Cyzicus (Euestheria) sp. 2
3. Cyzicus (Lioestheria) cf. regularis (Reed)
4. Monoleaia unicostata (Reed)
5. Undetermined conchostracans, poorly preserved

Figure 28. South America. Top: São Pascoal area, Brazil. *Cornia-Gabonestheria*-bearing units of Bed 15. Bottom: Rio do Rasto Formation, Upper Permian, Santa Catarina, Brazil, conchostracan-bearing beds. (Top: Profile and bed number after Professor H. Papp.)

CONCHOSTRACAN – BEARING BEDS BAURU FORMATION (U. CRETACEOUS)
SÃO PAULO , BRAZIL

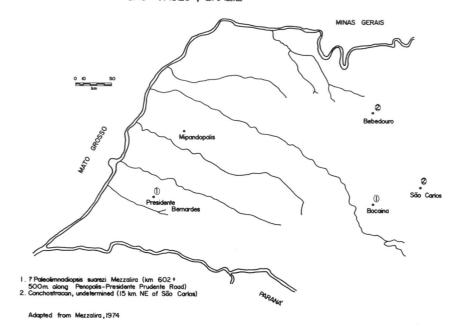

1. *? Paleolimnadiopsis suarezi* Mezzalira (km 602 +
   500 m. along Penopolis-Presidente Prudente Road)
2. Conchostracan, undetermined (15 km. NE of São Carlos)

Adapted from Mezzalira , 1974

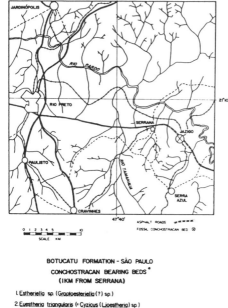

BOTUCATU FORMATION – SÃO PAULO
CONCHOSTRACAN BEARING BEDS *
(1 KM FROM SERRANA)

1. *Estheriella* sp. (*Graptoestheriella* (?) sp.)
2. *Euestheria triangularis* (= *Cyzicus* (*Lioestheria*) sp.)
3. *Lioestheria elliptica* (= *Cyzicus* sp.)
4. *Euestheria ribeiropretensis* (= *Cyzicus* (*Lioestheria*) sp.)
5. *Pseudestheria* sp (= *Cyzicus* (*Lioestheria*) sp.)

(* Map and faunal data after Souza, Sinelk, and Gonçalves.)

CONCHOSTRACANS BOTUCATU FORMATION
STATE OF MATO GROSSO , BRAZIL
(km. 51.5 - 51.8, JARDIM-BELLA VISTA ROAD)
(after Rocha-Campos and Farjallat ,1966)

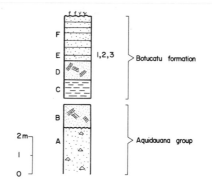

1. Paleolimnadia
2. (?) Echinestheria
3. Cyzicus , several species

F. Sandstone, fine to coarse intercalated with argillite
E. Conchostracan-bearing argillite
D. Cross bedded silt-to-sandstone

AREADO FORMATION , WESTERN MINAS GERAIS CONCHOSTRACAN–
BEARING OUTCROPS

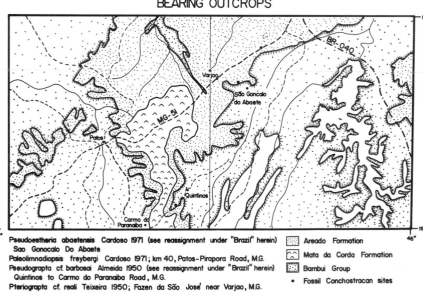

Pseudoestheria abaetensis Cardoso 1971 (see reassignment under "Brazil" herein)
    Sao Gonocalo Do Abaete
Paleolimnadiopsis freybergi Cardoso 1971; km 40, Patos-Pirapora Road, M.G.
Pseudograpta cf. *barbosai* Almeida 1950 (see reassignment under "Brazil" herein)
    Quintinos to Carmo do Paranaiba Road, M.G.
Pteriograpta cf. *reali* Teixeira 1950; Fazen da São José' near Varjao, M.G.

After Cardoso, 1971

Areado Formation
Mata da Corda Formation
Bambui Group
• Fossil Conchostracan sites

Figure 29. South America. Brazilian Mesozoic fossil conchostracan sites and species in three states. Upper: left, São Paulo, Bauru Formation; right, São Paulo, Botucatú Formation (de Souza and others, 1971). Lower: right, Mato Grosso, Botucatú Formation; left, Minas Gerais, Areado Formation. (*Pteriograpta cf reali* Teixeira 1950 should read *Palaeolimnadiopsis reali* (Teix., 1960); discussion in text.) (Cardoso, 1971, not Cardosa.)

22. *Acantholeaia regoi* Almeida, 1950 (= Almeida, 1950, text-Figs. 2–4, p. 3–6, Figs. 1–4). Syntypes, Departamento Geologia e Mineralogia No. 4073–4079, Universidad Federal São Paulo. Km 216 ±500 m along Rio Claro to São Carlos do Pinhal Road. (Cf. Mendes, 1954, p. 155.) Estrada Nova Group. Upper Permian.

22. ?*Acantholeaia* cf. *regoi* Almeida, 1950, Mendes 1954 (p. 163, Pl. 14, Fig. 12). Departamento Geologia Paleontologia do Universidad Federal São Paulo No. VII-130. Also, Pl. 31, Fig. 7 herein, based on Mendes photograph. It is uncertain that this is an *Acantholeaia*. Almeida's species displayed an anterior rib inset slightly from the anterior margin, a posterior rib, and a thickened dorsal margin from which a series of spinous projections arise. Only a posterior rib can be clearly seen in Mendes' photograph. Nor can it be determined whether the anterior rib (if it was actually present originally) was inset from the anterior margin. The dorsal margin characteristics also are not preserved on the Mendes specimen. Km 78 ± 600 m along Ponta-Grossa-Prudentopolis Road. Upper Permian.

Paraná

24. *Euestheria langei* Mendes, 1954 [= *Cyzicus (Euestheria) langei* (Mendes)]. Departamento Geologia Paleontologia, Universidad Federal São Paulo No. VII-123. Mendes, 1954, Pl. 12, Figs. 1 and 2. Km 115 + 200 m on the Ponta-Grossa-Guarapuáva Road. Rio do Rasto Formation, Morro Pelado facies. Permian.

24. *Monoleiolophus unicostatus* (Reed), Mendes 1954 [= *Monoleaia unicostata* (Reed), Tasch, 1969, p. R159; non *Monoleiolophus* Raymond 1946]. Mendes, 1954, p. 160–161, text-Fig. 3, Pl. 13, Figs. 6, 7; Pl. 14, Figs. 9–11. Rio do Rasto Formation, Morro Pelado facies, km 115 + 200 m on same road as *E. langei* above. Permian.

25. *Orthothemos regularis* (Reed) 1929, Mendes 1954 [= *Cyzicus (Lioestheria) regularis* (Reed)]. Departamento Mineralogia e Geologia Universidad Federal São Paulo No. 1397, Mendes, 1954, p. 158–159, text-Fig. 1. Santo António de Platina, town of Platina, about 1 km from town along river Boi Pintado. Rio do Rasto Formation, Morro Pelado facies. Permian.

12. *Unicarinatus octavioi* Cardoso 1965 (Holotype, Departamento Geologia e Mineralogia Universidad Federal São Paulo No. 4-648, near town of Quatigua, km 210 on Curitiba-Jacarezinho Road, Taquaral Member, Iratí Formation [Lower Permian?] or Itapetininga Formation. Cardoso, 1965, map, Fig. 2, p. 14 and genus description, p. 11–15.)

Monoleiolophus unicostatus (Reed) 1929, Mendes, 1954. [= *Monoleaia unicostata* (Reed), Tasch, 1969], p. R159. Repository: Universidad São Paulo. Located between Dorizon and Paulo Frotin, km 1.5 from Dorizon, Paraná. Possibly highest level of *Glossopteris* in the Paraná Basin). Permian.

Pernambuco

28. *Pseudograpta brauni* Cardoso, 1966 [= *Cyzicus (Lioestheria) brauni* (Cardoso)]. Lectotype, Departamento Geologia e Minera-

logia Universidad Federal São Paulo No. 4870; Syntypes; DGM 4866–4872. Cardoso, 1966, p. 46, text-Fig. 11, Pl. 1, Fig. 4. Candeias Formation, Varzea do Campinho, Ico. Lower Cretaceous.

27. *Palaeolimnadiopsis barbosai* Cardoso, 1966 [= *Macrolimnadiopsis barbosai* (Cardoso)]. Lectotype, Departamento Geologia e Mineralogia DGM 4860; designated here for figured specimen, one of five syntypes. Escola Federal Minas de Ouro Preto (= EFMOP), UFMG Cardoso, 1966, p. 78, Fig. 1; Pl. 1, Figs. 1, 2; p. 55–57. Aliança Formation, Mirandiba, km 555 on the National Road. Lower Upper Jurassic.

27. *Bairdestheria mawsoni* (Jones) var. *mirandibensis* Cardoso, 1966 [= *Cyzicus (Lioestheria) mirandibensis* (Cardoso) 1966, non *mawsoni* Jones]. Departamento Geologia e Mineralogia Nos. 4642-4647, EFMOP, UFMG. Cardoso, 1966, text-Fig. 10, Pl. 1, Figs. 5, 6. Same formation and locality as *"barbosai"* above. Lower Upper Jurassic. (Cardoso's variety is a valid species.)

26. *Notogripta costai* Cardoso, 1966 [= ?*Estheriina (Estheriina)? costai* (Cardoso)]. Cardoso, 1966, text-Fig. 8, Departamento Geologia e Mineralogia, UFMG (= DMG 4638), holotype* (incorrectly labeled *syntype* on the line drawing which differs markedly from the photograph (Cardoso, 1966, Pl. 3, Fig. 3) of a paratype. DGM-4641 incorrectly labeled syntype). The holotype suggests an estheriinid. BMG 4641 is badly eroded and crushed. Aliança Formation, Riacho do Machado, Petrolandia. Upper Jurassic.

Piaui

42. *Cornia semigibosa* (Cardoso) 1963
(Pl. 37, Fig. 2)

Echinestheria semigibosa Cardoso, 1962, p. 29–32, text-Fig. 4 only.

Original Diagnosis (translated and abbreviated): Small, subcircular valves, with straight dorsal margin whose length approximates the height of the valves; obese umbo placed anteriorly with a spinuous terminus; umbo convexity much greater than rest of valve. Three radial costellae (= apparent embryonic ribs) originate from base of umbonal spine and are directed posteriorly, then fade out high on the valve. (Emended): The apparent embryonic "ribs" do not occur in all available specimens of the species. Coarse, vertical ornamentation on growth lines.

Measurements: (mm) Figures for lectotype TC 90020 are followed by those of another syntype, DGP 7-989. (The following are new measurements.) L-1.2, 1.1; H-0.9, 0.7; Ch-0.5, 0.6; Cr-0.3, 0.3; Av-0.3, 0.3; Arr-0.4, 0.2; a-0.4, 0.2; b-0.3, 0.2; c-0.5, 0.4; H/L-.750, .636; Ch/L-.417, .545; Cr/L-.250, .273; Av/L-.250, .273; Arr/L-.333, .182; a/H-.444, .286; b/H-.333, .429; c/L-.417, .364.

---

*The confusion arises from dual labeling; line drawings and photograph figure noted above were both labeled "sintipo" but in the text measurements (Cardoso, 1966, p. 62) these are indicated as holotype and paratypes, respectively.

Material: Syntype, Departamento Geologia e Paleontologia Universidad Federal São Paulo syntype, DGP 7-989 (Cardoso, 1962; see Pl. 37, Fig. 2 herein). Fragmented; ventral portion absent; three faint, apparent embryonic ribs; Syntype DGP 7-990, umbonal spine but no embryonic ribs present (nor any signs of their original presence), and accordingly lack the main characteristics of Cardoso's species. TC 90020 (contributed by Cardoso) has an umbonal spine but no embryonic ribs (nor evidence of their original presence). This specimen is designated here the lectotype of the Cardoso species.

Locality: Motuca Formation (Upper Triassic), road from Floriano to Almirante, Piaui, Brazil.

Discussion: The only photograph among those contributed by Cardoso representing syntype DGP 7-989 apparently corresponds to his 1962 line drawing, Fig. 4. Syntype 7-990 and the lectotype (R 90020) lack the identifying characteristics of Cardoso's species. This writer did not see the embryonic ribs of the type because the specimens could not be located in the UFMG Museum during my visit. Because the available photograph does not display the "ribs" too clearly, I regard all but the segmented and shortest embryonic rib to require further data before identification. However, even if they are accepted as embryonic ribs, their restriction, or absence from other specimens assigned to this species, indicates that this feature cannot serve as an identifying characteristic of the species. Accordingly, this writer regards it as a deviant from the species norm [(Cf. *Leaia (Hemicycloleaia) mitchelli* (Etheridge, Jr.) from the Newcastle Coal Measures (Permian, Australia), for an equivalent example)].

It is clear that Cardoso's species does not belong to *Echinestheria,* which has a single, robust, very prominent spinous projection from an umbo, well inset from the anterior margin. Rather, the umbonal spine is a *Cornia*-type: it is minute, rises from a convex umbo, and occurs on two of Cardoso's specimens.

43. *Lioestheria florianensis* Cardoso, 1962 [= *Cyzicus (Lioestheria) florianensis* (Cardoso)]. Syntypes, Departamento Geologia e Paleontologia No. 7-996–7-999. Universidad Federal São Paulo. Cardoso, 1962, P. 33–35, text-Fig. 5, Pl. 1, Fig. 5. Mislabelled as *Lioestheria codoensis* Cardoso, in text-Fig. 5, but clearly stated in text as *"florianensis"* and Pl. 1, Fig. 5 legend. Motuca Formation, road from Floriano to Almirante. Upper Triassic.

39. *Cyzicus (Euestheria)* sp.
(Pl. 34, Fig. 8)

Lioestheria sp. Pinto and Purper, 1974, p. 308, Pl. 2, Fig. 6.

Diagnosis (slightly emended): Subovate valve with subterminal umbo; anterior margin steeply arched; posterior margin convex. Greatest height along a line through umbo to ventral margin; length of valve greater than length of dorsal margin. Ornamentation on growth bands, polygonal granules. Two cycles of growth-band spacing: upper sector wider spaced; lower more tightly spaced.

Measurements: (mm) New measurements based on photograph of specimen. No measurements provided by the indicated authors except magnification X20, (Pinto and Purper, 1974, Pl. 2, Fig. 6). L-2.10; H-1.50; Ch-1.47; Cr-0.35?; Av-0.15; Arr-0.25; a-0.65; b-0.40; c-0.80; H/L-.714; Ch/L-0.70; Cr/L-.167?; Av/L-0.71; Arr/L-.119; a/H-.433; b/H-.267; c/L.381.

Material: MP-M-39, Universidad Federal Rio Grande do Sul. Topotype, internal mold of right valve; dorsal margin partially covered from posterior to umbonal area; lower half of anterior margin covered; growth bands on umbo eroded.

Locality: Riacho Olhos d'Agua, Muzinho, 16 km northeast of Floriano Road, Floriano-Amarante, Piaui. Pastos Bons Formation. Upper Jurassic.

Discussion. Without better and additional specimens, it is not possible to compare the cyziciid described above. It is included here so that new finds will have a point of reference.

Macrolimnadiopsis Beurlen, 1954, emended;
Pinto and Purper 1974

Macrolimnadiopsis Beurlen, 1954, p. 5.
Macrolimnadiopsis Beurlen, Novojilov, 1958c, p. 103, text-Fig. 1.
Palaeolimnadiopsis Mendes 1960, p. 75–78.
Palaeolimnadiopsis Cardoso 1962, p. 27, 1966, p. 57.
Macrolimnadiopsis Beurlen, Pinto and Purper, 1974, p. 306–307.

Diagnosis: Large limnadiopsids with carapace rectangular to subrectangular, slight concavity on the upper part of the posterior margin; umbo does not surpass the dorsal margin; dorsal margin is in a contiguous straight line widely surpassing anteriorly the umbo and forming strong angles with the anterior and posterior margins; growth lines widely spaced backward and close together, in from the umbonal region; directed backward posteriorly and recurving abruptly forward at the junction of the dorsal margin but not projecting over it like a spine in recent *Limnadiopsis*; honeycombed sculpture; carapaces of males narrower than those of females.

Type: *Macrolimnadiopsis pauloi* Beurlen, 1954.

2. *Macrolimnadiopsis pauloi* Beurlen, 1954
(Pl. 33, Figs. 1, 2, 4, 6)

(Synonomy of species same as for genus above.)

Diagnosis: Beurlen, 1954 (slightly emended and abbreviated, after Pinto and Purper). Dimorphic, large valves with an almost subrectangular carapace that expands posteriorly; dorsal margin slightly concave over umbonal sector; umbo is small and apparently did not rise above the dorsal margin; growth band number variable; tightly spaced anterior to umbo, but wider spaced along dorsal margin to posterior margin; last ~10 growth bands recurved, resulting in a projecting dorsal margin terminus; growth bands generally more closely spaced ventrad. Ornamentation: contiguous granules on growth bands. Two forms of valves: the narrower is inferred to characterize males; the wider valves, females.

Measurements: (mm) Topotypes: MP-M-27 (female), MP-M-28 (female), MP-M-38 (male). Length and height by Pinto and Purper, 1974. Other measurements are new based on photographs supplied by Professor Pinto. The order of entries will be M-27, M-28, M-38. L-17.0, 14.0, 11.0; H-8.5, 7.0, 4.5; Ch-16.6?, 13.3, 10.6; Cr-2.0?, 2.2?, 1.4?; Av- --, 1.5, 1.9; Arr- --, --, --; a-2.8, 2.3, 1.3; b- --, --, --; c-11.1, 8.8, 6.7; H/L-.500, .500, .409; Ch/L-.976, .950, .964; Cr/L-.118, .157, .127; Av/L- --, .107, .173; Arr/L- --, --, --; a/H-.329, .329, .189; b/H- --, --, --; c/L-.653, .629, .609. Postero-dorsal and antero-dorsal angles measured on topotypes directly: postero-dorsal angle, 30°, 30°, 30°; antero-dorsal angle, --, 60°, 60°.

Material: Holotype, a broken, immature left valve. DGM 4244. Division de Geologia e Mineralogia do Departamento Nacional do Producão de Geologia e Mineralogia, Rio de Janeiro. Paratypes: DMG 4245–4248; topotypes: M.P. Universidad Federal Rio Grande do Sul, No. MP-M-27 to MP-M-38 (Pinto and Purper, 1974). These last include males, females, and juveniles.

Locality: Pastos Bons Formation. Riacho Olhos d'Agua, Muzinho, 15 km northeast of Floriano, road to Floriano-Amarante, Piaui. Upper Jurassic.

Discussion: The placement of this genus has long been puzzling. Beurlen figured an incomplete (broken) and immature valve of the type species *M. pauloi* Beurlen 1954, which led Mendes (1954) and this writer (Tasch, 1969, p. R162) to conclude that Beurlen's genus belonged under *Palaeolimnadiopsis.* However, the comparatively fine preservation of the plentiful topotype specimens collected by Pinto and Purper (1974) allows a much clearer overall grasp of the morphology of Beurlen's genus and species. As a result, this writer agrees with Pinto and Purper that *Macrolimnadiopsis* is a valid and very distinctive genus.

The Pinto and Purper collection of *Macrolimnadiopsis* differs markedly from that of the type species of *Palaeolimnadiopsis (Palaeolimnadiopsis) carpenteri* Raymond, which the writer studied at Harvard University repository, and duplicates of which were collected in the field (for detailed measurements see Tasch, 1962). The differences include: configuration—the elongate-almost subrectangular *"pauloi"* contrasts with the robust, sub-ovate *"carpenteri"*; umbonal placement and elevation—inset from anterior margin *"pauloi"* but terminal in *"carpenteri"*; the umbonal terminus is subdued and not above the dorsal margin *"pauloi,"* but prominent and above the dorsal margin in *"carpenteri"*; nature of recurvature—recurved last growth bands project beyond posterior margin *"pauloi,"* but do not project in *"carpenteri."*

2. *Pseudestheria* sp. 2 (= ?*Cyzicus* sp.) MP-M-41. Pinto and Purper, 1974, Pl. 2, Fig. 8. Pastos Bons Formation. Upper Jurassic.

2. *Pseudestheria* sp. 1 (MP-M-40). Pinto and Purper, 1974, Pl. 2, Fig. 7. Pastos Bons Formation. Upper Jurassic. (This species has growth bands that intercalate; the valves are both crushed and compacted. Because such intercalation is not possible in conchostracans, this species must be placed in the dubious catetory.)

2. *Asmussia* sp. (MP-M-42). Pinto and Purper, 1974, Pl. 2, Fig. 9. Pastos Bons Formation, Upper Jurassic; growth bands broken and overlap; no taxonomic clarification possible.

Rio Grande do Sul

20. *Estherites wianamattensis* (Mitchell) 1927, Pinto, 1956. [= Palaeolimnadia (Grandilimnadia) sp.]. MP-243, Museo de Paleontologia, Instituto de Ciencias Naturais, Universidad Rio Grande do Sul. Pinto, 1956, p. 79, Pl. 1, Fig. 3, *non "wianamattensis."* [See below for Mitchell's species, *Palaeolimnadia (Palaeolimnadia) wianamattensis.]* Santa Maria Formation, Upper Triassic.

20. ?*Estheriina* sp. [= *Estheriina (Nudusia)* MP-240, 242, same repository and university as *"wianamattensis"* above. Pinto, 1956, p. 79–80, Pl. 1, Fig. 2. Santa María Formation, Upper Triassic. [see also MP-M-99 Pl. 34, Fig. 5]

38. *Estheria* cf. *minuta* (von Zieten, 1833) [= *Cyzicus (Euestheria)* cf. *minuta* (von Zieten)]. Universidad Federal Rio Grande do Sul 4183. Katoo, 1971, p. 38–40, Pl. 4, Fig. 1, *non* Pl. 5, Figs. 1–5, "Belvedere" (see Appendix for site map). Upper Triassic.

38. *Euestheria* cf. *emmonsi* (Raymond) [= *Cyzicus (Euestheria)* sp. 1]. Katoo, 1971, p. 41, Pl. 7, Figs. 1–4. Universidad Federal Rio Grande do Sul 4187. "Belvedere." Upper Triassic.

39. *Lioestheria* sp. [= *Cyzicus (Lioestheria)* sp. 2]. Katoo, 1971, p. 34, Pl. 2, Figs. 1–5. Universidad Federal Rio Grande do Sul 4201. "Passo das Tropas." Upper Triassic.

39. *Pseudoasmussia* sp. Defretin-Le Franc [= *Cyzicus (Euestheria)* sp.]. Katoo, 1971, p. 44, Pl. 9, Fig. 4. Universidad Federal Rio Grande do Sul 4189. "Belvedere." Upper Triassic.

Family ASMUSSIDAE Pacht, 1849

(Emended and abbreviated): Valves of variable shape with straight dorsal margin, and umbo central to subcentral.

Genus *Pseudoasmussiata* Tasch (new name for new genus)

Pseudoasmussia Defretin-Le Franc 1969, p. 129 (*non Pseudoasmussia* Novojilov, 1954c). (Defretin-Le Franc (1969) used the name *Pseudoasmussia* and defined it as a new genus. This name was preoccupied by Novojilov (1954a). Tasch (1969) placed Novojilov's genus with polygonal ornamentation under synonomy of *Cyzicus (Euestheria).* However, Defretin-Le Franc's definition of *"Pseudoasmussia"* added a distinctive feature, i.e., the axis of the valve plunged posteriorly. Accordingly, it defines a valid new genus named above, *Pseudoasmussiata.)*

Original Diagnosis: Carapace legerement allongee aux angles carinaux nets et axe plongeant vers l'arrière.

Type: *Pseudoasmussia grasmücki* Defretin-Le Franc (= *Pseudoasmussiata grasmücki* Defretin-LeFranc).

38. *Pseudoasmussiata katooae* n. sp.
(Pl. 36, Fig. 5)

Name: Named for Ioco Katoo, who collected this and other fossil conchostracans and did her thesis on the Santa Maria Formation.

Diagnosis: Obliquely ovate valves, with short, straight dorsal margin; the umbo is raised slightly above the margin and inset from the anterior margin about one-fifth the length of the valve. An axis drawn through the middle of the umbo to the ventral margin plunges posteriorly. Anterior margin steeply arched; posterior margin rounded. Ornamentation of contiguous granules on growth bands.

Measurements: (mm) Based on Ioco Katoo's photograph and examination of the specimens at Universidad Federal Minas Gerais, courtesy of Professor Roberto Cardoso and Ioco Katoo. L-3.13; H-2.29; Ch-1.51; Cr-0.78; Av-0.78; Arr-0.78; a-0.90; b-1.02; c-1.51; H/L-.732; Ch/L-.482; Cr/L-.249; Av/L-.249; Arr/L-.249; Arr/L-.249; a/H-.393; b/H-.445; c/L-.482.

Material: Internal mold of right valve, Universidad Federal Rio Grande do Sul 4186. Photo courtesy of Ioco Katoo.

Locality: Uppermost Santa Maria Formation, Brazil. Defined as a fluvial facies and conchostracan-bearing. "Belvedere" (Station 4 on location map, Katoo, 1971, p. 24, and Figure 30 herein. Upper Triassic.

Discussion: This species is an excellent example of the genus *Pseudoasmussiata*. It differs from Defretin-Le Franc's type, also from the Triassic, (east Greenland) in that it has a greater angle of axial plunge, longer dorsal margin behind umbo, and more ovate than subcircular configuration.

38. *Pseudoasmussia*? sp. B (= *Pseudoasmussiata*? sp. B., Katoo, 1971.) Universidad Federal Rio Grande do Sul 4191. Ioco Katoo, 1971:45, pl. 10, Figs. 1, 2. "Belvedere." Upper Triassic.

39. *Echinestheria*? (= *Echinestheria*? Katoo, 1971). Universidad Federal Rio Grande do Sul 4192-B. Katoo, 1971, p. 47, Pl. 10, Figs. 3, 4. "Belvedere." Upper Triassic.

39. *Palaeolimnadia*? [= *Palaeolimnadia (Palaeolimnadia*?)* cf. *wianamattensis* (Mitchell)]. Katoo, 1971, Universidad Federal Rio Grande do Sul 4193, p. 47, Pl. 11, Fig. 1. (*Due to cover of ventral margin, there may be more growth bands than in *"wianamattensis."* If so, this species of Katoo would be assignable to subgenus *Grandilimnadia,* hence the question mark.) "Belvedere." Upper Triassic.

39. *Estheriina*? sp. (Too poorly preserved for identification.) Katoo, 1971, p. 46, Pl. 11, Fig. 2. Universidad Federal Rio Grande do Sul 4186. "Belvedere." Upper Triassic.

39. *Orthothemos*? sp. [= *Cyzicus (Euestheria)* sp. 2]. Katoo, 1971, p. 46, Pl. 11, Figs. 3, 4. Universidad Federal Rio Grande do Sul 4194, "Belvedere." Upper Triassic.

20. *Palaeolimnadiopsis* sp. 2 [No. CO-1, UFMG]. Occurs above a fish bed at north margin of Arroio das Tropas, 7.8 km from Santa Maria. Specimen studied and measured by Tasch through courtesy of Professor Cardoso and Ioco Katoo. Last few growth bands eroded along ventral margin. L-4.2 cms; H-1.7 or more cm; H/L-.400; $h_u/1_u$-.130 (u = umbo); posterior recurvature of numerous growth bands; an unusually large and elongate species. (Pl. 32, Fig. 3 herein.) Santa Maria Formation. Upper Triassic.

20. *Euestheria azambujai* Pinto 1956 [= *Cyzicus (Euestheria) azambujai* (Pinto), *non Pseudestheria* Katoo, 1971]. No. 238 MP of ICN (UFRGS). Passo das Tropas. Pinto, 1956, Pl. 1, Fig. 1 and text-Fig. 1. (See Pl. 35, Fig. 8 herein.) Santa Maria Formation. Upper Triassic.

71. *Cyzicus (Euestheria)* sp. 3. (Tasch Collection TC 90014, eroded mold of right valve. Porto Alegre-Santa María Road, from Porto Alegre, km 81–83. [See Bigarella, Pinto and Salamuni, Eds., 1967:118]. (Tasch field notes: estheriids poorly preserved at two levels.) Tasch Field Collection No. 20-2, Santa María Formation. Ratios: H/L-.625; Ch/L-.406; Cr/L-.313; Av/L-.313; Arr/L-.281; a/H-.450; b/H-.500; c/L-.563. These ratios do not match those of any of the other Santa María Formation euestheriids figured and described by Katoo (1971) or others from different sites. Santa María Formation. Upper Triassic.

Santa Catarina

4. *Orthothemos regularis* (Reed) 1929, Mendes, 1954. [= *Cyzicus (Lioestheria) regularis* (Reed)]. Reed, 1929, Figs. 1, 2, 2a only, non Figs. 3, 4, and 5. Approximately 1 km from railroad station at Poço Preto along the tracks. Middle Rio do Rasto Formation. Upper Permian.

[*Raymond (1946) erected the genus *Orthothemos* on the holotype *"Estheria draperi"* Jones. Tasch [1984, and see Africa section, this memoir) has shown that *"draperi"* belongs to *Cyzicus (Lioestheria).* Because Raymond's genus was based on a lioestheriid, it follows that the genus becomes a synonym of *Cyzicus (Lioestheria).]*

4. *Estheria subalata* Reed, 1939 [= *Palaeolimnadiopsis subalata* (Reed), Raymond, 1946.] (Cf. Mendes, 1954; Reed, 1929, Fig. 9–11.) Same locality and age as *"regularis"* above.

4. *Leaia pruvosti* Reed, 1929 [Leaia (Hemicycloleaia) pruvosti Reed]), Pl. 33, Fig. 8. Same locality and age as *"regularis"* above.

4. *Leaia curta* Reed, 1929 [= *Monoleaia unicostata* (Reed), Tasch, 1969, p. R159]. *Non Bileaia* Kobayashi 1954, *nom. dubium; non Monoleiolophus unicostatus* (Reed) to which Reed's species was assigned as a synonym (Mendes, 1954, p. 160, Pl. 13, Fig. 6, and text-Fig. 3). Mendes noted that the two dorsally superimposed valves created the illusion of two ribs but one is actually a dorsal margin. (See also Reed, 1929, Fig. 18.) Same locality and age as *"regularis"* above.

4. *Leaia unicostata* Reed, 1929 [= *Monoleaia unicostata* (Reed), Tasch, 1956]. *Non Monoleiolophus unicostatus* (Reed), Mendes, 1954. See Pl. 34, fig. 4 for photograph of Reed's type from Mendes, 1954. See also Reed, 1929, p. 12–13, Figs. 14–17. Same locality and age as *"regularis"* above. Another specimen collected by Tasch (Pl. 34, Fig. 2) is described below.

23. *Cyzicus langei* Mendes 1954 [= *Cyzicus (Euestheria) langei* Mendes]. Mendes, 1954, p. 158–159, Pl. 12, Figs. 1, 2. Poço

Preto on the Poço Preto-Porto União Road. (See also the State of Paraná for several localities.) Middle Rio do Rasto Formation. Upper Permian.

65. *Cyzicus (Lioestheria)* sp. (TC 60,000, Tasch Station 2, bed 4B, section below diastem.) Road along Highway Br 116, Serra Espigão, 34 km north of Santa Cecilia. (See Appendix herein for photograph of diastem outcrop.) Upper Rio do Rasto Formation. Upper Permian.

Genus *Gabonestheria* Novojilov 1958

Gabonestheria Novojilov, 1958c, p. 111–112, Pl. 2, Figs. 28, 29.

Large, robust spine on initial shell, situated in antero-d⟨ ⟩ l sector of the valve; sculpture reticulate (cf. Tasch, 1969).

Type: *Estheria (Pemphycyclus) gabonensis* Marlière, 1950.

66. *Gabonestheria brasiliensis* n. sp.
(Pl. 33, Figs. 3, 5)

Diagnosis: Valves of various configurations from ovate to subquadrate. In the antero-dorsal sector, the umbonal area is swollen above the level of the rest of the shell, despite flattening. Umbo is broad based and its dorsal terminus culminates in a spinous point; the spinous projection may occur slightly below the dorsal margin or extend beyond it in a flattened condition. (In life, the spinous ends must have been positioned at an angle relative to the valve sector from which they arose.) The entire umbo, including the spinous end, occupied approximately one-third of each valve. Growth lines generally from five to eight.

Measurements: (mm) Holotype followed by paratype TC 60502b. L-2.1, 2.5; H-1.5, 1.5; Ch-1.3?, 1.7?; Cr-0.3, 0.4; Av-0.3, 0.4; Arr-0.5, 0.4; a-0.5, 0.6; b-0.5, 0.7; c-0.7?, 1.2; H/L-.714, .692; Ch/L-.619?, .680?; Cr/L-.143, .154; Av/L-.143, .154; a/H-.313, .313; b/H-.333, .389; c/L-.333, .461; l_{sp} (length of spine)-0.9, 0.5.

Material: Tasch Collection. Internal mold of left valve. Holotype, TC 60500a; paratype TC 60502b, same as holotype; paratype TC 60501, internal mold of right valve. Sample n = 8, includes several incomplete specimens that display the species' chief characteristic.

Locality: Lower Rio do Rasto Formation, São Pascoal, approximately 400 km southeast of Curitiba; bed 3 (holotype), bed 5, lower (paratypes); both beds in red shale facies. Upper Permian.

Discussion: There are but three known species (two Permian; one Upper Triassic) to which the Brazil specimens can be compared: *G. gabonensis* (Marlière) 1950 from Gabon, Upper Triassic, *G. shandaica* Novojilov, 1970 from the USSR, Upper Permian, and *G. dickinsoni* Tasch, from the United States (Leonardian, Permian). The Brazil species differs from Marlière's gabonestheriid in having a shorter dorsal margin and generally a smaller height/length ratio. It is closest to this species in having a comparatively large antero-dorsal spinous umbo, although it is

larger and more pointed at the apex than the blunt-tipped Marlière species.

The new species differs from *G. shandaica* Nov. in having a lesser height/length ratio, a shorter dorsal margin, and valves that are more elongately ovate-to-subquadrate. *G. shandaica* is subcircular. The two species differ also in the spinous umbo, which is larger and more robust in the Brazil species and hence occupies a larger area of the anter-dorsal sector.

G. dickinsoni has a much more attenuated umbo in the antero-dorsal sector of the valve which results in a constricted terminal spine; the umbo occupies a smaller portion of the valve than it does in the new species. *G. dickinsoni* also has a higher upper limit for height/length ratio of the valves (.717–.737 vs. .692–.714 *"brasiliensis"*).

66. *Cyzicus (Euestheria)* sp. 1
(Pl. 32, Fig. 7

Diagnosis: Ovate valves with the umbo inset from the anterior margin and umbonal terminus rising slightly above the dorsal margin; the dorsal margin is short behind the umbo; margins: anterior rounded; posterior with convex bulge; ventral gently curved. Double growth lines. Ornamentation of contiguous granules on growth bands.

Measurements: (mm). L-3.3; H-2.6; Ch-1.3; Cr-1.0; Av-1.0; Arr-1.0; a-1.1; b-1.1; c-1.2; H/L-.788; Ch/L-.394; Cr/L-.303; Av/L-.303; Arr/L-.303; a/H-.432; b/H-.423; c/L-.364.

Material: TC 90520, internal mold of left valve; somewhat flattened and crushed behind umbo; terminus of umbo partly eroded. TC 90511, internal mold of right valve, antero-dorsal sector missing and dorsal margin behind umbo eroded and crushed.

Locality: Late-middle or upper Rio do Rasto Formation, Tasch Station 3, bed 3, 34 km north of Santa Cecilia; diastem section along Highway BR-116, Serra Espigão, Santa Catarina. Upper Permian.

Discussion: Specific identification is not possible due to incompleteness of the specimens. However, the Rio do Rasto species described by Reed (1929) and Mendes (1954), or Mesozoic forms elsewhere described by Katoo (1971) or Pinto (1956) do not fit the characteristics of the specimens described above.

66. *Cyzicus (Euestheria)* sp. 2
(Pl. 34, Fig. 1)

Diagnosis: Subovate valve with ovate umbo inset from and nearest to the anterior margin; dorsal margin slightly curved, possibly due to flattening; posterior margin with convex bulge; anterior margin steeply rounded; comparative height of posterior sector of valve, smallest; middle third, along a line through umbo to ventral margin, greatest. Ornamentation: contiguous granules.

Measurements: (mm) L-4.2; H-3.3; Ch-2.1; Cr-1.0; Av-1.0; Arr-0.9; a-1.7; b-0.8; c-1.9; H/L-.786; Ch/L-.667; Cr/L-.238; Av/L-

.238; Arr/L-.214; a/H-.515; b/H-.242; c/L-.452.

Material: TC 60000, external mold of right valve; crushed along medial sector, as well as flattened; eroded dorsad on umbo.

Locality: Upper Rio do Rasto Formation, Tasch Station 2, bed 4B. Upper Permian. [See *C. (Euestheria)* sp. 1 above for locality details and Fig. 28, lower panel for section.]

Discussion: No equivalent species occurs in the relevant literature of the Rio do Rasto Formation (i.e., by Pinto and students, Reed, or Mendes).

66. *Cyzicus (Lioestheria)* cf. *regularis* (Reed) 1929
(Pl. 31, Fig. 8; Pl. 32, Fig. 6)

Estheria regularis Reed, 1929, plate unnumbered, Figs. 1–5.

Diagnosis (emended): Dimorphic valves; subovate (male?) to elongate-ovate (female?), with small umbo inset from anterior margin; dorsal margin, short behind umbo and arched slightly in front of umbo; growth lines double. Ornamentation of minute pits on growth bands.

Measurements: (mm) TC 92774, male?, followed by TC 90503, female?, from the same bed. L-5.0, 5.6; H-3.8, 3.7; Ch-2.1, 2.3?; Cr-1.6, 1.9; Av-1.6, 1.9; Arr-1.3, 1.3; a-1.5, 1.5; b-1.7, 1.7; c-2.6, 2.5; H/L-.760, .661; Ch/L-.420, .411; Cr/L-.320, .339; Av/L-.320, .339; Arr/L-.260, .232; a/H-.395, .405; b/H-.340, .459; c/L-.500, .446.

Material: TC 92774, male valve?, internal mold of right valve; TC 90503, internal mold of left valve, female?

Locality: Upper Rio do Rasto Formation, Tasch Station 3, bed 4B. Upper Permian. [Same locality as *C. (Eu.)* sp. 1, above.]

Discussion: The above described specimens are closest to Reed's species *"regularis"* but now assignable to a different genus and subgenus. Reed's species came from Poço Preto (middle Rio do Rasto). It has an H/L-.720; H and L are the only measurements given. It also fits the general characteristics of umbonal placement, configuration, and ornamentation. Reed's Figure 5 (1929) indicates two different configurations of the valves of *"regularis."* The left valve in his figure is more elongate-ovate and probably represents the female valve (cf. 32, Fig. 6 herein; also see measurements for TC 90503 above).

Monoleaia Tasch 1956

Monoleaia Tasch, 1956, p. 1256, text-Fig. 10. Carapace subovate; imbricate or nodose, radial (= rib) extends from umbo to rounded posterior-ventral margin.

Type: *Leaia unicostata* Reed, 1929. *Monoleaia* Tasch and *Monoleiolophus* Raymond need to be distinguished because they may be confused. Raymond's material for *Monoleiolophus* (not based on Reed's type of *Leaia unicostata* R.), Museum of Comparative Zoology, Harvard, specimen 4752, was studied at the Museum by this writer and has the following characteristics: (1) an elongate lozenge configuration (vs. broadly subovate, *Monoleaia*); (2) the nodose rib fades out in both directions on the valve and

does not reach umbo (vs. rib begins on the posterior sector of the umbo and fades out ventrad on fifth growth band, *Monoleaia*); (3) rib has a marked anterior convexity (vs. rib straight, except gently curved where it crosses posterior sector of umbo, *Monoleaia*); (4) beta angle, approximately 40° (vs. 25° [Tasch] to 30° [Reed, 1929], *Monoleaia*); (5) length and height of valve measured by this writer yielded an H/L of .601; Raymond's figure was .615 (vs. 702 [Tasch], .705 [Reed] for *Monoleaia*). The above characteristics suffice to adequately distinguish the two genera.

4. *Monoleaia unicostata* (Reed) 1929
(Pl. 34, Figs. 2, 4)

Leaia unicostata Reed 1929, p. 12, 14; Fig. 14–17a.
Leaia curta Reed 1929, p. 12, Fig. 18 (cf. Mendes, 1954; see entry below, Mendes, 1954, p. 160, Pl. 13, Fig. 6). *Nom. dubium.*
Monoleiolophus unicostatus Raymond, 1946, p. 262, Pl. 3, Fig. 11.
Monoleiolophus unicostatus (Reed) Mendes, 1954. p. 160–161, Pl. 3, Fig. 6; Pl. 14, Figs. 9–11; text-Fig. 3.
Monoleaia unicostata (Reed) Tasch, 1956, p. 1248–1257.
Monoleaia uncostata (Reed) Tasch, 1969, p. R159, Figs. 55, 5a, b, c.

Diagnosis: Broadly subovate valve with a straight dorsal margin and a characteristic single nodose rib, having a beta angle of approximately 25°; the rib begins in the posterior sector of the umbo, then thickens and fades out ventrad on the fifth growth band above the ventral margin; the last nine growth bands are very tightly spaced, compared to the rest of the valve's bands. Ornamentation of reticulate pattern on growth bands.

Measurements: (mm) Two sets of figures are given for each parameter and ratio: the specimen described above, TC 60550, followed by Reed's species, DGM 1423 (studied at Universidad Federal São Paulo and also as figured by Mendes [1954, Pl. 14, Figs. 9, 10, 11]). L-4.7, 5.0; H-3.3, 3.1; Ch-2.7?, 3.1?; Cr-0.8, 1.0?; Av-0.8, --; Arr-1.1, 1.1; a-1.6, 1.3; b-1.1, 1.4; c-2.2, 2.6?; H/L-.702, .620; Ch/L-.702, .620; Cr/L-.170, --; Av/L-.170, --; Arr/L-.234, .220; a/H-.485, .419; b/H-.333, .452; c/L-.468, .520; beta angle-25°, 22° (however, in Mendes photograph, Collection No. 133, left valve has value of 30°).

Material: Internal mold of right valve, TC 60550, Station 3, bed 4. Flattening of the posterior margin midway led to a scalloped appearance for a portion of the margin. TC 90069, Station 2, bed 2b.

Locality: [See *Cyzicus (Euestheria)* sp. 1 above for locality data]. Tasch Station 3, bed 4, below diastem, and Station 2, bed 2B (TC 90069). Middle or Upper Rio do Rasto Formation. Upper Permian.

Discussion: Single-ribbed forms are now known from the Rio do Rasto Formation at several localities in Santa Caterina and Paraná: Reed's locality at Poço Preto, the Br 116 Serra Espigão, Santa Caterina (Tasch, above), Mendes locality at km 115–200 on the Ponto Grossa-Guarapuava Road from Ponto Grosso (Mendes, 1954, Fig. 3) and km 1.5 between Dorizon and Paulo Frotin. This last was collected by Dr. Mary Elizabeth Bernardes de Oliveira from Paraná, GP/3E-3169-2. This writer examined this specimen; it bears a single rib, curved as it crosses the entire umbo. It is definitely not *Monoleiolophus,* as labelled, and may be a new species.

The locality, Br 116, Santa Caterina, where the species described above had not been previously reported, occurs a few kilometres from the Rio do Rasto Formation/Botucatú Formation contact on the same highway. Published data are not available to determine whether the diastem section (34 km north of Santa Cecilia) is Middle or Upper Rio do Rasto. It might possibly be Upper-Middle Rio do Rasto as is the Poço Preto site of Reed—both have the same species of *Monoleaia.*

Genus *Estheriina* Jones, 1897a
Subgenus *Estheriina* Jones, 1897a

Estheriina-type valves with a more convex umbonal region bearing growth bands. Type: *Estheriina (Estheriina) bresiliensis* Jones, 1897a, Pl. 8, Fig. 5.

66. *Estheriina (Estheriina)* sp. 1
(Pl. 33, Fig. 7)

Diagnosis: Subovate valve with the umbo of the neanic portion of the valve elevated above the dorsal margin and distinctly separated from the adult portion of the valve; umbo subterminal; two distinct sets of growth bands; neanic valve with progressively wider spacing ventrad; adult portion, progressively closer spacing ventrad; the neanic valve has a smooth terminus and the entire valve has a postero-ventral obliquity.
Measurements: (mm) L-5.8; H-4.5; Ch-3.0?; Cr-1.1; Av-1.1; Arr-1.7; a-1.5; b-2.0; c-3.0; H/L-.776; Ch/L-.517; Cr/L-.190; Av/L-.190; Arr/L-.293; a/H-.333; b/H-.444; c/L-.517.
Material: TC 90067a, internal mold of left valve. The neanic valve has been slightly flattened on the anterior side; only the last one or two growth bands out of eleven were distorted. (For other conditions of preservation, see discussion below.)
Locality: Unit 1B of Papp's bed 15 (see Fig. 28, upper panel). Lower Rio do Rasto Formation, São Pascoal, Brazil. Upper Permian.
Discussion: This specimen has the telltale estheriinid characteristics of the appearance of the early valve being superimposed on the subsequent valve. Because it is badly eroded behind the neanic valve and elsewhere, it cannot be identified to species level, although the uneroded areas suffice for genus and subgenus identification. But it is recorded here because it is the first estheriinid reported from the Paleozoic Rio do Rasto Formation (Permian). "*Estheriina*" has generally been reported from beds of Mesozoic age.

Family VERTEXIIDAE Kobayashi, 1954
Genus *Cornia* Lutkevich, 1937

Valve shape varying from subovate to subrectangular; beak part subcentral to anterior; small spine or tubercle rising from umbo of initial valve; sculpture punctate.
Type: *Cornia papillaria* Lutkevich, 1937

66. *Cornia* sp.
(Pl. 35, Fig. 7)

Diagnosis: Subovate valve with umbo bearing a prominent subcentral, slightly oblique node that extends ventrad from very close to the dorsal margin to about one-fifth the height of the valve; length of valve greater than height; from a midpoint on the node, the distance to the anterior margin is less than to the posterior margin; both anterior and posterior margins rounded; dorsal margin arched; growth bands, 12–13 and widely spaced.
Measurements: (mm) L-4.4?; H-3.5; Ch-2.6?; Cr-1.3; Av-1.3; Arr-0.6?; a-1.8; b-1.6; c-1.7; H/L-.795; Ch/L-.591; Cr/L-.295; Av/L-.295; Arr/L-.136; a/H-.514; b/H-.457; c/L-.386.
Material: TC 90089, mold of left valve; also TC 90097.
Locality: Lower Rio do Rasto, Tasch unit 1-a, Papp bed 15. (See Fig. 28.) São Pascoal, Brazil.
Discussion: *Cornia* species has not previously been reported from the Rio do Rasto beds. It is of interest that in other sectors of the Gondwanaland continents (e.g., Lower Permian of Gabon, Lower Triassic of Northern Angola, Africa, as well as the Lower Triassic of India; Tasch, 1982a, and Fig. 15), *Gabonestheria* and *Cornia* species occur together in the same bed or section just as they do in the São Pascoal, Brazil (Upper Permian).

The mold is eroded on the anterior and posterior margins, making specific identification impossible. Nevertheless, the specimen is definitely assignable to *Cornia* by the presence of the umbonal node. Measurements that are given indicate magnitude rather than, for several parameters, unequivocal dimensions.

Family CYCLESTHERIIDAE Sars, 1899
Genus *Cyclestherioides* Raymond, 1946

Diagnosis: Cyclestheriidae with a beak (umbo) a bit further back than in modern *Cyclestheria.*
Type: *Estheria lenticularis* Mitchell, 1927

Subgenus *Cyclestherioides* Raymond, 1946

Diagnosis (emended): Minute, subcircular valves with subround, subterminal umbo directed anteriorly; distinctly unequal height and length; few widely spaced growth bands.
Type: Same as for genus.

65. *Cyclestherioides (Cyclestherioides) pintoi* n. sp
(Pl. 34, Fig. 3)

Name: This species is named in honor of I. D. Pinto. Both Professor Pinto and his students have made important contributions to our knowledge of fossil conchostracans in Brazil.

Diagnosis: Broadly and irregularly ovate valves with a slightly curved, short dorsal margin; umbo inset closer to anterior margin; postero-dorsal angle steeper than antero-dorsal; posterior margin convex, giving valves an eccentric posterior bias; greatest height of valve from umbo to ventral margin; bottoms of growth bands show beaded character (from middle of the valve and ventrad). These structures (in life) bore setae or spines.

Measurements: (mm) Holotype followed by paratype. L-7.1, 5.8; H-6.0, 4.7; Ch-2.5?, 1.7; Cr-2.5, 2.1; Av-2.5, 2.1; Arr-2.1, 2.0; a-2.0?, 2.0; b-2.4?, 2.3; c-4.2, 3.0; H/L-.845, .810; Ch/L-.352?, .293; Cr/L-.352, .362; Av/L-.352, .362; Arr/L-.296, .345; a/H-.333, .426; b/H-.400?, .489; c/L-.592, .517.

Material: TC 90333, holotype, external mold of left valve, with dorsal part of umbo eroded; valve crushed. TC 90003, paratype, external mold of right valve; valve crushed and terminus of umbo partially eroded. (Tasch field no. 17.9.)

Locality: For locality details, see Bigarella and others, 1967, p. 103, Fig. 21, road log, section 16-17. Estrada Nova Formation, Middle Permian.

Discussion: The new Estrada Nova Formation species is unique to the Permian of Brazil. It is unlike *Cyclestherioides (Cyclestheriodes) lenticularis* (Mitchell) from the Upper Permian, Newcastle Coal Measures of Australia as follows: different configuration (broadly and irregularly ovate vs. lenticular, n. sp.), umbonal placement (inset vs. subterminal, n. sp.), and dorsal margin (short and slightly curved vs. much shorter by one-half, n. sp.). The new species is also unlike a species described by Novojilov (1958c, p. 35, Pl. 3) under the name of *Cyclestheria krivickii* Nov. from the Permian (Tartarian) of the Sea of Laptev, USSR, which has an H/L of .330 vs. .845 for *"pintoi."*

65. *Cyzicus (Lioestheria) bigarellai* n. sp.
(Pl. 38, Fig. 1)

Name: The species is named for J. J. Bigarella, who was one of the organizers of the 1967 field trip which included a locality from which this writer collected the species described below.

Diagnosis: Compactly ovate valves with very short dorsal margin behind umbo; umbo robust, submedial, rises above dorsal margin in a low arched curve; greatest height along a line through center of umbo to ventral margin; posterior margin convex; anterior margin subdued round. Ornamentation of coarse hachure markings on growth bands which are pseudocancellate near ventral margin (i.e., due to tight spacing of growth bands bearing hachure markings).

Measurements: (mm) L-4.7; H-3.7; Ch-2.0; Cr-1.7; Av-1.9; Arr-1.4; a-1.6; b-1.1; c-1.5; H/L-.787; Ch/L-.425; Cr/L-.362; Av/L-.319; Arr/L-.298; a/H-.432; b/H-.297; c/L-.319.

Material: TC 90006, holotype (field no. 17.16), external mold of left valve; antero-dorsal sector slightly covered.

Locality: Estrada Nova Formation. Middle Permian. Same locality data as for *Cyclestherioides (Cyclestheriodes) pintoi* above.

Discussion: The four lioestheriid species [= *Cyzicus (Lioestheria)*] described and figured on Novojilov (1958c, p. 30-31) include *Lioestheria propinqua* Nov., 1958c from the Lower Triassic of the USSR Arctic; and three species from the Upper Permian (Tartarian) of the USSR Arctic: *Lioestheria nidymensis* Nov., 1958c; *Lioestheria toricata* (Nov.) 1946; and *Lioestheria evenkiensis* (Lutkevich) 1938. The new species differs from each of the above in its very short dorsal margin, submedial umbo (absent in all of the above Russian lioestheriids); umbo rises above dorsal margin in a less prominent form than in *"toricata"* (Nov.), and in its compact ovate configuration (absent in all the above).

São Paulo

Botucatú Formation (Lower Cretaceous)

11. *Graptoestheriella* cf. *fernandoi* Cardoso 1965 (see Pl. 36, Fig. 3 herein). A distinctive fragment figured herein; photograph provided by Pinto. Rifaina. Cardoso's holotype, DMG 4617, Rio de Janeiro, not figured here. Lower Cretaceous. (Note: A publication by A. de Souza, O. Sinelli, and N.M.M. Gonçalves [hereafter S.S. & G; 1971, p. 281] contains line sketches of several Botucatú conchostracans from the same locality, 26 km from Ribeirão Preto on the Ribeirão-Serrana-Cajuru Road. As Professor Pinto has indicated in his copy of de Souza, and others, 1971, the legends for Plates 1 and 2 should be interchanged. [I have a photocopy of Professor Pinto's copy.]

17. *Estheriella* sp. (= *Graptoestheriella?* sp. de Souza and others, 1971, Pl. 2, Fig. 1, incomplete left valve. For locality, see note above.)

17. *Euestheria triangularis* Gonçalves 1971 [= *Cyzicus (Lioestheria)* sp.]. One incomplete valve having hachure marks on growth bands. Available data inadequate for species identification in de Souza and others, 1971, Pl. 2, Fig. 4. Botucatú Formation. Lower Cretaceous.

17. *Lioestheria elliptica* Dunker, 1843, (= *Cyzicus* sp?). de Souza and others, 1971, Pl. 1, Fig. 1; cf. Jones, 1862, p. 103. Botucatú Formation. Lower Cretaceous.

17. *Euestheria ribeiropretensis* Gonçalves 1971 [= *Cyzicus (Lioestheria?)* sp.]. In de Souza and others, 1971, Pl. 1, Fig. 4.

17. *Pseudestheria* sp. S.S. & G 1971 [= *Cyzicus (Lioestheria)* sp.]. Punctate. *Pseudestheria* is a synonym of *Cyzicus* (cf. Tasch, 1961, p. R151). de Souza and others, 1971, Pl. 1, Fig. 2. Botucatú Formation. Lower Cretaceous.

18. *Pseudograpta?* barbosai Almeida 1950 [= *Cyzicus (Lioestheria) barbosai* (Almeida) 1950a]. Locality: 25 m before the "222 Km" sign on the road from Rio Claro to São Carlos. Lower Cretaceous.

18. *Palaeolimnadia petrii* Almeida 1950 [*Palaeolimnadia (Grandilimnadia) petrii* Almeida 1950]. Cf. Tasch *in* Tasch and Jones,

1979a, p. 11. Same road as above; 250 m before the "222 Km" sign. Lower Cretaceous. (See previous entry.)

18. *Bairdestheria mendesi* (Almeida), 1950 [= *Cyzicus (Lioestheria) mendesi* Almeida, 1950]. Same locality and age as *Palaeolimnadia petrii.*

59, 60. *Palaeolimnadiopsis suarezi* Messalira, 1974 (= ?*Palaeolimnadiopsis suarezi* Messalira, 1974). Messalira, 1974:20-21, pl. 2, figs. 9-11 (Nos. 757-1, 757-8). Km 600+ 500 m, on road between Penapolis-Presidente Prudente, Rio Feio, Iacri. Bauru Formation. Upper Cretaceous (Senonian).

72. Conchostracans undetermined, Messalira 1974 [= ?*Cyzicus (Lioestheria) sp.*]. Based on Messalira's brief description. 15 km northeast of São Carlos. Messalira, 1974, p. 121 (no figure given). Bauru Formation. Upper Cretaceous (Senonian).

Sergipe

16. *Estheriella brasiliensis* Oliveira 1953 [= *Graptoestheriella brasiliensis* (Oliveira) Cardoso, 1965]. Oliviera, 1953, holotype, Departamento Geologia e Mineralogia 3939, Rio de Janeiro; paratypes, Departamento Geologia Mineralogia 3944, 4056. Cardoso, 1965, Pl. 3, Figs. 1, 2. Serrinha, Bahía, Candeias, and Ilhas Formations. Lower Cretaceous (Wealden).

Chile

49. *Cyzicus (Euestheria) sp.* [= *Cyzicus (Euestheria)* cf. *forbesi* (Jones)]. Based on specimens collected by Professor Dr. A. von Hillebrandt, Technichische Universitat, Berlin. Locality: High Cordillera, below Jurassic section at Quebrado San Petrito-Yerbus Buenos, between lat 27°10′ to 27°25′S and long 69°20′ to 69°45′W. Upper Triassic.

Province of Aconcagua

61. *Cyzicus* sp. (Harrington, 1961, p. 174.) Central Chile, coastal region approximately 804.6 km south of Taltal. Upper Triassic.

Province of Antofagasta (Pular Formation)

58. *Cyzicus* sp. (Harrington, 1961, p. 177 and 194 [measured section]. Sierra de Almeida between Monturaqui Station and Quebrado de Pajonales. Cf. *Cyzicus* sp., Magellan Basin, southern Chile and Argentina.) Lower Cretaceous (Neocomian). *Cyzicus* sp. Harrington, 1961. (See entry for Argentina, Santa Cruz herein. If the "estheriid beds" continue from Argentina into the Magellan Basin area to the south and west in Chile, this cyziciid might be related to the Santa Cruz species described herein.) Lower Cretaceous (Neocomian).

Province of Aranca

52. *Estheria chilensis* Philippi, 1887 [= *Cyzicus (Euestheria) chilensis* (Philippi)]. Cf. *Estheria chilensis* Philippi, 1887; Jones,

1897b, (Section VIII), Pl. X, Figs. 4, 5. Lebu on the seacoast at lat 37°36′ or 39′S. Exact age undetermined, probably late Paleozoic. Exact locality and formation in doubt (Jones, 1897b, p. 290).

Colombia

75. *Estheriella*? (Will Maze Collection. C-2-79-3, a remnant of the narrow apron on the valve which apparently bears a few costellae preserved on a badly eroded left valve. Originally compared to *Estheriella taschi* Shah and Ghosh, which has an apron. However, that species has been reassigned [see *Cornutestheriella* under India section.] Bocas Formation, Santander massif, near Bucaramanga [see Fig. 30, upper Triassic–Lower Jurassic.)

Montebel

7. *Howellites columbianus* Bock, 1953 [= *Cyzicus (Lioestheria) columbianus* (Bock)]. Holotype 200118, Academy of Natural Sciences, Philadelphia, Pennsylvania. In gray-bluish, brittle shale near Montebel. Locality G1818, Imperial Oil Company, Toronto, Canada, Rhaetic or Lower Jurassic.

Quebrado de los Indios

68. *Estheria* sp. Trumpy 1943 (= *Cyzicus* sp.). Trumpy, 1943, p. 1299, 1304, Pl. 1, Fig. 3. Quebrado de los Indios, approximately 32 km southeast of Fundacion, southwest of the Santa Marta Mountains. Triassic.

Guyana (formerly British Guiana)

Takutu River

5. *Pseudestheria* sp., Ball [= *Cyzicus (Lioestheria) sp.*]. Ball, British Museum (Natural History), Letham Shales. Right bank of Takutu River at St. Ignatius Mission. Permian or Triassic?

Peru

Arica (Department of Arequipa)

53. *Cyzicus (Euestheria) aricensis* (Jones)
(Pl. 36, Fig. 4)

Estheria aricensis Jones, 1897b, p. 264–265, Pl. X, Figs. 1–3.
Dadaydedeesia? aricensis (Jones) Raymond 1946, p. 261.
Euestheria aricensis (Jones) Kobayashi 1954, p. 107.
Cyzicus (Euestheria) aricensis (Jones) Morris 1980, p. 34, Pl. 1, Figs. 1 and 2.

Diagnosis: (After Jones, 1897b, p. 264 abridged.) "Obliquely suborbicular valves . . . with the longest diameter" about "60° to the hingeline . . . The umbo is in the anterior third of the cardinal margin and rather prominent above the hingeline . . . "In one

116

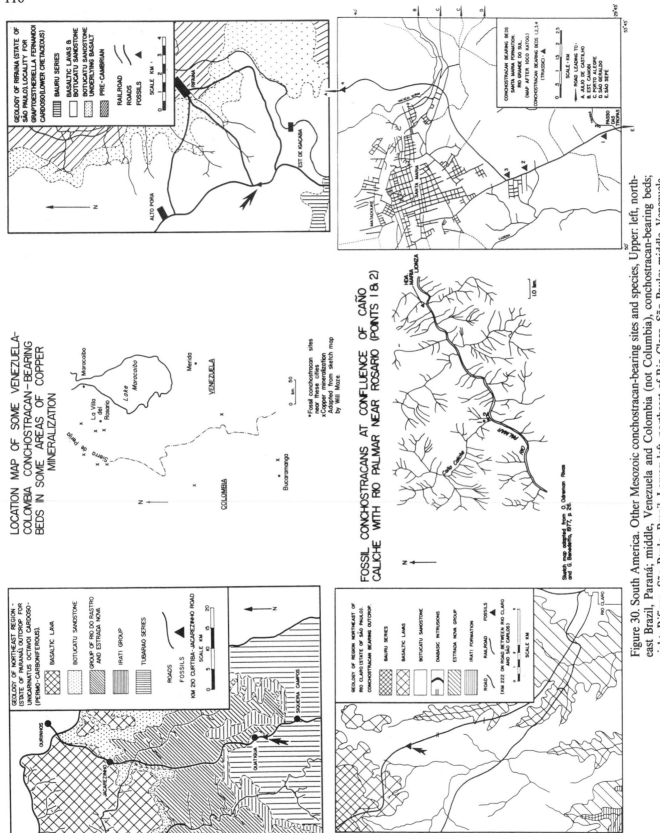

Figure 30. South America. Other Mesozoic conchostracan-bearing sites and species, Upper: left, north-east Brazil, Paraná; middle, Venezuela and Colombia (not Columbia), conchostracan-bearing beds; right, Rifaina, São Paulo, Brazil. Lower: left, northeast of Rio Claro, São Paulo; middle, Venezuela, conchostracan-bearing sites; right, Santa Maria Formation, Rio Grande do Sul, Brazil, fossil conchostra-can localities (Ioco Katoo, 1971).

large hollow mould . . . vertical burrelets or small transverse ridgelets are visible in the broader interspaces".

(Emended, based on study of the types.) Subovate valves with umbo close to the anterior margin; the terminus of the umbo is raised and projects slightly above the dorsal margin. Anterior and posterior margins rounded; height of posterior third greater than that of the anterior sector, and confers an obliquity to the valve. Ornamentation of contiguous granules on growth bands. (The small transverse ridgelets are due to erosion.)

Measurements: (mm) Based on lectotype (f-2). L-9.8; H-7.8; Ch-5.2; Cr-2.8; Av-2.8; Arr-1.8; a-3.2; b-3.6; c-5.8; H/L-.796; Ch/L-.296; Cr/L-.286; Av/L-.286; Arr/L-.184; a/H-.410; b/H-.461; c/L-.596.

Material: Syntype, British Museum I-6814a, b; I-6815. Slab I-6814a, collected by David Forbes, has several badly eroded specimens: three molds, f-1, 3, and 5 and an impression of f-4. Specimen f-1 preserves the euestheriid ornamentation; i.e., contiguous granules. Slab I-6815 contains the best preserved specimens of the syntype (f-2), and one of these, a complete mold of a left valve, is designated here as the lectotype (ICZN 1961, p. 79).

In addition, there are three or four other external molds partially preserved on the slab; right and left valves are represented. Slab I-6814b contains three impressions and two incomplete molds of *C. (Eu.) aricensis*, in addition to several incomplete impressions. One of these, (I-6814b, f-3) corresponds to Jones' illustration (1897b, Pl. X, Fig. 3) and displays granular ornamentation.

Locality: Arica, Department of Arequipa, southern Peru, lat 18°25′S; long 70°15′W. Bed F, Moro de Arica. Carboniferous.

Discussion: The syntype (f2) slab shows one complete but badly eroded mold (lectotype) and numerous other eroded molds, mainly incomplete. Accordingly, only measurements for the lectotype are given. Another specimen of the syntypes (I-6814a, f-3) appears to possess hachure-type markings on the growth bands, but this is a pseudo-ornamentation due to erosion.

Raymond (1946) and Kobayashi (1954) accepted Jones' original drawings in making their reassignments. Morris (1980), who had access to the syntypes, did not figure the lectotype given herein (Pl. 36, Fig. 4), but figured only f-3 on slab 6814b.

Uruguay

69. *Cyzicus (Lioestheria) ferrando*, Herbst, 1985. Herbst and Ferrando, 1985, p. 29–47, p. 35, Pl. 1, Figs. 1–10. (Paraná.) Tacuarembó Formation (lower part). Route 5, km 356, en route to Tacuarembó. Uruguay. Upper Triassic. (For fragment only, Pl. 37, Figs. 5, 6 herein).

Venezuela

5. *Isaura olsoni* Bock 1953 [= *Cyzicus (Lioestheria) olsoni* (Bock)]. Holotype 200108, Academy of Natural Sciences, Philadelphia, Pennsylvania, in black laminated shale, near Merida. Core from Imperial Oil Company, Toronto, Canada. Rhaetic?

Sierra de Perija

74. *Cyzicus (Euestheria)* sp. 1 (Will Maze Collection, 231D). (Pl. 38, Fig. 2 herein). Tinacoa Formation, near Villa del Rosario, Sierra de Perija, northwest Venezuela. This species is distinct from *Cyzicus (Lioestheria) mazei* n. sp. described herein and *Cyzicus (Lioestheria) olsoni* (Bock). It has polygonal ornamentation (unlike *"mazei"* and *"olsoni"*) and an umbo that has a tapered nipple-type terminus, that rises above the dorsal margin. ?Lower Jurassic.

Posidonia ovata Lea, 1856 [*Cyzicus (Euestheria) aff. ovata* (Lea), Lea 1856, p. 77]. Lea's species came from near Phoenixville, Pennsylvania. Upper Triassic. MPI-046 redescribed by Rivas and Benedetto, 1977, p. 20. Locality: confluence of Cano Caliche with Rio Palmar. Points 1 and 2 on map (Rivas and Benedetto, 1977, and Fig. 30, lower, middle). Tinacoa Formation, Edo, Estado Zulia, Venezuela. Jurassic.

Isaura olsoni Bock, 1953 [= *Cyzicus (Euestheria)* cf *olsoni* (Bock)]. Bock, 1953, p. 66, Pl. 12, Fig. 1–6. Bock's species came from near Merida, Venezuela. MPI-046, redescribed by Rivas and Benedetto, 1977, p. 21. Tinacoa Formation, Venezuela. Jurassic.

Howellites columbianus Bock, 1953 [= *Cyzicus (Lioestheria) columbiana* (Bock)]. Bock, p. 72, Pl. 11, Figs. 5–7, 12. Bock's species came from Montebel, Colombia. MPI-046, redescribed by Rivas and Benedetto, 1977, p. 21. Tinacoa Formation, Venezuela. Jurassic.

74. *Howellites columbianus* Bock, 1953 [= *Cyzicus (Lioestheria) cf columbianus* (Bock). Bock, 1953, p. 72, Pl. 11, Figs. 5–7, 16. His species came from Montebel, Colombia. Rhaetic or slightly younger. Rivas and Benedetto, 1977. Will Maze Collection; Tinacoa Formation, near Villa del Rosario, Venezuela. ?Lower Jurassic.

Pseudestheria pricei Cardoso, 1966 [= *Cyzicus (Lioestheria)* cf. *pricei* (Cardoso)] Cardoso, 1966, p. 71–72, Pl. 3, Fig. 1, text-Fig 16. Candeias Formation, Bahía, Brazil, Departamento de Geologia e Mineralogia 4865. Lower Cretaceous. Will Maze Collection; Pl. 39, Fig. 3 herein. Four specimens (WM 230 A-D). La Quinta Formation, Venezuela. ?Lower Cretaceous.

74. *Cyzicus (Euestheria)* sp. 2 (Pl. 38, Fig. 8 herein). Coquina of shells, crushed internal mold of right valve with fragments of inner layers of original valve. Ornamentation: granular. H/L-.775. Will Maze Collection WM-231D. Tinacoa Formation? Venezuela.

74. *Cyzicus (Lioestheria) mazei* n. sp.
(Pl. 39, Figs. 1, 2)

Name: This species is named for Will Maze. As a Ph.D. candidate at Princeton University, in the course of field research on copper mineralization, he collected the conchostracan fossils described here, and others noted elsewhere in this section and the section titled "Colombia".

Diagnosis: Compact subovate valves with umbo inset, but close to the anterior margin; dorsal margin short; steep descent from umbonal flank to anterior margin; posterior margin rounded; ventral margin broadly and gently convex. Ornamentation of coarse, hachure-type markings on growth bands.

Measurements: (mm) (In each instance, the syntype 231 B is the first figure for comparative purposes, followed by *C. (Lio.) olsoni* (Bock), on the basis of measurements determined from Bock's (1953, Pl. 12, Fig. 2) "male" photograph. L-3.7, 1.5; H-2.9, 1.3; Ch-1.9, 1.0; Cr-0.8, 0.3; Av-0.9, 0.3; Arr-0.8, 0.6; a-1.2, 0.6; b-1.1, 0.6; c-?, 0.6; H/L-.784, .866*; Ch/L-.486, .526; Cr/L-.216, .158; Av/L-.243, .159; Arr/L-.216, .316; a/H-.414, .333; b/H-.379, .400; c/L- --, .526. (*Based on Bock's [1953, p. 69] figures for H-1.3, L-1.5 mm, which yields H/L-.866 ["male" valve]; and H-1.4, L-2.2 mm ["female" valve] which yields H/L-.636.)

Locality: La Villa del Rosario, northwestern Venezuela (Will Maze site 213 B). Tinacoa? Formation, Sierra de Perija. ?Lower Jurassic.

Material: Will Maze Collection No. 231 B, syntype: one piece, with the figured specimen (Pl. 39, Figs. 1, 2). WM 231 B (1), a left valve with inner portion of the original valve, poorly preserved and crushed; other valves (WM 231 A) with many juveniles, are all fragmentary.

Discussion: *Cyzicus (Lioestheria) olsoni* (Bock) 1953 is the only Venezuelan species to which *"mazei"* may be compared. Bock's species has two configurations, subovate and subcircular (which he considered evidence for sexual dimorphism). These respective configurations yield a difference in length of the dorsal margin (long in the "female" and short in the "male"), with corresponding difference between male and female in placement of the umbo (inset from dorsal margin male vs. subterminal female). In the new species described above, the dorsal margin is short; there is no detectable dimorphism and the ratios differ. H/L is less than in Bock's male specimen but greater than in the female specimen. Similarly, Cr/L, Av/L, and a/H are greater in *"mazei"*, whereas Arr/L, Ch/L, and b/H are smaller than in *"olsoni"*.

Revised systematic taxonomy of Mitchell's types

INTRODUCTION

John Mitchell. When John Mitchell, an educator from Newcastle, New South Wales, Australia, studied Australian fossil conchostracans in the second decade of the twentieth century (1925, 1927), very little was known about the fossil record of these forms. Beyond the publications of Jones and very few others, there was scant reference material on fossil conchostracans, and much of that not generally available.

Thus, Mitchell's taxonomy, which had several shortcomings, must be regarded as a pioneering effort which effectively contributed to the worldwide expansion of the infant field of conchostracan paleontology. Further, Australia, being a critical component of the Gondwanaland configuration, had and has special relevance for the systematics of Permian ribbed and nonribbed fossil conchostracans of Permian and Triassic age, respectively. Mitchell's species also serve as a reference for collections from Antarctica, Africa, southern Africa, India, and South America, as well as Russian, European, and North American fossil conchostracans.

Technical Shortcomings. (Generic and Specific Assignments.) The assignment of nonribbed conchostracan species from the Newcastle Coal Measures and the Australian Triassic to the genus *Estheria* (a preoccupied name), or ribbed types to the genus *Leaia* resulted in endless reassignments of Mitchell's genera and species by Raymond initially (1946), and later by Kobayashi (1965), Novojilov (1956a, 1958c, etc.), and others.

Photography. The problem with these reassignments is that the respective paleontologists relied solely on Mitchell's poor photographs as well as his descriptions and limited measurements (with occasionally additional measurements of the photographs to determine angles, as in the case of Novojilov, for example). Because Mitchell's photographs had been retouched with ink lines and white paint, there was inadvertent distortion, but these were accepted as the actual fossil data by Raymond, Kobayashi, and others. None of these investigators studied the original types and hence many reassignments of Mitchell's species and genera were based on inaccurate data derived from retouched photographs.

Measurements. At best, Mitchell's measurements were inadequate; therefore, more complex measurements are used throughout this revision. These are based on restudy of Mitchell's types, which are deposited in the Australian Museum, and measurements made on rubber imprints of the types. Where measurements were made on photographs of the types, these were new photographs, not Mitchell's published figures. These latter measurements, while few, were needed in some instances because of difficulties in measuring the types themselves.

Angles. The problem of measuring angles on ribbed conchostracans is important to consider. Measurements made on the rotating stage of a petrographic microscope are the most accurate way to measure alpha, beta, and other angles. Measurements by protractors of any kind will differ by as much as 8° from the petrographic microscope determinations.

Overlapping Valve Parameters. Many valve parameters in ribbed and nonribbed conchostracans overlap and therefore it is desirable to have an enlarged array of choices in measurement, instead of a restricted few and the more subjective evaluation commonly employed. In this memoir, including the revision of Mitchell's types, *decisions on specific and generic assignments were based chiefly on the extended array of compared measurements.* This procedure is recommended to all conchostracan workers. (See Fig. 18 for all measured parameters and angles.)

This writer's laboratory assistants measured all ribbed conchostracan angles from rubber imprints of Mitchell's types. Each repeated the measurement three times and averaged. Operator variation is expressed in the error indicated after each angle measurement. Parameters on Mitchell's nonribbed types were measured on new enlarged photographs, except where rubber imprints were available. Mitchell's types, including a few subsequently reported, and apparently misplaced, had been studied on three different visits to the Australian Museum, and with one or two exceptions, detailed observations recorded on valve characters.

Mitchell's Genera and Species. Many Mitchell species and genera are based on single specimens. This is a practice all workers, including this writer, have been compelled to use on those occasions when the distinctive morphology so indicated. However, the assignment of fragmentary specimens to a new species, as Mitchell did several times, is a practice to be avoided.

Having collected from the Newcastle Coal Measures (see Appendix) as well as from the Australian Triassic, this writer can appreciate the limited number of well-preserved specimens that are most suitable for taxonomic treatment. (Cf. section on Australian Conchostracans.)

The concept of the species as a population showing considerable variability was not very much appreciated, if at all, in Mitchell's day. In view of our present understanding, it is desirable to leave as "species undetermined" all instances of incomplete single specimens; i.e., incomplete in critical sectors of the valve. This procedure has been followed in the systematic sections of the general text, but not for Mitchell's species that are incomplete, because they have a long taxonomic history (synonomy), and more complete material has been reported by other workers.

Family LEAIIDAE Raymond, 1946

Genus *Leaia* Jones, 1862

Carapace outline variable from quadrate to semicircular; valves bearing two radial ribs (indistinct in some species) and a third rib may be present where dorsal margin thickens (cf. Pruvost, 1920).

Type: *Cypricardia leidyi* Lea, 1855

Subgenus *Leaia* Jones, 1862

Rectangular, subquadrate to subelliptical leaiids. (Tasch, in Tasch and Jones, 1979a, p. 10.)
Type: *Cypricardia leidyi* Lea, 1855; = *Leaia (Leaia) leidyi* (Lea).

Subgenus *Hemicycloleaia* Raymond, 1946

Bicarinate to tricarinate leaiids with circular to ovate configuration. (Tasch, 1979a, p. 12, cf. Pruvost, 1914.)

Type: *Hemicycloleaia laevis* Raymond, 1946. (= *Leaia (Hemicycloleaia) laevis*) Ray.)

Leaia (Hemicycloleaia) mitchelli Etheridge, Jr.
(Pl. 40, Figs. 1, 2)

Leaia mitchelli Etheridge, Jr., 1892, p. 307. (Two freehand drawings, lacking a scale.)
Leaia mitchelli Etheridge, Jr., Mitchell, 1925, p. 440, Pl. XLI, Figs. 1, 2; Pl. XLIII, Fig. 19.
Trileaia mitchelli (Etheridge, Jr.) Kobayashi, 1954, p. 145, 161.
Hemicycloleaia compta (Mitchell) Novojilov, 1956a, p. 32, Pl. IV, Figs. 5–7. (Mitchell's 1925, Plate XLIII, Fig. 19 only.)
Mimoleaia mitchelli (Etheridge, Jr.) Novojilov, 1956a, Table 61, p. 63, Pl. XI, Fig. 3. (Mitchell's 1925, Plate XLI, Figs. 1 and 2.)
Mimoleaia mitchelli (Etheridge, Jr.) Molin and Novojilov, 1965, Pl. VI, Figs. 1–3.
Diagnosis (emended): Subovate valves with two major carinae that cross the subterminal umbo, and sometimes the presence of a short, incipient carina close to the dorsal margin; anterior rib

slightly concave but straightens ventrad, and fades out at the second or third from the last growth band; posterior carina also curved dorsad on upper part of umbo and then continues straight, ending at the third from the last growth band; the short, incipient rib (see arrow, Pl. 41, Fig. 1), which is not present on all members of this species, originates near the dorsal margin slightly beyond the main posterior rib and deviates slightly from the dorsal margin.

A more broadly ovate form (alpha-87°, beta-30° ±2°) (Mitchell, 1925; Pl. 43, Fig. 19, Pl. 41, Fig. 2, herein) is interpreted to be a dimorph.
Measurements: (mm) L-5.7; H-4.0; Ch-4.2; Cr-1.9; Av-0.8; Arr-1.0; a-1.7; b-1.5; c-3.0; H/L-.702; Cr/L-.333; Av/L-.140; Arr/L-.175; Ch/L-.737; a/H-.425; b/H-.375; c/L-.526; d_1**0.5 ±0.1; d_2-3.8 ±0.2; alpha-81°; beta-32° ±2.0°; delta*-5°; gamma*-9°. (*delta can vary ±2.0°; gamma-1.0°–3.0°. **d_1, d_2 =distance between points where carina extended, touches, or would touch dorsal [d_1] or ventral [d_2] margins.)
Material: One left valve, Australian Museum F25487A (Mitchell, 1925, Pl. XLI, Fig. 1), holotype (by subsequent designation herein). F25426, dimorph (see above). (There are a large number of specimens assignable to this species in the Tasch collection and many also in the Australian Museum Mitchell Collection.)
Locality: At Charlestown between Newcastle and Lake Macquarie (Etheridge, Jr., 1892). Permian (Tartarian).
Discussion: The critical features of *Leaia (H.) mitchelli* are the angle measurements, the extent and curvature of the major carina, and the various ratios. It may be observed that previous workers did not study the actual types, but relied on Mitchell's poor published photographs. Mitchell's figured specimen (his Pl. XLIII, Fig. 19, and see new photograph, Pl. 41, Fig. 1 herein) is thought to be a dimorph of *"mitchelli"* and is not assignable to *Hemicycloleaia compta* (Mitchell) as Novojilov (1956a) concluded, because important metrical and morphological differences separate *"compta"* and *"mitchelli,"* despite the closeness of alpha and beta angles. These include: for *"mitchelli,"* a smaller H/L, Cr/L, Av/L, Ch/L, a/H, and Arr/L (.175 vs. 0.71 *"compta"*); b/H and c/L are greater. In addition, the eta angle of the curved posterior rib is 0° (vs. 4° *"compta"*).

Kobayashi (1954) erected *Trileaia* on Mitchell's *Leaia belmontensis* (see below, Pl. 42, Fig. 3). This assignment was based on the erroneous conclusion that there were three carinae which it does not have. (For full description see below.) That fact leaves *L. belmontensis* M. without a type, and it is placed in synonymy of *"mitchelli"* on the basis of a new set of measurements.

Because all *"mitchelli"* specimens at the Australian Museum, exclusive of the holotype and those from several Newcastle Coal Measures localities in this writer's collection, lack an incipient rib, we can treat this feature as a deviation from the species norm and, as such, does not merit taxonomic recognition.

Novojilov's assignment of *Mimoleaia* is also unacceptable because that genus is a synonym of *Leaia* (Tasch, 1958, p. 1099).

Besides assignment of *"Leaia mitchelli* (Etheridge, Jr.)," Permian, (Tartarian) to *Mimoleaia* Novojilov, 1956a (= *Leaia*),

Novojilov also placed "*Leaia ovata* M." under this genus (Novojilov, 1956a, Pl. 12, Figs. 3 and 8). Mitchell's Australian leaiids thus became relatives of older USSR species from the basin of the River Viatki (Permian, upper Kazanian), the Kouznetsk Basin (Lower Permian), and the Karagand Basin (Upper Carboniferous–Stephanian). This Russian-Australian relationship during the Permian–Carboniferous ties in with other data that relate USSR, China, Australia, and Antarctica (for greater detail, see section below on dispersal).

Leaia (Hemicycloleaia) mitchelli (?)
(Pl. 40, Fig. 5)

Leaia mitchelli Etheridge, Jr., 1892, p. 307.
Leaia quadriradiata Mitchell 1925, p. 441, Pl. XLI, Fig. 7.
Brachioleaia quadriradiata (Mitchell) Novojilov 1952, p. 4, Fig. 3.
Quadriradiata quadriradiata (Mitchell) 1925 Kobayashi 1954, p. 144, text-Fig. 30k.

Diagnosis (emended): Subovate valves with two major carinae that cross the subterminal umbo and do not meet; there are one or two apparent incipient or minor carinae—a rare occurrence. A questionable "anterior minor carina" is short and deviates slightly from the anterior margin; it begins near the origin of the major anterior carina. The posterior incipient or minor carina is somewhat larger and diverges slightly from the dorsal margin; it occurs above the major posterior carina which it approaches near to the point of origin of the latter. The major posterior carina (right valve) ends at the fifth growth band from the ventral margin, and the major anterior carina fades out at about the fourth growth band above the ventral margin. The dorsal margin thickens posteriorly.

Measurements: (mm) "Type;" L-6.5; H-4.0; Ch-4.7; Cr-2.3; Av-1.1; Arr-0.5; a-1.8; b-1.5; c-3.5; H/L-.615; Cr/L-.354; Av/L-.169; Arr/L-.077; Ch/L-.723; a/H-.450; b/H-.375; c/L-.538; alpha-80° ±2.0°; beta-31° ±1.0°; delta-5.0° ±1.0°; gamma-9.0°; d_1-0.2; d_2-2.3 ±0.9; growth band number, 20 or more.

Locality: Chert quarry near Belmont, New South Wales.
Material: Right valve overlying left valve (internal mold), F 25465A.

Discussion: There are two specimens labelled *Leaia quadriradiata* Mitchell in the Australian Museum Collection numbered F 25465: one, a single valve (not figured; Mitchell, 1925) labelled "holotype"; the other, with the right valve overlapping the left valve, labelled "type" (the figured specimen in Mitchell, 1925). Reference to the last-named specimen will be to F 25465A, and to the single valve, F 25465B.

Restudy of F 25465A indicated the presence of a minor carina that deviated from the dorsal margin and was considerably thicker and larger than the other reported anterior minor carina, which is very faint, shorter, and about as thick as a single hair. It may well be a mere crease (a not infrequent occurrence observed in fossil conchostracan valves). When Mitchell whitened both the posterior and anterior sector of the valve in his published photograph for emphasis (1925, Pl. XLI, Fig. 7), he unfortunately gave greater thickness and prominence to the anterior incipient carina than it actually displays. From Mitchell's photograph, Kobayashi and Novojilov independently erected new genera based on this exaggeration.

Tasch (1958) indicated that *Brachioleaia* Novojilov, 1952 was a valid genus that had priority over Kobayashi's (1954) *Quadriradiata*. However, the validity of *Brachioleaia* as a genus distinct from *Leaia* is now questionable. This is due to a detailed study of Mitchell's type discussed above, and restudy of published data of the only other species of *Brachioleaia; B. ruasiensis,* described by Novojilov in 1952 from Lutkevich's (1941) figured specimen of *Leaia kargalensis* Lutk. Permian (Kazanian), Tartar Republic.

As noted by Mitchell (1925, p. 441–442), *L. quadriradiata* has a "contour of the valve much the same as those of *L. mitchelli*" and differs chiefly in the presence of the anterior and posterior minor carinae. As shown by the measurements above, it is also metrically close to "*mitchelli.*" The presence of two additional embryonic carinae in "*quadriradiata*" cannot be proven to be either bona fide or regular features of the valve based on a single specimen. In this writer's Newcastle Coal Measures collection, *L. (H.) mitchelli* was abundant; however, no specimens attributable to *Leaia quadriradiata* were present (see Figs. 9 and 10). Evidence indicates that the so-called incipient carinae, if bona fide features, occur occasionally in "*mitchelli*" and, accordingly, do not merit taxonomic recognition.

Leaia (Hemicycloleaia) mitchelli Etheridge, Jr.
(Pl. 41, Fig. 1)

Leaia mitchelli Etheridge, Jr., 1893, p. 307–309.
Leaia intermediata Mitchell 1925, p. 440–441, Pl. 41, Figs. 3, 4.
Trileaia intermedia (sic) (Mitchell) Kobayashi 1954, p. 145, 158.
Hemicycloleaia intermediata (Mitchell) Novojilov, 1956a, p. 34, Pl. IV, Figs. 9, 10 (included *L. intermediata* Mitchell 1925, Pl. 41, Fig. 3 and *L. elliptica,* Mitchell 1925, Pl. 43, Fig. 10. See subsequent entry for *elliptica*).

Diagnosis (emended): Subovate valves with two long carinae and a brief, curved, beaded posterior structure that Mitchell interpreted to be a "third rib;" dorsal margin straight, anterior margin convex, and the posterior margin much less so; ventral margin slightly curved. Anterior of valve of greater height than the posterior. Gently arcuate anterior carina thickens ventrad and fades out in the middle of the last growth band; the straight posterior carina does not thicken as much, and thins ventrad; it also terminates midway on the last growth band. Growth bands, 11 or more (umbonal growth bands unclear due to erosion).

Measurements: (mm) L-6.0; H-4.1; Ch-4.4; Cr-2.0; Av-1.0; Arr-0.6; a-1.9; b-1.5; c-2.7; H/L-.683; Ch/L-.733; Cr/L-.333; Av/L-.166; Arr/L-.100; a/H-.463; b/H-.366; c/L-.450; alpha-79°; beta-29° ±2.0°; delta-7°; gamma-10° ±1.0°.

Material: F 25459A, holotype, left valve.

Locality: Chert quarry near Belmont, New South Wales Permian (Tartarian).

Discussion: Mitchell (1925, p. 441) observed that in outline and in most other respects, *L. intermediata* "resembles *L. mitchelli.*" A check of the newly measured holotypes of each confirms this determination. The distinctive difference that Mitchell noted was the presence of a third rib. Kobayashi (1954) assigned *L. intermediata*, solely on the basis of Mitchell's photographs, to a new genus, *Trileaia;* set up with Mitchell's *L. belmontensis* as the type. This last is bicarinate and was misidentified by Kobayashi as a tricarinate form.

Mitchell's photograph of the holotype (1925, Pl. 41, Fig. 3) was whitened in such a manner that, inadvertently, the feeble, brief structure, the so-called "third rib", and the upper portion of an adjacent valve crease appeared to be a unitary feature, creating the impression of a much thicker and more extensive structure than actually is present on the valve (cf. Pl. 41, Fig. 1, below—a new photograph of the holotype—to Mitchell's figure cited above). As the new photograph brings out, the actual structure, the incipient rib, can only be occasional in occurrence.

It therefore follows that the above observations and data require reassignment of *L. intermediata* to *L. mitchelli;* it also has an occasional incipient carina.

Leaia (Hemicycloleaia) mitchelli Etheridge, Jr.
(Formerly *Leaia belmontensis*)
(Pl. 41, Fig. 4)

Leaia mitchelli Etheridge, Jr., 1892, p. 307.
Leaia belmontensis Mitchell 1925, p. 445, Pl. 42, Fig. 15.
Trileaia belmontensis (Mitchell) Kobayashi 1954, p. 144, text-Fig. 30-1.
Hemicycloleaia belmontensis (Mitchell) Novojilov 1956a, p. 33, Pl. VI, Fig. 8.

Diagnosis (emended): Subovate, bicarinate valve with a very slightly thickened straight dorsal margin.

Both carinae originate dorsad on the umbo but do not meet; anterior carina gently curved; posterior carina straight; anterior carina terminates on the second from the last growth band; the posterior carina terminates on the third from the last.

Measurements: (mm) L-4.9; H-3.3; Ch-3.5; Cr-1.2; Av-0.6; Arr-0.3; a-1.4; b-1.2; c-2.7; H/L-.673; Ch/L-.714; Cr/L-.245; Av/L-.122; Arr/L-.061; a/H-.424; b/H-.364; c/L-.551; d_1-0.2 ±0.1; d_2-3.3 ±0.1; alpha-81° ±1.0°; beta-33° ±1.0°.

Material: F 25429, not the holotype; "right" valve, according to Mitchell, but actually a fragment of left valve overlain by a complete right valve of *"mitchelli"* creates the visual impression that the true dorsal margin of the bicarinate valve (= *"mitchelli"*) is a third rib of the "expanded" valve. The underlying valve is at a lower level on the slab than the *"mitchelli"* valve, above it.

Locality: Chert quarries, Belmont, New South Wales.

Discussion: Kobayashi (1954), working from Mitchell's poor and retouched photograph, erected a new genus, *Trileaia*, with *L.*

belmontensis as the holotype. Because this genus is based on Mitchell's original misinterpretation, and because none of Mitchell's leaiids that Kobayashi assigned to it are tricarinate, the genus is invalid.

Novojilov (1956a) correctly rejected the *Trileaia* assignment and placed *belmontensis* under the genus *Hemicycloleaia,* subsequently reassigned as a subgenus of *Leaia* by Tasch in Tasch and Jones, 1979a). However, the measurements and particularly the ratios cited above, not previously available, when compared to those of *L. (H.) mitchelli* clearly indicate that the bicarinate valve described above belongs to this last-named species. Incidentally, on the same bedding plane (F 25429 slab, Australian Museum) another valve of *L. (H.) mitchelli* occurs close to the specimen described above. Accordingly, all previous assignments of *"belmontensis"* become synonyms of *"mitchelli."*

Leaia (Hemicycloleaia) collinsi Mitchell 1925
(Pl. 40, Fig. 3)

Leaia collinsi Mitchell 1925, p. 446, Pl. 43, Fig. 16.
Leaia collinsi Mitchell, Kobayashi, 1954, p. 155.
Australoleaia collinsi (Mitchell) Novojilov, 1954b, p. 1241–1244; Novojilov, 1956a, p. 69, Pl. XIII, Fig. 4 (*Paris Centre d'Études et Documentation Paléontologique.* Translation 1581).

Diagnosis (emended): Subovate, bicarinate valves with straight dorsal margin that is prominent and thickened posteriorly; umbo terminal; both carinae cross umbo but do not meet, although they are quite close on the umbo; carinae curve gently anteriorly to the second or third growth band following a straight line thereafter to the ventral margin.

Measurements: (mm) L-6.3; H-4.3; Ch-5.0; Cr-1.2; Av-0.5; Arr-1.0; a-1.6; b-1.7; c-1.7; H/L-.683; Cr/L-.190; Av/L-.079; Arr/L-.158; Ch/L-.794; b/H-.385; c/L-.270; d_1-0.2; d_2-4.0; alpha-69°; beta-34° ±1.0; delta-3.0° ±1.0°; gamma-8.0°.

Material: Holotype F 25421, right valve.

Locality: Chert quarries near Belmont. Permian (Tartarian).

Discussion: *L. (H.) collinsi* was originally designated *L. mitchelli* (as indicated on Australian Museum labels accompanying the type), from which it differs markedly, as Mitchell realized when he assigned it subsequently to a new species *"collinsi."* The dorsal margin of *"collinsi"* is thickened posteriorly but does not curve. Mitchell did not give angles for the carinae (but these are provided by measurements above).

Leaia (Hemicycloleaia) pincombei Mitchell 1925
(Pl. 40, Fig. 4)

Leaia pincombei Mitchell 1925, p. 443, Pl. 42, Fig. 9.
Leaia pincombei Mitchell, Kobayashi, 1954, p. 164.
Hemicycloleaia pincombei (Mitchell) Novojilov, 1956a, p. 32, Pl. IV, Fig. 8.

Diagnosis (emended): Subovate, bicarinate valves with terminal umbo; dorsal margin slightly sinuate posteriorly. Both carinae cross antero-dorsal part of umbo but remain separated; anterior carina slightly concave and reaches ventral margin; posterior carina straight throughout and ends at about the fourth to fifth growth band from the ventral margin.

Measurements: (mm), L-6.6; H-5.0; Ch-5.3; Cr-2.0; Av-0.7; Arr-0.6; a-1.7; b-2.0; c-3.2; H/L-.758; Cr/-.303; Av/L-.106; Arr/-.091; Ch/L-.803; a/H-.340; b/h-.400; c/L-.485; d_1-0.5 ±0.1; d_2-4.1 ±0.1; alpha-70° ±2.0°; beta-30° ±1.0°; delta-4.0°; gamma-6.0° ±1.0; growth band number, 17 or more.

Material: Holotype, right valve, F 25469.

Locality: A short distance above the old Cardiff seam (i.e., Awaba Tuff Member) northwest shore of Belmont Bay. Permian (Tartarian).

Discussion: Mitchell's criteria for this species included the number of growth lines, which are a nonspecific factor; however, they denote sexual dimorphism in some living species, individual life span in fossils, and can indicate ecological data on factors affecting individual growth (food supply, shrinking water levels of the habitat). He also reported straightness of the anterior carina which actually is gently concave posteriorly. The shape of the valves was recorded as subscaphoidal (i.e., boat shaped) which exaggerates the slight, partial sinuate curvature of the dorsal margin and overlooks the overall subovate configuration of the valves. Mitchell's figure (1925, Pl. 42, Fig. 9) has inked-in continuations of the growth bands and darkening of others posteriorly.

Genus *Cycloleaia* Novojilov, 1952

Shape discoidal; valves with unequal radial ribs originating from blunt umbo, far apart distally.

Type: *Leaia discoidea* Mitchell, 1925.

Cycloleaia discoidea (Mitchell) Novojilov, 1952
(Pl. 40, Fig. 6)

Leaia discoidea Mitchell, 1925, p. 441, Pl. 41, Fig. 6; Pl. 43, Fig. 22.

Cycloleaia discoidea (Mitchell) Novojilov, 1952, p. 1361–1372, Fig. 2a; Novojilov, 1960, p. 249,

Cycloleaia discoidea (Mitchell) Novojilov, 1952. Tasch, 1958, p. 1100.

Discoleaia discoidalis sic (Mitchell) Kobayashi, 1954, p. 141, text-Fig. 30f.

Diagnosis (emended): Longitudinally ovate, bicarinate valves with short dorsal margin; postero-dorsal angle steep, antero-dorsal and ventral margins rounded. Both carinae cross a flat umbonal area; anterior carina gently curved, fades out at fourth growth band above the ventral margin, while the straight posterior carina continues to the next to last growth band. Growth bands, 14 or more.

Measurements: (mm) L-5.5; H-4.3; Ch-3.0; Cr-2.3; Av-1.2; Arr-1.6; a-2.2; b-2.0; c-3.0; H/L-.782; Cr/L-.418; Av/L-.218; Arr/L-.291; Ch/L-.545; a/H-.512; b/H-.465; c/L-.545' sigma*-49° ±2.0°; d_1-0.6 ±0.1; d_2-3.4 ± 0.3. (*Due to difficulty in accurately measuring angles in specimens like that of the holotype of *Cycloleaia discoidea,* an angle sigma was measured by two inside tangents to the two carinae projected to the point of intersection dorsad. Cross hairs are aligned along one tangent before measurement is taken, using a petrographic microscope stage.)

Material: Holotype F 25419 (Mitchell, 1925, Pl. 43, Fig. 22) by subsequent designation herein, left valve, exterior mold.

Locality: Chert quarries near Belmont, New South Wales.

Discussion: Once again, the confusion in taxonomic placement that can arise when the establishment of new genera relies solely on published figures is apparent. Novojilov's genus has priority over Kobayashi's. However, the Mitchell figure as depicted by Novojilov (1952, Fig. 2a) as a line drawing had an umbo raised above the dorsal margin and the two carinae meet on the umbo—a morphology which does not occur.

Subsequent Designation of Holotype: Kobayashi and Novojilov erected their genera on two different aspects of the species as figured by Mitchell. Kobayashi used Mitchell's Plate 43, Figure 22 because it was the more complete specimen. Novojilov used Plate 41, Figure 6—a specimen which has been lost and is no longer available for reference. Accordingly, a new type is selected here by subsequent designation from Mitchell's suite, because Novojilov's genus has priority, and because he did cite and illustrate (though incorrectly) Mitchell's original type.

Genus *Leaia* Jones, 1862
Subgenus *Leaia* Jones, 1862

Rectangular, subquadrate to subelliptical leaiids. (Tasch *in* Tasch and Jones, 1979a, p. 11.

Type: *Cypricardia leidyi* Lea, 1855 [= *Leaia (Leaia) leidyi* (Lea)].

Leaia (Leaia) oblonga Mitchell, 1925
(Pl. 40, Fig. 7)

Leaia oblonga Mitchell, 1925, p. 441, (*non* Pl. 41, Fig. 5); Pl. 43, Fig. 18.

Leaia oblongata (sic) (Mitchell) Kobayashi, 1954, p. 109, 163; not figured.

Siberioleaia oblonga (Mitchell) Novojilov, 1956a, p. 27, Fig. 18, Pl. III, Fig. 6 (ex Mitchell, 1925, Pl. 41, Fig. 5 only).

Diagnosis (emended): Suboblong, bicarinate valves divided into three unequal sectors (smallest, anterior; largest, median); valve length twice the height. Dorsal margin thickened posteriorly with slight concavity along thickened portion; posterior margin convex; ventral margin eroded; anterior margin rounded. Both ribs cross umbo but do not meet. Posterior rib straight, fades out at next to last growth band, considerably larger than anterior rib, slightly concave, and extends to about the tenth growth band (due to the eroded ventral margin, its terminus cannot be determined). Two large, raised loops surround and/or merge with the anterior carina (Pl. 40, Fig. 7). These may represent a pathologic condition or trace fossils.

Material: F 25423, holotype (Mitchell, 1925, Pl. 43, Fig. 18 only, and Pl. 40, Fig. 7 herein), right valve with eroded ventral margin. (This writer agrees with Novojilov that only F25423 can represent the species Mitchell described.)

Locality: Chert quarries near Belmont, New South Wales. Permian (Tartarian).

Measurements: (mm) Holotype (F 25423). L-7.8; H-?; Ch-6.0; Cr-2.4; Av-1.3; Arr-?; a-1.2; b-?; c-?; H/L-?; Ch/L-.153; Cr/L-.308; a/H-?; b/H-?; c/L-?; d_1-0.2; d_2-3.6 ±0.2; alpha-78° ±2.0°; beta-29° ±2.0°.

Discussion: Novojilov (1956a) reassigned Mitchell's species to *Siberioleaia,* a synonym of *Cycloleaia* (Tasch, 1958, p. 1101). It now seems that there was no acceptable reason to take this species out of the genus *Leaia.*

Leaia (Leaia) oblongoidea n. sp.
(Pl. 40, Fig. 8; Pl. 24, Fig. 1)

Leaia oblonga Mitchell, 1925, p. 441, Pl. 41, Fig. 3 (non Pl. 43, Fig. 18 = *L. (L.) oblonga* Mitchell).

Diagnosis: Elongate, rectangular, bicarinate valves with slightly curved dorsal margin. Anterior rib gently arcuate, terminates on fourth from last growth band; posterior rib straight, same termination ventrad as an anterior rib, but of greater overall length. Greatest height in anterior third.

Measurements: (mm). First listed number in each instance is the holotype; the second is the paratype: L-6.6, 5.8; H-3.9, 3.3; Ch-5.3, 4.6; Cr-1.5, 1.3; Av-0.7, 0.7; Arr-0.1, 0.5; a-1.5, 1.5; b-1.5, 1.3; c-3.2, 2.8; H/L-.591, .569; Cr/L-.227, .224; Av/L-.106, .017; Ch/L-.803, .793; a/H-.385, .455; b/H-.385, .394; c/L-.485, .483; d_1-0.3, 0.3; d_2-4.1 ±0.2, 3.8 ±0.2; alpha-76°, 76° ±2.0; beta-31°, 32° ±2.0.

Material: Holotype, right valve (Mitchell, 1927, Pl. 43, Fig. 18), Australian Museum F 25420; paratype, right valve, TC 10608 (Pl. 24, Fig. 1 herein).

Locality: Holotype, chert quarry near Belmont, New South Wales, Paratype, Tasch Station 4, bed 3. Newcastle Coal Measures. Permian (Tartarian). (See Figs. 10 and 11 for locality and distribution.)

Discussion: The new species is close to but distinguishable from Mitchell's *Leaia (Leaia) oblonga* (restr.) by its elongate rectangular configuration, slightly concave dorsal margin, and the large posterior sector of the valve.

Leaia (Hemicycloleaia) paraleidyi Mitchell
(Pl. 42, Fig. 8)

Leaia paraleidyi Mitchell, 1925, p. 442, Pl. 41, Fig. 8.
Leaia paraleidyi Mitchell, Kobayashi, 1954, p. 163.

Leaianella paraleidyi (Mitchell) Novojilov, 1956a, p. 58, Pl. X, Fig. 2; Molin and Novojilov, 1965, Pl. VII, Figs. 5, 6.

Diagnosis (emended): Subovate, bicarinate valves with straight dorsal margin that thickens posteriorly; umbo, subterminal, both carinae cross umbo but do not meet; anterior rib slightly arcuate, thickened ventrad, and terminates on next to last growth band; posterior rib, while thinner than anterior rib, also thickens ventrad, slightly curves on umbo but straight thereafter, and terminates on next to last growth band.

Measurements: (mm). L-6.0; H-4.1; Ch-4.8; Cr-2.0; Av-0.5; Arr-0.4; a-1.5; b-1.1; c-2.4; H/L-.683; Ch/L-.800; Cr/L-.333; Av/L-.083; Arr/L-.066; a/H-.366; b/H-.268; c/L-.585; d_1-8.0; d_2-3.9 ±0.2; alpha 73° ±2.0°; beta-31° ±1.0°; delta-5.0° ±1.0°; gamma-9.0° ±1.0°.

Material: F 25454, holotype, left valve (external mold).

Locality: Chert quarries near Belmont, Lake Macquarie (a short distance above the old Cardiff coal seam, i.e., the Awaba Tuff member of the Eleebana Formation).

Discussion: Mitchell's retouched photograph (1925, Pl. 41, Fig. 8) is misleading when compared to the new photograph of the holotype (Pl. 42, Fig. 8, below). Growth bands and dorsal margin were inked in, creating a variety of optical illusions; i.e., that the "radials originate in a point" (Mitchell, 1925, p. 442), which they do not, even though they are quite close to each other. The retouching also creates the impression that the configuration is subquadrate, which it is not, and that there was a "slight emargination" anteriorly, which does not occur. This illusory configuration is due to a piece having been broken and subsequent erosion.

Novojilov (1956a), relying on the above illusory feature, assigned Mitchell's species to the new genus *Leaianella,* that belongs under synonymy of *Leaia* (Tasch, 1958). Comparison of the dimensions of *L. (H.) mitchelli* shows that they are close and/or identical in the angles beta, delta, and gamma, and in most valve measurements. *L. (H.) mitchelli* had a slightly smaller valve and larger ratios, except for c/L. The unique metrical difference is the distinctive alpha angle: 73° ±2.0° in *paraleidyi,* compared to 81° ±2.0°. Also, the carinae in *paraleidyi* and *mitchelli* have different ventrad terminations. Thus, rather than being close to *Leaia leidyi* Lea, as Mitchell thought, *paraleidyi* is a near relative of *mitchelli* and, as remarked earlier, most probably derived from the *mitchelli* population. (Both species occurred together at several Tasch localities, with *mitchelli* abundant.)

As with other Mitchell hemicycloleaiids, a Russian–Australian connection is indicated, particularly during Permian time. Thus, two species from the Kouznetsk Basin (Lower Permian), *Leaianella* (= *Leaia*) *kaltanensis* Nov., 1956a, is a bicarinate, subovate form and *Leaianella* (= *Leaia*) *tenera* Nov., 1956a is also. Both of the indicated species are assignable to *Leaia (Hemicycloleaia)* and related to the geologically younger Permian (Tartarian) Australian *Leaianella* (= *Leaia*) *paraleidyi* (Mitchell) (Novojilov, 1956a, p. 58), on both the generic and subgeneric levels.

Leaia (Hemicycloleaia) elliptica Mitchell
(Pl. 41, Fig. 7)

Leaia elliptica Mitchell, 1925, Pl. 42, Fig. 10.
Leaia elliptica Mitchell, Kobayashi, 1954, p. 166.
Hemicycloleaia intermediata (Mitchell) Novojilov, 1956a, p. 34.

Diagnosis (emended): Elliptical bicarinate valves with comparatively wide separation of the two ribs on the umbo. The anterior rib is slightly curved and thickens ventrad; it terminates on the second from last growth band; the posterior rib is straight and terminates on the second or third from the last growth band. The median sector between the carinae is the largest; the posterior margin approaches the configuration of the inner curve of a parabola.

Measurements: (mm) L-5.3; H-3.2; Ch-3.7; Cr-2.0; Av-0.7; Arr-0.1; a-1.5; b-1.0; c-2.1; H/L-.566; Ch/L-.698; Cr/L-.377; Av/L-.132; Arr/L-.113; a/H-.500; b/H-.333; c/L-.392; d_1-0.7; d_2-3.2; alpha-83° ±0.3°; beta-24° ±2.0°.

Material: F25463, holotype, left valve, eroded and compressed. (For other specimens of this species, see text, Newcastle Coal Measures, Permian.)

Locality: Chert quarries near Belmont, Lake Macquarie, New South Wales (Awaba Tuff Member of the Eleebana Formation).

Discussion: Novojilov's assignment of this Mitchell species to *H. intermediata* is unacceptable because the elliptical configuration is not a distinguishing feature and parameters differ. It now appears that *L. (H.) intermediata* is attributable to *Leaia (H.) mitchelli.* (See above discussion on diagnosis on *"mitchelli".*)

Leaia (Hemicycloleaia) latissima Mitchell)
(Pl. 41, Fig. 2)

Leaia latissima Mitchell, 1925, Pl. 42, Fig. 11.
Leaia latissima Mitchell, Kobayashi, 1954, p. 31, 159.
Hemicycloleaia compta (Mitchell) Novojilov, 1956a, p. 33, Pl. IV, Figs. 5-7.

Diagnosis (emended): Ovate (except for short, straight dorsal margin) bicarinate valves; both carinae cross umbo but do not meet; anterior carina thickens ventrad and terminates on the third from the last growth band; the posterior carina curves very slightly on the uppermost sector of the umbo and then proceeds uncurved and without thickening to the third from the last growth band, where it terminates.

Measurements: (mm) L-8.0; H-6.5; Ch-5.8; Cr-3.2; Av-1.1; Arr-1.1; a-3.2; b-2.0; c-4.0; H/L-.725; Ch/L-.725; Cr/L-.400; Av/L-.138; Arr/L-.138; a/H-.492; b/H-.308; c/L.500; d_1-0.9 ±0.1; d_2-5.3 ±0.3; alpha-72°; beta-30°.

Material: F 25466, holotype, right valve (external mold).

Locality: A short distance above the coal seam outcropping near sea level on the north shore of Belmont Bay (this is the Awaba Tuff Member, Eleebana Formation). Permian (Tartarian).

Discussion: Novojilov (1956a) assigned *Leaia compta* Mitchell to *Hemicycloleaia latissima* Mitchell. These two species have several distinctive differences in measured parameters (*latissima* is cited first, *compta* second): Av/L-.138 vs. .214, respectively; Arr/L-.138 vs. 0.71; d_1-1.0 vs. 0.4; alpha angle 72° ±0.0 vs. 87° ±2.0. In addition, there are visible morphological differences: *compta* has a curved, posterior rib throughout its extent that is lacking in *latissima;* that rib also terminates higher above the ventral margin than does its equivalent in *latissima; compta* has a seimcircular configuration, an umbonal length greater than height and larger than that found in other ribbed specimens described by Mitchell. By contrast, *latissima* is ovate in configuration and has a longer dorsal margin than *compta*. These factors suffice to reject Novojilov's reassignment of *compta.*

Leaia (Hemicycloleaia) sulcata (Kobayashi)
(Pl. 41, Fig. 3)

Leaia sp. undet. Mitchell, 1925, p. 446, Pl. 43, Fig. 20.
Trileaia sulcata Kobayashi, 1954, p. 144–145.
Hemicycloleaia sulcata (Kobayshi) Novojilov, 1956a, p. 39, Pl. VI, Fig. 10.

Diagnosis (emended): Elongated, ovate, bicarinate valves with straight dorsal margin; anterior margin more steeply convex than posterior; ventral margin gently curved; anterior rib slightly arcuate, thickens ventrad, and fades out on the fourth growth band above the ventral margin; posterior rib much thinner than anterior rib, straight for most of its course even where it crosses the umbo; an exception is the slight pseudocurvature midway, due to valve flattening and compaction posterior to its occurrence; ends on the third from last growth band. These two ribs cross umbo but do not meet dorsad.

Measurements: (mm) L-7.3; H-4.5; Ch-5.0; Cr-2.6; Av-1.1; Arr-1.0; a-1.7; b-1.8; c-2.5; H/L-.616; Ch/L-.685; Cr/L-.356; Av/L-.151; Arr/L-.137; a/H-.378; b/H-.400; c/L-.343; d_1-0.6 ±1.0; d_2-4.5 ±0.4; alpha 77°; beta-31° ±2.0°.

Material: F 25461, holotype, left valve, margins eroded; anterior and ventral sectors of valve flattened posteriorly.

Locality: Chert quarry near Belmont. Permian (Tartarian).

Discussion: Kobayashi (1954, p. 145) misinterpreted a shallow fold near the dorsal margin in Mitchell's figured holotype as a third rib. Novojilov (1956a) apparently recognized this error also, because he placed Kobayashi's *Trileaia* under synonymy of *Hemicycloleaia.* This last-named genus is now regarded as a subgenus of *Leaia* (Tasch *in* Tasch and Jones, 1979c, p. 11).

The species described above is metrically distinct from all other Mitchell types, and also in its elongated ovate configuration. It differs from *L. (H.) etheridgei* (Kobayashi) in the greater width of the sector between its two ribs, which are also farther apart on the umbo (0.6 vs. 0.2 mm); it is also more elongate, as evidenced in the smaller H/L ratio for *L. (H.) sulcata* (Mitchell) (.616 vs. .652); in addition, *"sulcata"* has a smaller alpha angle (77° vs. 86° ±3.0°) and a smaller beta angle (31° ±2.0° vs. 44° ±3.0°).

Leaia (Hemicycloleaia) compta (Mitchell)
(Pl. 41, Fig. 5)

Leaia compta Mitchell, 1925, p. 444, Pl. 42, Fig. 14.
Leaia compta Mitchell, Kobayashi, 1954, p. 108, 115.
Hemicycloleaia compta (Mitchell) Novojilov, 1956a, p. 35, Pl. VI, Figs. 5–7.

Diagnosis (emended): Semicircular, bicarinate valves with a straight, slightly thickened, dorsal margin; a rather prominent ovate umbo. Both ribs curve anteriorly across the umbo, the anterior rib more sharply than the posterior; the two ribs do not meet at point of origin; the anterior rib thickens ventrad and terminates very close to the last growth band; the much thinner posterior rib is gently arched along its entire extent and terminates at about the eighth growth band.

Measurements: (mm) L-7.0; H-5.0; Ch-5.5; Cr-3.0; Av-1.5; Arr-0.5; a-2.2; b-1.7; c-3.3; H/L-.714; Ch/L-.786; Cr/L-.429; Av/L-.214; Arr/L-0.71; a/H-.440; b/H-.340; c/L-.471; d_1-0.5; d_2-4.2 ±0.3; number of growth bands: 12–14?; alpha-87° ±2.0°; beta-34° ±2.0°; sigma-54°; eta-4°.

Material: F25424, holotype, left valve, gauged and eroded on the ventral-anterior sector, flattened and compacted on the ventral-posterior sector.

Locality: Approximately 122 m below the Belmont Chert beds, just above the Dirty Coal seam, Merewether Beach. Permian (Tartarian).

Discussion: As noted under the discussion of *Leaia (Hemicycloleaia) mitchelli* and *L. (H.) latissima*, both of which Novojilov (1956a) assigned to *Hemicycloleaia compta*, these are not co-specific. Mitchell (1925) noted the prominent umbo of *L. (H.) compta* M. (which he thought to be tumid, but in reality is not), referred to a straight posterior rib (which, in reality, is bent like an archer's bow), and emphasized the convexity of the valve. He further described the ribs as uniform throughout (actually, the anterior rib thickens ventrad).

The outstanding characteristics, aside from configuration, are the prominent umbo and curved, posterior rib, which yields an angle eta of 4°, in contrast to an eta angle of 0° in *"mitchelli."* The combined characteristics noted suffice to distinguish *compta* from *mitchelli*.

Leaia gondwanella Tasch [= *Leaia (Hemicycloleaia) gondwanella* T., 1965] from the Permian of the Antarctic Ohio Range was found to resemble two bicarinate species, *L. pruvosti* Reed from the Brazilian Permian Rio do Rasto Formation, and *Leaia (Hemicycloleaia) compta* Mitchell from the Australian Permian Newcastle Coal Measures (Tasch *in* Doumani and Tasch, 1965). The Antarctic and Brazilian species both overlap in H/L at the higher part of the range (.714 *"compta"* vs. .590–.790), or mid-range d_2 (4.2 ±0.3 *"compta"* vs. 3.2–6.0). For about 60% of the "gondwanella" sample, alpha angle is greater (87° ±2° *"compta"* vs. 90° "gondwanella"); however, the Antarctic species has a broader range, (75°–103°). Thus, overlap with the indicated

Brazilian and Australian species occurs only in the lower part of the range.

Mitchell's *"compta"* has a smaller beta angle, larger sigma angle (54° vs. 38°–48°), and larger eta angle (4° vs. 0°). In addition, *"compta"* has a thickened dorsal margin absent in *"gondwanella,"* and a different configuration (semicircular vs. subcvate *"gondwanella"*).

Leaia (Hemicycloleaia) etheridgei (Kobayashi)
Pl. 41, Fig. 6)

Leaia sp. undet. Mitchell, 1925, p. 446, Pl. 43, Fig. 21.
Trileaia etheridgei Kobayashi, 1954, p. 144, 157.
Hemicycloleaia etheridgei (Kobayashi) Novojilov, 1956a, p. 39, Pl. VI, Fig. 9.

Diagnosis (emended): Ovate, bicarinate valves except for straight, dorsal margin. Both carinae orginate approximately midway on umbo and do not meet but are very close together; both thicken posteriorly; anterior rib terminates on third from last growth band; posterior rib terminates on the fourth from last; both carinae are straight on umbo but curve posteriorly, the anterior carina more than the posterior. The sector beyond the posterior rib to the dorsal margin is larger than usual. Growth bands 19 or more.

Measurements: (mm) L-4.6; H-3.0; Ch-3.2; Cr-1.3; Av-0.8; Arr-0.8; a-1.2; b-1.0; c-2.2; H/L-.652; Ch/L-.696; Cr/L-.238; Av/L-.174; Arr/L-.174; a/H-.400; b/H-.333; c/L-.478; d_1-0.2; d_2-2.5; alpha-86° ±3.0°; beta-44° ±3.0°.

Material: F 25427, holotype, right valve, external mold, bearing two rib grooves and a raised area (Mitchell's "third rib").

Locality: Chert quarries near Belmont, New South Wales. Upper Permian (Tartarian).

Discussion: Mitchell, and Kobayashi (1954), using Mitchell's description and poor photograph, interpreted an impression made by a superimposed apparent soft part of another leaiid as a third rib. However, as Novojilov (1956a) apparently realized, it was impossible to have grooves representing imprints of the two ribs as well as a raised "rib" on the same valve. The raised structure called "rib" is segmented and tapers at the lower end. This suggests it may represent a detached piece of the antenna of this specimen or of another leaiid impressed on the soft sediment. At any rate, the species is *not* tricarinate.

The distinctive characteristics of this species are the closeness of the two ribs, as well as their origination on the midportion of the umbo, and the larger than usual posterior valve section beyond the posterior rib.

Leaia (Hemicycloleaia) ovata Mitchell
(Pl. 41, Fig. 8)

Leaia ovata Mitchell, 1925, p. 443, p. 42, Fig. 12.
Leaia ovata Mitchell, Kobayashi, 1954, p. 163.
Mimoleaia ovata (Mitchell) Novojilov, 1956a, p. 43, Pl. XI, Fig. 8.

Diagnosis (emended): Subovate, bicarinate valves; posterior rib slightly curved across umbo, otherwise gently convex to where it terminates ventrad on next to last growth band. Anterior rib maintains gentle curvature from over the umbo to second from last growth band. The two ribs do not meet on the umbo. Margins: dorsal straight; anterior rounded; posterior gently convex; and ventral moderately convex.

Measurements: (mm) L-3.0; H-4.2; Ch-4.7; Cr-1.8; Av-0.5; Arr-1.3; a-2.1; b-1.7; c-2.9; H/L-.600; Ch/L-.671; Cr/L-.257; Av/L-.166; Arr/L-.186; a/H-.500; b/H-.405; c/L-.414; d_1-0.4; d_2-4.0; a-88°; b-30°.

Material: F 25456, holotype, mold of left valve (not "paratype" as labelled on Australian Museum Card, because Mitchell described and figured only a single specimen). A new photograph is figured here (Pl. 41, Fig. 8). Mitchell's figure of *ovata* was retouched with ink lines that distorted the umbo and masked the crinkled posterior sector. The original valve was eroded along the postero-ventral and postero-dorsal margins.

Locality: Near Belmont, New South Wales, not much above the coal seam outcropping nearly at sea level on the north shore of Belmont Bay (i.e., Awaba Tuff Member, Eleebana Formation). Upper Permian (Tartarian).

Discussion: Kobayashi (1954) accepted Mitchell's designation of this species but Novojilov (1956a) reassigned it to a genus *Mimoleaia*—subsequently reassigned as a subgenus of *Leaia* (Tasch, 1958, p. 1099). The difficulties that arise in measuring the alpha and beta angles on Mitchell's photos are indicated in Novojilov (1956a, p. 74); failure to note the curvature of ribs and to measure from dorsal margin to tangents to such curved ribs led to angle values that do not correspond to those of Mitchell's type. No distinguishing angularity exists such as is required in Novojilov's description of *Mimoleaia*. The valve margins are eroded.

Bond (1955, p. 95) compared one of his two leaiids from the Middle Madumabisa Shale (Lower Beaufort, Upper Permian, Zimbabwe), *Hemicycloleaia sessami* Bond [= *Leaia (Hemicycloleaia)*] and *Leaia ovata* Mitchell [= *Leaia (Hemicycloleaia)*]. He noted a "resemblance in size and shape" but "a marked difference in density and fineness of lirae" (growth lines); i.e., fewer and stronger in *"ovata"* than in *"sessami."* Fortunately, detailed, comparative measurements not previously available now permit a metrical comparison of the Zimbabwe and Australian species.

Bond's *"sessami"*, when compared to Mitchell's *"ovata"* is close in H/L, c/L, Arr/L, alpha angle (80°–86° vs. 88°), and beta angle that overlaps (30°–39° vs. 30° *"ovata"*), but differs in having a greater Cr/L (.349 vs. .257 *"ovata"*), d_1 (1.06–1.07 vs. 0.40 *"ovata"*), d_2 (6.06–6.15 vs. 4.0 *"ovata"*), Av/L, and a smaller Ch/L (.566 vs. .671 *"ovata"*), a/H (.305–.325 vs., .520 *"ovata"*), and b/H (.253 vs. .405 *"ovata"*). The anterior rib in *"sessami"* is generally straight whereas in *"ovata"* it maintains a gentle curvature from over the umbo to its termination; the posterior rib in *"sessami"* is also straight, but in *"ovata"* it is slightly curved across the umbo and gently convex to its terminus.

Family CYZICIDAE Stebbing, 1910

Emended: Valve outline variable from ovate and elliptical to subrectangular, with numerous growth lines. Ornamentation distinct and variable, ranging from polygonal pattern on growth bands to longitudinal striae that may anastamose.

Genus *Cyzicus* Audouin, 1837
Subgenus *Lioestheria (sensu stricto)*, Depéret and Mazeran 1912, emended

Generally ovate to subovate valves with growth bands on the umbo and lacking a node, broad based umbonal spine, a curved spine, or a "radial element," or any kind of recurved posterior margin. Ornamentation punctate or with vertical markings (hachure marks) on growth bands.

Type: *Estheria (Lioestheria) lallyensis* Depéret and Mazeran, 1912 (Fig. 1).

Cyzicus (Lioestheria) fictacoghlani n. sp.
[Pl. 25, Fig. 5 (See text for photograph)]

Estheria coghlani Cox, 1881, p. 276 *(nomen nudem)*.
Estheria coghlani Cox, Mitchell, 1927, p. 10, Pl. 2, Fig. 3 *(non 5 and 6)*.

Diagnosis: Subovate valve with subterminal umbo bearing growth bands; a line through the middle of the umbo to the ventral margin would denote the highest portion of the valve; valve height decreases anteriorly. Margins: anterior rounded; posterior convex; ventral gently curved. Ornamentation: hachure lines on growth bands; first six growth bands deeply incised and widely spaced; total number of growth bands, 15.

Measurements: (mm) (MMF 3113) L-3.2; H-2.5; Ch-2.0; Cr-1.4; Av-0.6; Arr-0.2; a-1.0; b-1.1; c-1.5; H/L-.781; Ch/L-.625; Cr/L-.438; Av/L-.188; Arr/L-.063; a/H-.400; b/H-.440; c/L-.469.

Material: Holotype, Geological Survey of New South Wales (Mining Museum), MMF 3113 (= Mitchell, 1927, Pl. 2, Fig. 3 only). Many badly eroded and incomplete valves of this species labelled "*E. coghlani*" and "type" are on deposit at the Australia Museum. One of these slabs, F 35720, Dent's Bore, bears poorly preserved specimens, one of which may be related to MMF 3113. (For *Cornia coghlani* Etheridge, Jr., see Pl. 43; Fig. 1 herein.)

Locality: Cremorne Bore No. 2, depth ~674 m. Narrabeen Group (Early Triassic).

Discussion: The three specimens assigned by Mitchell (1927, Pl. 2, Figs. 3, 4, 5) to *Estheria coghlani* have now been reassigned as follows: Figure 4 (Mitchell, 1927) assigned to *Cornia coghlani* (Etheridge, Jr.) on the basis of Etheridge's (1888, Figs. 1–4) recognition of the presence of an umbonal node and the actual occurrence of such a node on forms studied in Mitchell's Collection at the Australian Museum and the Mining Museum; Figure 5 has been reassigned twice, first by Novojilov (1958c) to *Belgolim-*

nadiopsis australensis Nov. and by the present author to *Cyzicus (Lioestheria) australensis* (Novojilov) (Tasch *in* Tasch and Jones, 1979a, p. 16); Figure 3, redescribed above, is assigned to the new species *C. (L.) fictacoghlani;* i.e., "false *coghlani*" after study of a mold of MMF 3113.

Cyzicus (Euestheria) obliqua (Mitchell)
(Pl. 43, Fig. 2)

Estheria obliqua Mitchell, 1927, p. 109, Pl. 4, Fig. 1.
Pseudestheria obliqua (Mitchell) Raymond, 1946, p. 253.
Euestheria obliqua (Mitchell) Kobayashi, 1954, p. 109, 163.
Pseudestheria obliqua (Mitchell) 1926 (sic.) Novojilov, 1958, p. 33, Fig. 30; Pl. 3, Fig. 33 (p. 76).

Diagnosis (emended): Transversely subovate valves with straight dorsal margain, subterminal umbo; greatest height of valve along median line; anterior height less than posterior. Margins: Anterior sharply convex; posterior rounded and appears truncated; ventral gently curved. Remnant of polygonal ornamentation on growth bands.

Measurements: (mm) (Measurements taken directly from the holotype were L-6.0, H-3.5. Other measurements taken from an enlarged new photograph.) Ch-3.6; Cr-4.1; Av-1.3; Arr-1.4; a-1.3; b-1.5; c-3.6; H/L-.583; Ch/L-.600; Cr/L-.683; Av/L-.217; Arr/L-.233; a/H-.217; b/H-.428; c/L-.600; growth bands approximately 15. (Because of delicate preservation, a rubber replica was not available for measurements.)

Material: F 25481, holotype, left valve that overlies embedded right valve; eroded along antero-ventral and postero-ventral margins; posterior quarter of valve compressed; compression of dorsal margin and umbonal area creates the erroneous impression of a short dorsal margin and a prominent umbo.

Locality: Merewether Beach, one or more metres below the Dirty Coal Seam, New South Wales (= Tasch Station 14). Upper Permian (Tartarian).

Discussion: Raymond's (1946) assignment to *Pseudestheria* (= *Cyzicus*), reiterated by Novojilov (1958c) is unacceptable because Mitchell's specimen is not only a cyziciid, but it has euestheriid polygonal ornamentation. Kobayashi's assignment to *Euestheria* on the basis of sculpture unknown (without having seen or studied the type) was a lucky guess. Accordingly, Mitchell's specimen is assigned to *Cyzicus (Euestheria) obliqua* (Mitchell) as described above.

Novojilov (1958c, p. 33, Fig. 30, Pl. 2, Fig. 33) recorded a Russian Permian (Tartarian) specimen of *Pseudestheria obliqua* (Mitchell) from the subsurface south of Gulf of Katanga, Cape Ilia, USSR (Arctic). This was not the only Australian species noted by Novojilov [cf. *C. (E.) ipsvicensis* (Mitchell) and *C. (Eu.) novocastrensis* (Mitchell) described below].

Cyzicus (Euestheria) ipsviciensis (Mitchell)
Pl. 42, Fig. 6

Estheria mangaliensis Etheridge, Jr. (nec Jones 1862) *in* Jack and Etheridge, Jr., 1892, p. 387 (no figure).

Estheria ipsviciensis Mitchell, 1927, p. 107, Pl. 3, Figs. 1–4.
Pseudestheria ipsviciensis (Mitchell) Raymond, 1946, p. 255.
Euestheria ipsviciensis (Mitchell) Kobayashi, 1945, p. 110, 158.
Pseudestheria ipsviciensis (Mitchell) Novojilov and Kapel'ka, 1968, p. 126, Pl. B, Fig. 3.

Diagnosis (emended): Laterally subovate valves with subterminal umbo and straight dorsal margin which meets the rounded posterior margin obliquely; anterior margin rounded to the umbo; ventral margin slightly arcuate; polygonal ornamentation present on some sectors of the valve on growth bands (as seen on a rubber replica).

Measurements: (mm) (Lectotype) L-5.2; H-3.5; Ch-3.2; Cr-2.0; Av-1.0; Arr-1.4; a-1.5; b-1.2; c-2.6; H/L-.673; Ch/L-.615; Cr/L-.385; Av/L-.192; Arr/L-.269; a/H-.429; b/H-.343; c/L-.500; growth bands 20 or more.

Material: F 25488, Australian Museum, Sydney, lectotype, right valve, umbo compressed. (Left and right valves, Mitchell, 1927, Pl. 3, Figs. 2–3; of these, the right valve, Fig 3 [= "type," Geological Museum of Brisbane] and left valve, Fig. 4 [= No. 27, Geological Museum of Brisbane].)

Locality: Denmark Hill, Ipswich, Queensland. Upper Triassic.

Discussion: Etheridge, Jr. (1892) described but did not figure this species. The type (see data under material above) referred to by Mitchell (1927, p. 107, 108) is one of several specimens of the syntype of this species. In the studies by Raymond, Kobayashi, and Novojilov and Kapel'ka, there was no concern with the "type" for this species; the first and last-named authors referred to Mitchell, 1927, Figures 1–4. Accordingly, specimen F 25488 is here designated the lectotype of the species described above, because it was the first mentioned and figured in Mitchell, 1927.

Furthermore, Raymond's genus *Pseudestheria* (long since assigned to *Cyzicus*) was established to embrace estheriids of uncertain disposition which lacked sculpture except for punctation. However, a rubber replica of the lectotype prepared by the Australian Museum from F 25488 shows clear polygonal sculpture preserved on part of the valve. Mitchell made no mention of this feature. Other investigators also did not examine F 25488.

Novojilov and Kapel'ka (1968) attributed a specimen found in a lower Middle Jurassic core from East Siberia (USSR), to a species originally described from the Upper Triassic Wianamatta Series, Denmark Hill, Queensland, Australia (Mitchell, 1927), *Cyzicus (Euestheria) ipsvicensis* (Mitchell). Despite the Lower Jurassic time gap between the Australian and Russian occurrences, other data set forth in this memoir and in this writer's publications over many years (see References cited) have pointed to a Russian-Australian and Antarctic connection for fossil insects as well as conchostracans (Tasch, 1970b, 1981a). (See below for other of Mitchell's species, that were recorded by Soviet investigators, from the Russian Arctic.)

Genus Cyzicus Audouin, 1837
(Subgenus Euestheria Depéret and Mazeran, 1912, emended.)

Carapace generally ovate (or subovate), but with wide variation in shape and size. Characterized by a pattern of minute polygons

(granules) on growth bands, not arranged vertically or parallel.
Type: *Posidonia minuta von Zieten, 1833.*

Cyzicus (Euestheria) novocastrensis (Mitchell)
(Pl. 42, Fig. 5)

Estheria novocastrensis Mitchell, 1927, p. 109, Pl. 3, Figs. 5, 6.
Pseudestheria novacastrensis (sic.) (Mitchell) Raymond, 1946, p. 252.
Pseudestheria novacastrensis (sic.) (Mitchell), Novojilov (1958c, p. 33). See Novojilov, *in* Molin and Novojilov, 1965, p. 24, Fig. 28.
Euestheria (?) *novocastrensis* (Mitchell) Kobayashi, 1954, p. 109, 162.

Diagnosis (emended): Robust broadly subovate valves with straight dorsal margin and umbo inset from the anterior margin about one-fourth the length of the valve; anterior margin steeply inclined from umbo, grading into a low amplitude convexity; posterior margin of greater convexity than anterior. Remnants of polygonal ornamentation visible on growth bands near the posterior and ventral margins.
Measurements: (mm) Holotype (F 25469). L-7.1; H-5.4; Ch-5.0; Cr-2.8; Av-1.1; Arr-1.0; a-1.9; b-1.7; c-2.8; H/L-.761; Ch/L-.704; Cr/L-.394; Av/L-.155; Arr/L-.141; a/H-.352; b/H-.315; c/L-.394.
Material: F 25469, holotype (by subsequent designation here, based on first figured specimen), right valve, crushed along antero-ventral sector and crinkled on face of the valve; F 25470, paratype, mold of right valve. These new type indications are the reverse of the information on the type-specimen labels in the Australian Museum.
Locality: Approximately 1m below the Dirty Coal seam (Newcastle Coal Measures), Merewether Beach; a short distance southwest of Newcastle sewerage, between high and low tide. Associated with *Glossopteris* flora. Upper Permian (Tartarian). (See Fig. 9, Tasch Station 14.)
Discussion: *Pseudestheria* is a synonym of *Cyzicus* and *Lioestheria,* a subgenus. Raymond (1946) noted that *Pseudestheria* and *Euestheria* were very similar except for punctate ornamentation on *Pseudestheria,* as contrasted with polygonal ornamentation in *Euestheria.* Because polygonal ornamentation occurs in the species described above, it is clearly assignable to *Cyzicus (Euestheria)* (Tasch, 1969).

The distinctive character of this species is the steep incline from the umbo anteriorly, the robust subovate configuration, and the large size.

Novojilov (1958, p. 30–34, Pl. 3, Fig. 34) recorded a specimen attributed to Mitchell's *"Pseudestheria novacastrensis* (M.) [= *Cyzicus (Euestheria)*] that came from littoral deposits of the Sea of Laptev, Upper Permian, Tartarian, as well as the Lower Toungouska deposits. Along with *Cyzicus (Euestheria) obliqua* (M.) of the same age, there is also a Lower Jurassic form discussed above found in the Siberian Arctic (USSR) and attributed to Mitchell's species *Cyzicus (Eu.) ipsvicensis* (M.) described above.

Cyzicus (Euestheria) belmontensis (Mitchell)
(Pl. 43, Fig. 3)

Estheria belmontensis Mitchell, 1927, p. 110, p. 4, Fig. 5.
Pseudestheria belmontensis (Mitchell) Raymond, 1946, p. 252.
Lioestheria (?) *belmontensis* (Mitchell) Kobayashi, 1954, p. 108, 109, 154.
Glyptoasmussia belmontensis (Mitchell) Novojilov, (1958c, p. 38). See Molin and Novojilov, 1965, p. 30, Fig. 36.

Diagnosis (emended): Robust, ovate valves with straight dorsal margin, subterminal umbo; height of anterior sector behind umbo is greatest; valves decrease in height posteriorly. Margins: anterior sharply convex; posterior rounded; ventral gently convex; polygonal ornamentation on growth bands. Growth bands, 17 or more.
Measurements: (mm) L-5.0; H-4.0; Ch-3.5; Cr-1.3; Av-0.6; Arr-1.0; a-1.3; b-1.4; c-3.0; H/L-.794; Ch/L-.686; Cr/L-.255; Av/L-.118; Arr/L-.196; a/H-.325; b/H-.350; c/L-.588.
Material: F 25473, holotype, right valve, partly covered dorsal margin, slightly eroded postero-dorsal sector and peripherally on ventral and anterior margin.
Locality: Chert quarries near Belmont, Kahibah Parish, Northumberland County, New South Wales. Upper Permian (Tartarian).
Discussion: Mitchell (1927, p. 110) suggested "strong resemblance" to *Estheria ipsviciensis* [= *Cyzicus (Euestheria)*], but *ipsviciensis* is more elongate and less ovate in configuration, as can be seen in the comparative H/L-.794 *(belmontensis)* and .673 *(ipsviciensis).*

Raymond's (1946) *Pseudestheria* and Kobayashi's (1954) *Lioestheria?* assignments are equally unacceptable because relict polygonal ornamentation is preserved on growth bands of the type of *"belmontensis."*

Novojilov's assignment to *Glyptoasmussia* (1958c) was based on the impression gained from Mitchell's figure of *belmontensis,* that it had an asmussid, straight dorsal margin in front of and behind the naupliid portion of the valve. (cf. Novojilov, 1958c, Pl. 3, Fig. 37, with Pl. 43, Fig. 3, this paper.) In actuality, it curves sharply in front of the umbo.

?Cyzicus (Lioestheria) sp.
(Pl. 43, Fig. 4)

Estheria? bellambiensis Mitchell 1927, p. 111–112, Pl. 4, Figs. 7, 8.
Lioestheria bellambiensis (Mitchell) Kobayashi, 1954, p. 109, 154.
Concherisma bellambiensis (Mitchell) Novojilov, 1958c, p. 46, Pl. 4, Figs. 51, 52. Molin and Novojilov, 1965, p. 38, Fig. 47.

Diagnosis (emended): Usually elongate, ovate valves with subterminal umbo bearing hachure-like marks between growth lines (seen on lectotype F 25485). Margins: uncertain, variable sets of growth bands; i.e., wide to tight spacing.

Measurements: (mm) Because the lectotype and Mitchell's published figures of this species are both covered and eroded, it is not possible to provide a complete set of the usual measurements. What follows are the only useable figures: Mitchell—L-10; H-6; H/L-.60 (F 25479); Tasch—L-10; H-5+; H/L-.50 (F 25479); Tasch—L-9.5; H-?; H/L-? (F 25485).

Material: F 25485, the reverse side of F 25479 (= Mitchell's 1927 Pl. 4, Fig. 7), is figured herein (Pl. 43, Fig. 4). The reason for figuring this specimen is due to the fact that Mitchell's Pl. 4, Fig. 8 (= F 26384) is retouched and undecipherable. Furthermore, the original is missing. A search of the drawers carrying Mitchell's collections at the Australian Museum failed to produce this specimen.

Locality: North side of railroad between South Bulli Colliery and staiths of Bellambi roadstead (Mitchell, 1927, p. 11).

Discussion: The exact classification of this species is uncertain at best, as Mitchell also observed. The hachure-like markings suggest a lioestheriid, as Kobayashi (1954, p. 109) speculated without seeing the type material, and is confirmed here. The configuration of the shell and nature of the umbo are also unclear due to cover and erosion of portions of the preserved valves. Novojilov's reassignment to *Concherisma* is also unwarranted because he used published photographs which, as noted, were inked over originally; this led to a subjective and misleading interpretation of the actual type material. Mitchell's *"bellambiensis"* is reviewed here to clarify the record.

Cyzicus (Euestheria) trigonellaris? (Mitchell)
(Pl. 42, Fig. 2)

Estheria trigonellaris Mitchell 1927, p. 109, Pl. 4, Fig. 6.
Pseudestheria trigonellaris (Mitchell) Raymond, 1946, p. 252.
Euestheria (?) *trigonellaris* (Mitchell) Kobayshi, 1954, p. 108, 109, 159.

Diagnosis (emended): Eccentrically subovate valves; umbo inset one-third valve length; posterior third of valve of greater height than anterior. Margins: dorsal partially covered; anterior eroded, sharply convex; posterior gently convex; ventral gently concave upward and foreshortened anteriorly.

Measurements: (mm) L-6.2; H-4.3; Ch-4.7; Cr-1.9; Av-1.4; Arr-0.7; a-1.7; b-1.7; c-3.0; H/L-.694; Ch/L-.759; Cr/L-.306; Av/L-.226; Arr/L-.113; a/H-.395; b/H-.395; c/L-.484; growth band number - 14.

Material: F 25484, holotype, external mold of right valve.

Locality: Approximately 1 m below the Dirty Coal seam (Newcastle Coal Measures), Merewether Beach, New South Wales. Upper Permian (Tartarian).

Discussion: Valve outline was erroneously interpreted as "subtriangular" (Mitchell, 1927, p. 109); actually, the holotype is clearly eroded on the ventral and anterior margins and the dorsal margin is partially covered. Restored to an original configuration, the valve is clearly subovate with an eccentric tendency posteriorly.

Raymond (1946) placed this form in *Pseudestheria* (= *Cyzicus*) due to the lack of reticulate ornamentation on growth bands. Kobayashi (1954, p. 109) tentatively grouped *trigonellaris* with several other Mitchell species (*novocastrensis, obliqua,* and *lata*) under *Euestheria* (?) because the sculpture was unknown in all of these. However, the holotype of *novocastrensis* actually reveals polygonal ornamentation. At any rate, assignment of this species is at best tentative, due to the poor preservation.

Because *trigonellaris* occurs on the same horizon and at the same locality as *novocastrensis,* a comparison of the respective species is indicated. The valves of each show distinctive configurations (eccentrically subovate vs. broadly ovate; cf. Pl. 42, Fig. 2 with Fig. 5).

Metrically, *"trigonellaris"* differs from *"novocastrensis"* in having a smaller H/L (.694 vs. .761); Cr/L (.306 vs. .394), larger Ch/L (.759 vs. .704); Av/L, b/H, and c/L (.484 vs. .394). Configuration and umbonal position (inset one-third vs. inset one-fourth) alone could suffice to distinguish these two species, if preservation of *"trigonellaris"* were better.

Cyzicus (Euestheria?) lata (Mitchell)
(Pl. 42, Fig. 3)

Estheria lata Mitchell, 1927, p. 110, Pl. 3, Figs. 8, 9.
Pseudestheria lata (Mitchell) Raymond, 1946, p. 252.
Euestheria? lata (Mitchell) Kobayashi, 1954, p. 108, 109, 159.
Cyclestheria mitchelliana Novojilov, 1957, in Novojilov, 1958c, p. 35. Novojilov, 1960, Fig. 475.

Diagnosis (emended): Obliquely subovate valves; subterminal umbo; greatest height posterior to the median line; posterior height greater than anterior; dorsal margin slightly curved. Ornamentation: faint pattern of granules. Growth bands 22 or more.

Measurements: (mm) L-6.6; H-4.7; Ch-4.0; Cr-3.0; Av-1.0; Arr-1.8; a-2.0; b-2.0; c-3.0; H/L-.712; Ch/L-.606; Cr/L-.455; Av/L-.152; Arr/L-.273; a/H-.425; b/H-.425; c/L-.455; growth bands 22 or more.

Material: F 25472, lectotype (designated here), left valve (= Mitchell, 1927, Pl. 3, Fig. 8).

Locality: Just below the Dirty Coal seam near Newcastle sewage outlet. Merewether Beach. (See stratigraphic data and text under Australia.) Upper Permian (Tartarian).

Discussion: Raymond (1946) assigned Mitchell's species to *Pseudestheria.* As with other of Mitchell's species, that was solely because of the lack of a reticulate ornamentation on growth bands. Kobayashi (1954) also, but tentatively, assigned *lata* to *Euestheria?* because ornamentation was unknown.

Novojilov (1958c) described a new species based on Mitchell's poor photographs because he interpreted the shape of the valves of *lata* as differing from living *Cyclestheria* "uniquely", i.e., thus supposedly have oblong valves rather than circular valves, and a great number of growth bands. The configuration of Mitchell's species is distinctive. An equivalent number of growth bands is found on many species and therefore cannot be definitive for *lata* or any other species because these merely reflect life span conditioned by the duration of the water body and temperature. Tighter and closer spacing of growth lines are more useful because they reflect cycles of growth indicative of a drying or filling basin and variable food supply.

It is difficult to assign this species with great confidence. I will tentatively follow Kobayashi and assign it to *Cyzicus (Euestheria?)* on the basis of the faint granular pattern on growth bands. On the same slab (F 25472) there are two specimens labelled "syntypes." One is the present lectotype and the second is so poorly preserved that it adds little additional data.

Genus *Cornia* Lutkevich, 1937

Emended slightly: Valve shape varies from subovate to subrectangular; umbo subcentral to anterior in position: small spine or tubercle (node) rises from initial valve but may be flattened on umbo. Ornamentation: punctate or with hachure markings on growth bands.
Type: *Cornia papillaria* Lutkevich, 1937.

Cornia coghlani (Etheridge, Jr.)
(Pl. 43, Fig. 1)

Estheria coghlani Cox 1881, p. 276 (indication by reference to Charles Lyell, p. 450, Fig. 490).
Estheria coghlani Cox, Etheridge, Jr., 1888, p. 6, Pl. 1, Figs. 1–10 (drawings based on Cox's specimens; Pl. 1, Figs. 1–4, *non* Fig. 5 = Cornia).
Estheria coghlani Cox, Mitchell, 1927, p. 106, Fig. 4 (*non* Figs. 3 and 5 = *Cyzicus (Lioestheria) australensis* Novojilov, 1958).
Palaeolimnadia coghlani (Etheridge, Jr.) Raymond, 1946, p. 264.
Euestheria (?) coghlani (Etheridge, Jr.) Kobayashi, 1954, p. 110.
Cornia coghlani (Mitchell), Tasch 1979, *in* Tasch and Jones, 1979a, p. 16. Correction herein: *Cornia coghlani* (Etheridge, Jr.).

Diagnosis (emended): Small, obliquely ovate valves; subdued arcuate and short dorsal margin; subterminal umbo; umbonal node or spine almost centered. Margins: antero-dorsal rounded; postero-dorsal less rounded; ventral broadly concave.
Measurements: (mm) (From photograph of the lectotype F 25724.) L-2.0; H-1.5; Ch-?; Cr-0.74; Av-0.18; Arr-0.37; a-0.74; b-0.74; c-0.92; H/L-.750; Ch/L-?; Cr/L-.370; Av/L-.09; Arr/L-.185; a/H-.493; b/H-.493; c/L-.460; antero-dorsal angle -44.5° ±3°; posteror-dorsal angle-28° ±1°. Growth band number - 11.
Material: F 25724, lectotype (designated here), right valve, Mitchell, 1927, Pl. 2, Fig. 4 only, *non* Figs. 3 or 5; F 35720, F 35723. These three specimens constitute the syntypes.
Locality: Holt Sutherland Estate, Georges River. Narrabeen Group, Sydney Basin, New South Wales. Lower Triassic.
Discussion: Of the three specimens that Mitchell designated *"Estheria coghlani,"* only one (1927, Pl. 2, Fig. 4) is *Cornia coghlani.* However, there are two other specimens at the Australian Museum with the *coghlani* label that are also corniids: "F 35720,

depth 1483 feet" and "F 35723, depth 1932 feet; Dents Creek Bore."

As discussed elsewhere (Tasch *in* Tasch and Jones, 1979a, Pl. 16), the typical specimens of *"Estheria coghlani* Etheridge, Jr." are corniids with an umbonal node.

Family CYCLESTHERIIDAE Sars, 1899

Carapace laterally compressed with few and indistinct growth lines. Shell almost circular, with beaks far forward.

Genus *Cyclestherioides* Raymond, 1946

Cyclestheriidae with the beak (umbo) a bit further back than in the modern *Cyclestheria.*
Type: *Estheria lenticularis* Mitchell

Emended: Minute, subcircular valves with subround, subterminal umbo directed anteriorly; distinctly unequal height and length; few widely spaced growth bands.
Type: Same as genus.

Cyclestherioides (Cyclestherioides) lenticularis (Mitchell)
(Pl. 42, Fig. 4)

Estheria lenticularis Mitchell, 1927, p. 109, Pl. 3, Fig. 7.
Cyclestherioides lenticularis (Mitchell) Raymond, 1946, p. 275, Pl. 5, Fig. 3.
Cyclestherioides lenticularis (Mitchell) Raymond, 1946, Kobayashi, 1954, p. 134.
Cyclestherioides lenticularis (Mitchell) Novojilov, 1954c, p. 19, 22. *C. lenticularis,* Novojilov, 1960, Fig. 463.
Rhabdostichus lenticularis (Mitchell) Tasch, 1969, p. R155.

Diagnosis (emended): Minute, subcircular valves with subround, subterminal umbo directed anteriorly; few widely spaced growth bands.
Measurements: (mm) L-1.7; H-1.3; Ch-1.0; Cr-0.8; Av-0.3; Arr-0.2; a-0.6; b-0.6; c-1.0; H/L-.765; Ch/L-.588; Cr/L-.471; Av/L-.176; Arr/L-.118; a/H-.462; b/H-.462; c/L-.588; growth band number - 5.
Material: Holotype F 25481, left valve.
Locality: A short distance below the Dirty Coal seam (= Dudley seam), Merewether Beach, New South Wales. Upper Permian (Tartarian).
Discussion: Mitchell's holotype and only specimen, by its diminutive size and growth band spacing, appears to be a naupliid. Raymond assigned it to *Cyclestherioides* because it was reminiscent of living *Cyclestheria.* The latter, described by soft part anatomy, has a circular valve with a small umbo that is so far forward (i.e., anterior) on the valve that it is recumbent on the upper portion of the anterior margin. Raymond (1946) doubted that fossil valves were identical to living ones of this genus in every respect (including recumbent umbo), and made a more

conservative decision to assign it to a new genus, *Cyclestherioides*, rather that to *Cyclestheria* Sars.

Mitchell's species *"lenticularis"* has relatives in the Permian of Brazil and the Triassic of India: *C. (Cyclestherioides) pintoi* n. sp., (Brazil) and *C. (C.) machkandensis* n. sp., (India). Mitchell's *"lenticularis"* differs from the India species in having a greater H/L, Ch/L, c/L (.580 vs. .311–.400), and a smaller Av/L (.176 vs. .217–.300) and Arr/L; a/H and b/H overlap. *"lenticularis"* differs from the Brazilian species *"pintoi"* in having a smaller H/L (.765 vs. .810–.845), Av/L, Arr/L, and a larger Ch/L (.588 vs. .293–.350), Cr/L, and a/H; b/H and c/L overlap.

Genus *Palaeolimnadia* Raymond, 1946

Relatively long oval carapace, large, smooth umbonal region, and few growth bands. (Raymond, 1946, abridged.)

Type: *Estheria wianamattensis* Mitchell, 1927.

Subgenus *Palaeolimnadia* Raymond

A prominent naupliid carapace; swollen umbonal area occupies a considerable area of the valves; growth bands 3 to 6. Valve form ratio of height to length 0.82.

Type: *Palaeolimnadia (Palaeolimnadia) wianamattensis* (Mitchell 1927).

Palaeolimnadia (Palaeolimnadia) glabra (Mitchell)
(Pl. 42, Fig. 7)

Estheria glabra Mitchell 1927, p. 110–111, p. 4, Figs. 2, 3.
Palaeolimnadia glabra (Mitchell) Raymond, 1946, p. 264.
Estheriina glabra (Mitchell) Kobayashi, 1954, p. 157.
Limnadia glabra (Mitchell) Novojilov *in* Molin and Novojilov, 1965, p. 54, Pl. 3, Figs. 7, 8 (no description). *L. glabra* (Mitchell) Novojilov *in* Molin and Novojilov, 1965, p. 74, text-Figs. 28, 29.

Diagnosis (emended): Subovate valves with large, ovate umbonal area, slightly oblique anteriorly and occupying a considerable portion of the valve; few growth bands.
Measurements: (mm) F 25475, holotype. L-3.0; H-2.0; Ch-2.0; Cr-1.2; Av-0.7; Arr-0.4; a-1.0; b-1.0; c-1.6; H/L-.666; Ch/L-.666; Cr/L-0.40; Av/L-.233; Arr/L-.133; a/H-.500; b/H-.500; c/L-.533; growth band number - 3–4 or more; 1_u-2.0; h_u-1.3; $h_u/1_u$-.650 (h_u, 1_u—height and length of umbo.)
Material: F 25475, holotype (designated here; = Mitchell, 1927, Pl. 4, Fig. 2) mold of right valve. The other figured specimen (Mitchell, 1927) was F 25486 (paratype designated here), a left valve with postero-ventral sector missing. The illustration of this specimen (Mitchell, 1927, Pl. 4, Fig. 3), like that of the holotype, was inked in.
Locality: Chert quarries near Belmont (Kahibah Parish), New South Wales. Same horizon as *Leaia (Hemicycloleaia) belmon-*

tensis and associated with *Glossopteris* species and insects. Upper Permian (Tartarian).
Discussion: Comparing *Palaeolimnadia (Palaeolimnadia) wianamattensis* with *P. (P.) glabra,* the valve eccentricity and greater size of the umbonal area in the latter are distinctive. The comparative measurements are close, or overlap, with a few exceptions: Av/L-.233–.296 *(glabra);* .364 *(wianamattensis);* $h_u/1_u$-.636–.650 *(glabra);* .571 *(wianamattensis).*

Novojilov *in* Molin and Novojilov, 1965, considered this species a limnadiid. This assignment is unacceptable because Novojilov's (1970) placement of fossil species in modern *Limnadia* includes those having small as well as large umbos, variable configurations, and even more significant, mostly straight dorsal margins, whereas modern limnadiid species have prominent umbos, are generally broadly ovate, and have markedly arched dorsal margins (Daday de Deés, 1925). Raymond (1946) noted this last-cited difference in a brief discussion of his choice of genus *Palaeolimnadia* Ray. 1946 for fossil species.

Also unwarranted is Kobayashi's (1954) assignment to *Estheriina,* because the umbonal area is not more convex than the rest of the valve.

Palaeolimnadia (Palaeolimnadia) wianamattensis (Mitchell)
(Pl. 42, Fig. 1)

Estheria wianamattensis Mitchell 1927, p. 108, Pl. 2, Fig. 8 (*non* Fig. 7).
Palaeolimnadia wianamattensis (Mitchell) Raymond, 1946, p. 263–264, Pl. 3, Fig. 8 (*non* Fig. 7).
Estherites wianamattensis (Mitchell) Kobayashi, 1954, p. 30, 110, 168.
Kontikia wianamattensis (Mitchell) 1926 *(sic.)* Novojilov, 1958c, p. 87, Fig. 2.
Palaeolimnadia (Palaeolimnadia) wianamattensis (Mitchell), Tasch, 1979c, Pl. 41–42, Pl. 6, Fig. 8 (*non* Fig. 7); Tasch *in* Tasch and Jones, 1979a, p. 18.

Diagnosis: Small ovate to subovate valves, with a comparatively large, smooth subcentral umbo relative to rest of the valve; umbo of variable shape (ovate to elliptical) and size, but always the dominant feature of the carapace; few growth bands.
Measurements: (mm) L-3.3; H-2.5; Ch-0.69; Cr-1.6; Av-1.2; Arr-0.6; a-1.1; b-1.1; c-1.7; H/L-.757; Ch/L-0.69; Cr/L-.485; Av/L-.364; Arr/L-.182; a/H-.440; b/H-.440; c/L-.515; h_u-0.9; 1_u-1.5 ±0.1; $h_u/1_u$-.600; growth band number, 4 to 6.
Material: F 25490, lectotype (see Novojilov, 1958c; Tasch *in* Tasch and Jones, 1979a), internal mold of right valve.
Locality: Near Glenlee Homestead, from a cutting on the Great Southern Railway, Wianamatta Group (Middle to Upper Triassic).
Discussion: Two northern Angola (Africa) taxa from the Upper Triassic Cassanje III Series have close equivalents in the Lower and Middle Triassic, respectively, in Australia: *Estheriina*

(Nudusia) cf rewanensis Tasch 1979, and *Palaeolimnadia (Palaeo-limnadia) cf wianamattensis* Mitchell, 1927 is another. Compar-ing the Angola *"wianamattensis"* to the Australian type indicates: H/L (.846 Angola vs. .757 Australia); other ratios, a/H, b/H, Ch/L, vary by .01 to .07 except for Cr/L (.308 Angola vs. .485 Australia). Umbonal prominence and position on the carapace are also similar, as well as shared small size and presence of few growth bands.

Genus *Palaeolimnadia* Raymond, 1946
(Subgenus *Grandilimnadia* (Tasch, 1979a, p. 18)

Ovate to subovate valves with comparatively large, smooth umbo of variable position (from medial to subterminal). Umbo occu-pies a smaller shell volume than in *Palaeolimnadia (Palaeolim-nadia)*. Growth bands 4 to 20 or more.

Type: *Estheria wianamattensis* Mitchell, 1927 (= *Palaeolimnadia (Grandilimnadia) mitchelli* Tasch, 1979a).

Palaeolimnadia (Grandilimnadia) glenleensis (Mitchell)
(Pl. 43, Fig. 5)

Estheria glenleensis Mitchell, 1927, p. 108, Pl. 2, Fig. 6.
Palaeolimnadia glenleensis (Mitchell) Raymond, 1946, p. 265.
Estheriina glenleensis (Mitchell) Kobayashi, 1954, p. 110, 157.
Limnadia (Palaeolimnadia) wianamattensis (Mitchell), 1926 *(sic.),* Novojilov, 1970, p. 108, Figs. 82, 83 (Fig. 83 only, *non* 82).

Diagnosis (emended): Small, transversely ovate valves with a comparatively large subterminal umbo, inset slightly less than one-third the length of the valve from the anterior margin; short straight dorsal margin; anterior margin steeply convex; posterior margin rounded; ventral gently curved. Growth-band number variable (to seven).
Measurements: (mm) (Lectotype) L-3.0; H-2.0 or more; Ch-2.1; Cr-1.1; Av-0.4; Arr-0.3; a-0.7; b-0.7; c-1.4; Ch/L-.700; Cr/L-.367; Av/L-.133; Arr/L-.133; Arr/L-.100; a/H-.250; b/H-.350; c/L-.467.
Material: Because Mitchell (1927) figured a single specimen and it is lost, it is necessary to take the remaining specimens of this species in the type suite as syntypes and to select a lectotype, F 25482. This specimen is a mold of a left valve. The anterior sector of the dorsal margin, including a portion of the umbo, is covered or eroded, as is the antero-ventral margin. Two other syntypes are F. 25411 and F. 25478.
Locality: Cutting Great Southern Railway near Glenlee Home-stead, Wianamatta Group. Middle to Upper Triassic.
Discussion: Raymond (1946) and Novojilov (1970) agreed that Mitchell's species is a palaeolimnadiid. Novojilov recognized it as pertaining to the subgenus *Palaeolimnadia,* which he thought to be a subdivision of living *Limnadia.* This is a very questionable assignment because many of the forms which Novojilov (1970)

assigned to the genus itself are palaeolimnadiids. To also have *Palaeolimnadia* as a subgenus of *Limnadia* is redundant. For more details see discussion above for *Palaeolimnadia (P.) glabra* (M.).

The distinguishing characteristic of *P. (G.) glenleensis* (M.) is the comparatively large subterminal umbo which sets it apart from *Cyzicus (Lioestheria) fictacoghlani* (Mitchell, 1927, Pl. 2, *Fig. 3) and from C. (Lioestheria) australensis* (Novojilov, 1958c; Mitchell, 1927, Pl. 2, Fig. 5). The smaller umbonal size and its position, the greater number of growth bands, and more impor-tantly, the transversely ovate valve configuration, distinguishes it from *Palaeolimnadia (Palaeolimnadia) wianamattensis.*

Mitchell (1927) inked in his photo (Pl. 2, Fig. 6) and thereby distorted the true nature of the valve. For example, there is no mild sag in the dorsal margin, nor is the beak prominent, as it appears in his illustration.

Two Antarctic conchostracans are related to *Palaeolimnadia (Grandilimnadia) glenleensis* (Mitchell) from the Australian Middle Triassic. (The species range in Australia is Upper Per-mian, and Middle to Upper Triassic.) The Antarctic Lower Juras-sic specimens were collected at Storm Peak, Upper Flow, bed 7 (TC 9857a, Pl. 13, Fig. 8 herein) and Upper Flow, bed 6, lower half (TC 9033, Pl. 19, Fig. 1 herein). The ratios are close (H/L .666 Australia vs. .667-.731 Antarctica) or overlap (c/L .467 Australia vs. .423-.500 Antarctica). Umbonal position as well as configuration are equivalent; most other ratios are larger in the Antarctic specimens.

Palaeolimnadia? (Grandilimnadia?) linguiformis (Mitchell)

Estheria linguiformis Mitchell, 1927, p. 111, Pl. 4, Fig. 4.
Palaeolimnadia? linguiformis (Mitchell) Raymond, 1946, p. 264.
Estheriina linguiformis (Mitchell) Kobayashi, 1954, p. 42, 160.

Diagnosis: (Because the holotype—the only specimen—is "lost," one must rely solely on Mitchell's inked-in photograph and de-scription. Without new observations and measurements, nothing conclusive can be said.) Mitchell's diagnosis (abbreviated slight-ly): "Carapace obliquely flabellate . . . Dorsal margin, straight and long; anterior margin, short and rounded; posterior one, wide and gently rounded; beak, anterior, inconspicuous; concentric striae, about eighteen in number, obliquely directed towards the posterior-ventral margin, fine and compacted near the umbo and towards the postero-ventral margins . . ." (Mitchell, 1927, p. 111).
Measurements: (mm) (Mitchell's figures.) L-5.0; H-4.0; H/L-.800.
Comment: The valve is obviously compressed because the sixth and eighth growth bands below the umbo converge into a single growth band (Mitchell, 1927, Pl. 4, Fig. 4). Thus, the shape cannot be as Mitchell interpreted it, as further evidenced by ob-vious antero-ventral erosion along the margin. It follows that the further details as to growth-band disposition on the valve in Mitchell's description cannot stand, nor can any other details,

until a reconstruction of the valve prior to compression and erosion is considered. It would not be productive to make this attempt without the holotype.

Material: Holotype, Number ?, internal mold of left valve.

Locality: Chert (tuff) quarries near Belmont, New South Wales.

Discussion: Raymond's uncertain assignment to *Palaeolimnadia* is tentatively accepted, and this writer's current assignment to the subgenus *Grandilimnadia* is likewise tentative. These are used solely to remove the names *Estheria* (preoccupied) and *Estheriina,* each of which is inappropriate.

Taxonomic Clarification of Papers by Kozur and others Bearing on Taxa in this Memoir

INTRODUCTION

From the data set forth below, it will be seen that Kozur and others (1981), Kozur and Sittig (1981), and Martens (1982) have presented no convincing evidence (and, in fact, have misinterpreted the available evidence) to validate their emendation of *Lioestheria* Depéret and Mazeran, 1912. Their wholesale transfer of valid genera under the aegis of *Lioestheria* emended is a violation of the rules governing such genera and stems from failure to use proper procedures to correct the original taxonomic assignment of Depéret and Mazeran. To emend the taxonomy, the procedure should be based on *all* specimens that have growth bands on the umbo and/or hachure markings on growth bands. Further, to establish *Lioestheria sensu stricto* would retain the original authors' intent.

It may be noted that several conchostracan genera are known to occur in the same water body today and also in the fossil record of such assemblages (Tasch, 1969). These several distinct genera would obviously be fossilized together. (Note: This writer requested of M. Pacaud at the museum housing Depéret and Mazeran types, loan of these slabs and/or a rubber imprint of them. Regrettably, no response was forthcoming.)

Cyzicus (Lioestheria) sensu stricto, Depéret and Mazeran, emended.

Estheria (Lioestheria) Depéret and Mazeran, 1912, p. 172–173 (in part).
Lioestheria Depéret and Mazeran emended, Kozur and others, 1981, p. 1439. (Only forms on the original slab that show growth bands on the umbo and/or hachure-type ornamentation on the growth bands.)
Cyzicus (Lioestheria) Depéret and Mazeran, Tasch, 1969, p. R151.

Description: Generally ovate to subovate valves with growth bands on the umbo; lacks a node (*Cornia* type), a broad-based umbonal spine, (*Gabonestheria*-type), a curved spine (*Curvacornutus*-type), multiple spines (Vertexia), or a "radial element" of any kind, a recurved posterior margin (*Palaeolimnadiopsis*-type) or a combination of two of these characteristics (*Pemphilimnadiopsis*-type); bears ornamentation of punctae or

parallel vertical markings (coarse or fine hachure markings) on growth bands. Umbo usually is subterminal to slightly inset. Growth band number variable.

Type: *Estheria (Lioestheria) lallyensis* (Number 1789, Depéret and Mazeran, 1912, Fig. 1).

Comment: The indicated type of *C. (Lioestheria)* Depéret and Mazeran, sensu stricto is a turreted specimen on the original slab; the turret is a pseudo-structure, a result of compression of the umbonal region of the valve. That specimen bears clearly displayed growth bands on the umbo (in the basal middle sector of the slab; Pl. 5, Fig. 1; and Pl. 44, Figs. 8, 9 herein). Included are all paratypes lacking the features indicated above in the emended subgeneric description and/or bearing hachure markings on growth bands or punctae.

Cyzicus (Lioestheria) lallyensis, sensu stricto, Depéret and Mazeran, emended. (Pl. 44, Figs. 8, 9)

Estheria (Lioestheria) lallyensis, Depéret and Mazeran, 1912, Pl. 5, Fig. 1.
Lioestheria lallyensis Depéret and Mazeran, 1912, Raymond, 1946, p. 231 (in part).
Lioestheria lallyensis Depéret and Mazeran, 1912, emended. Kozur and others, 1981, p. 1438–1441 (in part).
Estheria paupera Fritsch, 1901 (=*Estheria (Lioestheria) lallyensis* Depéret and Mazeran, 1912; Kozur and Sittig, 1981, p. 16. Not a lioestheriid sensu stricto.) (See Pl. 44, Fig. 6 herein.)

Diagnosis: Ovate to subovate valves with umbo subterminal to inset from the anterior margin: growth bands on umbo and rest of valve number from 20–22. No valve structures other than growth bands and their ornamentation of hachure markings or punctae.

Discussion: The photograph of Depéret and Mazeran (1912, Pl. 5, Fig. 1), with a turreted umbo due to compression of a left valve, bears growth bands on the umbo and hence does not fit the Kozur and others (1981) emended definition of the species. If more than one valve on the same side as the "turreted specimen" can be shown to have remnants of umbonal growth bands and/or lioestheriid ornamentation (as suggested above, there probably are), then the type *sensu stricto* should become the lectotype. All other related valves would become additional syntypes. Kozur and others (1981) plate, Fig. 2, which is the only figured paratype, does not conform to their emended definition and has lioestheriid characters (sensu stricto) such as growth bands on the umbo and hachure markings on growth bands.

Background and Clarification: In 1912, Depéret and Mazeran described two conchostracan subgenera: *Estheria (Lioestheria)* Depéret and Mazeran, type species *(E. Lio.) lallyensis* Depéret and Mazeran (Pl. 5, Fig. 1), and *Estheria (Euestheria)* Depéret and Mazeran, type species *Estheria (Euestheria) minuta* Alberti Depéret and Mazeran (Pl. 5, Figs. 2, 3) (cf. Feys, 1960; Warth, 1969), both from Lower Permian strata of Lally near Autun, France. Raymond (1946) raised (Lioestheria) to genus rank and retained the same type species. *Euestheria* was also raised to genus rank and, because its type species was found to be a *nomen nudem,* a new type species had to be selected: *Posidonia minuta* von Zieten, 1833 (cf. Beyrich, 1857; Bronn, 1850). The two genera were once again restored as subgenera, but *Cyzicus,* the name of a living genus, was used to replace *Estheria* (Tasch, 1969). It should be emphasized that both of the above subgenera have growth bands on the umbo and have distinctive ornamentation: hachure marking or punctae (lioestheriids), and polygonal ornamentation (contiguous granules) on growth bands (euestheriids; see Pl. 44, Fig. 7), as originally described and figured by Depéret and Mazeran (1912) and restated by Raymond (1946). It is these specific valve characteristics that have been ignored by Kozur and others in their drastic revision of a huge portion of accumulated systematics.

Lioestheria lallyensis Depéret and Mazeran, 1912 emended Kozur and others, 1981.

In 1981, Kozur and others restudied Depéret and Mazeran's types on deposit at the Museum d'Histoire Naturelle, Autun, France, and reached the conclusion that among nineteen specimens on the original slab (front and back), there was not a single specimen that showed any of the characteristics described by the original authors. Instead, they reported finding only specimens with a large umbo, all lacking growth bands on the umbo which bore a "hemispherical bulge" and/or a "radial element" (see Pl. 44, Figs. 1 and 2 and Figs. 21, 1, and 3 herein). From these observations, Kozur and others decided to discard the subgenus and species of Depéret and Mazeran by emending it to include the above-stated characteristics, and eliminated features the original authors used to describe their new subgenus (genus of Raymond, 1946). Their emended genus and species now embraced a host of distinct genera and species: *Cornia, Gabonestheria, Palaeolimnadia, Palaeolimnadiopsis, and Vertexia,* among others. (Cf. Kozur, 1980; Holub and Kozur, 1981; Martens, 1984.)

For *Lioestheria* emended, Kozur and others (1981) designated a new holotype (not seen on the figured slab of Depéret

and Mazeran but reputed to appear on that slab). It had a so-called node and "radial element" (MHN No. 1789, Kozur and others, 1981, Fig. 1/2 and plate, Figure 1 [non 2]; i.e., *Lioestheria lallyensis* Depéret and Mazeran, 1912, emended). Kozur and Sittig (1981, p. 16) designated "*Estheria paupera* Fritsch 1901" (=*Estheria (Lioestheria) lallyensis* Depéret and Mazeran, emended) as the holotype of *Lioestheria,* emended. Thus, in the same year, two holotypes were designated for the same genus, one with the so-called "radial element" and smooth umbo (No. 1789 above) and the other (Fritsch's "paupera"), an apparent gabonestheriid with a smooth, flattened umbo having a tapered terminus and positioned in the antero-dorsal sector of the valve (Fritsch, 1901, Tafel 161, Fig. 5).

Other data were assembled by Martens (1982), who studied the Czechoslovakian Obora Collection. He, as one of the collaborators on the emendation of *Lioestheria* again referred to *Lioestheria lallyensis* D. & M. emended as the type in describing his Obora Collection.

MARTENS OBORA EVIDENCE

In his 1982 paper, Martens figured the Permian (Rotleigendes) (cf. Kozur, 1980), specimens of *Lioestheria lallyensis* Depéret and Mazeran, 1912, emended, from Obora, Czechoslovakia, ČSSR, on several plates of photographs (Pls. 13–15) and line drawings. Analysis of these figured specimens revealed that Plate 13 contained illustrations of genus *Cornia* and *Gabonestheria,* only. Plate 14 contained *Gabonestheria, Palaeolimnadia* and *Cornia,* with a fossilized biramous second antenna arising from a scape (Martens, 1982, Fig. 4) and positioned obliquely from the anterior end of the dorsal margin across the umbonal area (cf. Pl. 44, Fig. 3 and 6, and Fig. 21,5 herein). This anatomical structure was interpreted to be a "radial element" by Martens. Plate 15 had figures of *Palaeolimnadiopsis* (cf. Kozur, 1980), *Cornia,* and *Palaeolimnadia* (cf. Holub and Kozur, 1981).

Clearly, Martens, following the logic of *Lioestheria* Depéret and Mazeran, emended, included several valid genera under *Lioestheria.* That concept is rejected here as leading to a crazy-quilt genus, and because it is in complete disregard of the rules regarding the fixity of a valid genus. The validity of Depéret and Mazeran's genus will be considered further below.

VALIDITY OF *LIOESTHERIA* DEPÉRET AND MAZERAN

1. Does any specimen of the nineteen available on the original slab of Depéret and Mazeran bear growth bands on the umbo? Despite very poor photographs in Depéret and Mazeran, 1912, one can discern a telltale specimen at the bottom of the slab (Pl. 44, Fig. 8 and 9 herein) which has a turreted umbo due to compression. That umbo clearly displays three growth bands that appear white. To this one can add Kozur and others 1981 photograph (Pl. 44, Fig. 2 herein), which shows a crushed umbo

with several partially preserved growth lines in white. (Cf. Pl. 44, Figs. 4 and 5, for *Euestheria* with growth bands on the umbo.)

Another of Kozur and others photographs (Pl., Fig. 1; and see Pl. 44, Fig. 1 herein) reveals apparent growth bands, very faint, but detectable when one rotates the page bearing the photograph 90° to the right and a magnifying glass is used. At least three growth bands underlay the so-called "radial element." Thus, three specimens on the original slab display umbonal growth bands. That eliminates any genus with a smooth umbo such as *Palaeolimnadia.*

2. The so-called "radial element" (Pl. 44, Fig. 1 and Figs. 21,5 herein) on the umbo is a single, segmented or beaded, curved structure, which could represent a second antenna (cf. Martens, 1982, Pl. 14, Fig. 4; and Pl. 44, Figs. 3 and 6 herein); i.e., a single strand, or a fecal string. Thus, the two figures that Kozur and others (1981) intended to depict the new characters (i.e., smooth umbo and "radial element" of *Lioestheria* Depéret and Mazeran, emended) actually negate the interpretation given

and that negation becomes supporting evidence for the validity of the original subgenus of Depéret and Mazeran (raised to genus by Raymond, 1946), as restricted above.

3. The only other information (Kozur and others, 1981, p. 1440–1441) about the remaining seventeen paratypes is "juvenile forms also frequently occur with none to four growth bands which are even particularly characteristic of the species" (Pl. 44, Figs. 2 and 9 herein, which show growth bands on the umbo). Curiously, nothing is said about the so-called "radial element" or the hemispheric bulge (node) that juveniles ought to display on their "smooth" umbos, these being characteristics of the larval valve. Martens (1982), for example, has the following legend (1982, Pl. 14, Fig. 4) with reference to a specimen attributed to *Lioestheria lallyensis* Depéret and Mazeran, emended, "mit radialen Element auf der larvelen Schale" but that, as shown, is a biramous second antenna and not the shell structure the triad of authors termed "radial element."

Paleozoic/Mesozoic Dispersal of Conchostracan Taxa in the Southern Continents

FACTORS IN DISPERSAL AND SURVIVAL

Salinity Tolerance. Living branchiopod crustaceans, conchostracans included, are nonmarine creatures, that generally inhabit ephemeral small water situations. Evidence that this writer has gathered over several decades indicates that fossil conchostracans in life were also nonmarine. This writer's experiments in hatching eggs of living conchostracans in waters of varying degrees of salinity demonstrated that these eggs will not hatch in regular or synthetic salt water. One can refer to the Antarctic Ohio Range Permian ecosystem with conchostracans, where increases in salinity terminated, the system (Tasch, 1977a, Table 2; Tasch and Gafford, 1968, on paleosalinity). The fossil record of some estuarine forms, other reports of fossilization of conchostracans with marine fauna or of living conchostracans temporarily immersed in water enriched by salt spray, all indicate a limited tolerance of slightly brackish water; and that higher salinity leads to a comatose state (Gislén, 1936).

Conchostracan Eggs. In most instances, the water bodies occupied by living conchostracans seasonally dry up by evaporation, or the strand line of a larger body of water lowers, leaving relict pools or ponds that have conchostracan eggs in, or on, the substrate; these will hatch when the usual precipitation and/or drainage refills the water body.

Egg Transport. Appreciation of the dispersal evidence set forth below requires awareness that conchostracans cannot, and from associated biota and sedimentary contexts in which we find the fossils, probably never could, have spread by any marine route since the original transition from marine to nonmarine habitats in Carboniferous time (Tasch, 1963a). All other transport possibilities have also been explored (Tasch, 1971).

Transport by Birds. There were no Permian-Carboniferous birds. This writer has remarked elsewhere that although today we have circumpolar birds that fly from other continents to breed in the Antarctic, not one has yet brought any conchostracan eggs to even a single one of the many Antarctic lakes. These lakes are barren of conchostracans among their other branchiopod fauna (cladocerans, anostracans). In addition, there is no evidence of any circumpolar birds reaching Antarctica, if any existed before the Cretaceous. However, there is one report of a bird's feather in Lower Cretaceous (Vallangian–Aptian) beds of Victoria, Australia (Talent and others, 1966). This Cretaceous fossil was found in a section that yielded a conchostracan reported in this memoir.

Transport by "Noah's Ark" (floating logs). This writer has found no fossil evidence (nor is there any for living forms) that conchostracans chose/choose fallen logs that eventually reach water and occupy them prior to their flotation, as habitats. Although the Antarctic Jurassic sequence contained petrified wood (log beds), the conchostracan fauna lived when different conditions prevailed. The opposite of floating-log transport is suggested by the abundance of palynomorphs in some conchostracan beds, which denotes wind transport. (See Antarctica, Storm Peak section, this memoir.)

Furthermore, study of the evidence set forth below will make it clear that it would require a remarkably fortuitous sequence of currents to deliver logs bearing eggs of the *same* conchostracan genus and/or species to each of the several southern continents, or to four or three of them, positioned relative to each other as they are today.

Aerial Transport. This is presently, and presumably was in the geologic past, an important mode of dispersal via windblown conchostracan egg-bearing substrates that had dried out. The eggs of the magnitude of approximately 0.13 mm would move as fine dust along with the sedimentary fines when wind swept. However, while such transport over small distances on some floodplains, for example, or by "stepping stones" from one neighboring shallow pool or pond to another, is a viable mode of dispersal (Tasch *in* Tasch and Jones, 1979c), it is not a viable mode of transport for distances that presently separate continents. Why not?

Aerial transport of contemporary insects over long distances is well established for the Southern Hemisphere, but the transported insects, when they arrive at another continent, are found to have been desiccated enroute (Gressitt, 1961). Furthermore, insects are mainly aerial creatures (exclusive of aquatic larval forms) and being picked up by a wind is always likely. On the other hand, conchostracans are not wind transported, and they are unable to survive out of water. Their eggs, however, are so transported where dried-out small water situations prevail. (See above for "Salinity Tolerance.")

DISPERSAL EVIDENCE

The bioprograms of four conchostracan genera were dispersed to all five continents of the Southern Hemisphere during portions of Paleozoic/Mesozoic time: *Cornia, Estheriina, Palaeolimnadia,* and *Cyclestherioides,* to each of four southern continents: *Leaia* (all except India); *Pseudoasmussiata* (all except Australia); and to each of three continents: *Palaeolimnadiopsis* and related genera (South America, Africa, Australia); *Estheriella* and related genera; as well as *Gabonestheria* (South America, Africa, India). *Cyzicus* is excluded in the discussion that follows because it is ubiquitous both in horizontal spread and geological (vertical) persistence. However, individual species, where relevant, will be reviewed.

All the evidence cited below is either based on species described or catalogued in this Memoir.

Cornia. In India, East Bokaro Coal Field, west Bengal, a borehole slice 250 m above the Triassic (Panchet Formation)-/Permian (Raniganj Formation) boundary contained *Cornia panchetella* n. sp. This species was also collected from the village of Mangli. This writer placed the Mangli horizon yielding the corniid as equivalent to the 92 m level above the *Glossopteris/-Schizoneura* datum in the Raniganj Basin.

South American corniid fossils include *Cornia semigibosa* (Cardoso, 1962) (formerly *Echinestheria*) from the Brazilian Upper Triassic Motuca Formation, and a *Cornia* species from the Upper Permian, lower Rio do Rasto Formation of Brazil, São Pascoal (Santa Caterina).

Cornia angolata n. sp. (among other conchostracans) was found in Oesterlen's northern Angola, Africa, collection, Upper Triassic (Cassanje I Series), and *Cornia haughtoni* Tasch was uncovered from the Upper Triassic Cave Sandstone of Lesotho.

The Australian Permian and Triassic have several corniids: *Cornia* sp. (Australian Museum No. 51987), Upper Permian, Sirius Creek, Queensland; the Lower Triassic of the Sydney Basin contains abundant *Cornia coghlani* (Etheridge, Jr.) from George's River and also Port Hacking, Heathcote, Narrabeen and Liverpool—all near Sydney; and at Tuggeruh Lake at a depth of 209 or more m.

Only the Jurassic of Antarctica (Blizzard Heights and its environs) yielded corniids at Tasch Station 0, bed 7 (*Cornia* sp. 1) and David Elliot's site, D.E. 25.58 c (*Cornia* sp. 2).

Other than the Upper Permian corniid in Australia and the species in South America (Brazil), all other Southern Hemisphere corniids now known (exclusive of Antarctica) are chiefly Triassic. Of interest is the late dispersal during Jurassic time of corniid eggs to Antarctica. The Australian mainland has no reported Jurassic conchostracan-bearing beds and Tasmania's Triassic has not yielded any corniids (Tasch, 1975). Furthermore, no conchostracan-bearing beds have been discovered thus far in the Antarctic Triassic. (This writer has examined several Triassic sites during the 1969–70 Antarctic season and found them to be dominated by plant fossils and/or coal deposits.) Accordingly, the probable original source of the Antarctic Jurassic corniids is narrowed down to either the Upper Triassic of Angola or the Lower to Upper Triassic of India.

Estheriina. (This genus has two subgenera, *Estheriina [=E]* with growth bands on an umbo that is generally raised and distinct from the rest of the valve, and *Nudusia [=N]*, lacking growth bands on the umbo.) The Upper Triassic of Angola, Africa, Cassanje I Series, had *Estheriina (N.)* cf. *rewanensis* Tasch (see Fig. 20A for section data).

Several estheriinids are known from the Australian column: *Estheriina (N.) blina* Tasch in the Blina Shale, Lower Triassic (well cuttings, Canning Basin) (Tasch *in* Tasch and Jones, 1979a), and two estheriinids from the Bowen Basin; *Estheriina (N.) rewanensis* Tasch and *Estheriina (N.) circula* Tasch from the base of the Rewan Group, Lower Triassic. (Tasch *in* Tasch and Jones, 1979c.)

In the Antarctic Jurassic at Blizzard Heights, several species of estheriinids occur: *Estheriina (N.)* sp. 1 (BH, Station 1, bed 8); *Estheriina (E.)* sp. 1 (BH, Station 1, bed 4); *Estheriina (N.)* sp. 2 (D.E, site 23.59b); and *Estheriina (N.) brevimargina* Tasch (D.E, site 33.56A). Both Upper and Lower Flow at Storm Peak had estheriinid fossils in the interbed: *Estheriina* (E.) sp. 2 (Lower Flow, Station 0, bed 2); *Estheriina (N.)* sp. 3 (Upper Flow, loose), and *Estheriina* (N.) *stormpeakensis* Tasch (Upper Flow, beds 1, 6, and 10). This last-named species is also found at Blizzard Heights, Station 1, beds 5, 7, and 9A, and Station 0, beds 7 and 8.

India has distinctive Lower Jurassic estheriinids, all from the Kotá Formation. Included are: *Estheriina (N.) indijurassica* Tasch (Station K-2, bed 2); *Estheriina (N.) adilabadensis* Tasch (Station K-1, bed 8A); *Estheriina (E.) pranhitaensis* Tasch (Station K-7, bed 6); and *Estheriina (N.) bullata* Tasch (Station K-2, bed 3).

South America (Brazil) has a broad spectrum of estheriinids ranging from Permian to Cretaceous time: *Estheriina (E)* sp. 1 (Lower Rio do Rasto Formation, São Paulo, Santa Caterina, Upper Permian), *Estheriina* (N.) sp. (Santa Maria Formation, Rio Grande do Sul, Lower Triassic). No Jurassic forms are presently known. Jones described three species from the Lower Cretaceous of Bahía: *Estheriina (E.) bresiliensis* (genus type), *Estheriina (E.) astartoides,* and *Estheriina* (E.) expansa.

There are several items in the spread of the estheriinid bioprogram that merit further comment.

1. South America is the only southern continent having Permian estheriinids—the geologically oldest known representatives of the genus.

2. Accordingly, dispersal of estheriinid eggs from the Upper Permian site (State of Santa Caterina) to the Lower Triassic (State of Rio Grande do Sul) seems possible, although heterochronous. As noted above, the Canning Basin of Western Australia and the Bowen Basin of eastern Australia also had Lower Triassic estheriinids. Because the Blina Shale of the Canning Basin is geologically younger than the base of the Rewan Formation of the Bowen Basin, dispersal would have been from east to west (see Tasch *in* Tasch and Jones, 1979c, p. 37–38). That

leaves the question of how the estheriinids reached the Bowen Basin.

3. The Upper Triassic of Angola could have received dispersed estheriinids from South America.

4. The Upper Triassic of Africa (Angola) is a possible original source of the Antarctic's Lower Jurassic estheriinids, but data on a Lower Jurassic link is presently lacking.

5. India, with its several Lower Jurassic estheriinid species, is a more viable option than Angola for the source of Antarctic estheriinids (see Frontispiece—"Gondwana Fossil Conchostracan Distribution"). It is worth noting that India and Antarctica, respectively, have the only Lower Jurassic estheriinids in the reconstructed continental assemblage.

6. Dispersal of Antarctic Jurassic estheriinids presumably was the ultimate source of Brazilian forms of Lower Cretaceous age, although the time hiatus still needs to be bridged.

Palaeolimnadia. Australia's Upper Permian and Lower to Upper Triassic have yielded several palaeolimnadid species. *Palaeolimnadia (G.)* cf. *glenleensis* Mitchell comes from the Upper Permian Blackwater Group of the Bowen Basin. Mitchell (1927) originally described the species from the Wianamatta Group (Middle to Upper Triassic). The same species has also been found in the Arcadia Formation, Upper Rewan Group, Lower Triassic, the base of which is directly over the Blackwater Group. The Hawkesbury Sandstone, directly below the Wianamatta Group, also has a palaeolimnadid species, as do the Upper Triassic beds of Ipswich, Queensland. Mitchell described *Palaeolimnadia (P.) wianamattensis* from the Wianamatta Group, which this writer found to be widely disseminated in the Permian (Tartarian) Newcastle Coal Measures (Croudace Bay Formation). (See Fig. 26.)

Other palaeolimnadids from Australia include a Lower Triassic species from the Blina Shale (Canning Basin), *Palaeolimnadia (G.) medaensis* Tasch, and from the Joseph Bonaparte Gulf, *Palaeolimnadia (G.) profunda* Tasch; the upper Lower Triassic of Garie Beach, eastern Australia, yielded *Palaeolimnadia (G.) arcadiensis* n. sp. In the KNocklofty Formation of Tasmania (Lower Triassic), equivalent to the Blina Shale, two palaeolimnadid species were found, and from the Ross Sandstone, an older bed of the Tasmanian Lower Triassic, *Paleolimnadia poatinis* Tasch was described.

In Africa, *Palaeolimnadia (P.)* cf. *wianamattensis* Mitchell (Middle to Upper Triassic, Wianamatta Group) also occurs in the Angola Phyllopod Beds. *Palaeolimnadia (Grandilimnadia)* species also are known from the same Angola beds and in the Cave Sandstone Formation, Upper Triassic of Lesotho. Another genus and species described in this memoir, *Afrolimnadia siberiensis* n.g., n. sp., belongs to the palaeolimnadid clan and was found at Siberia, Republic of South Africa (Tasch Station 1A, bed 4).

Palaeolimnadia (Grandilimnadia) sp. is known from the Lower Jurassic Kotá Formation of India (Tasch Station K-2, bed 2). The writer also found palaeolimnadid species in the Lower Triassic Panchet Formation subsurface (Andal area, west Bengal) as follows: Site RNM-2, 47.5—50.0 m below the surface (=bts);

site RNM-3, 47.0 m bts; site RNM-4, 40.0 m bts. Others were found in the East Bokaro Coal Field, Panchet Formation, at site EBP-4, 250.0 m bts. (See Figs. 15, 16.)

From Bolivia, the Vitiacua Formation (Permian or Triassic Aguaraque Range) has a palaeolimnadid species in a subsurface slice. Another palaeolimnadid species, *P. (Grandilimnadia) petrii* Almeida (1950), is known from São Paulo State, Brazil in the Botucatú Formation (Lower Cretaceous); a palaeolimnadid species from the same formation has been reported from Mato Grosso State.

Antarctic palaeolimnadids include *Palaeolimnadia (Palaeolimnadia)* sp., found at Mauger Nunatak (D.E site 62-10-1A, Lower Jurassic), *Palaeolimnadia (Grandilimnadia) minutula* Tasch (Upper Flow, bed 6), as well as *Palaeolimnadia (Grandilimnadia)* cf. *glenleensis* Mitchell (Upper Flow, beds 2, 7, and 8)—both species from this writer's Storm Peak collection.

Cyclestherioides. This genus was dispersed to five continents. The geologically oldest fossils of this genus in the southern continents came from South America (Brazil, Estrada Nova Formation, Middle Permian). In Africa (Tanzania), Janensch (1933) described two species of conchostracans from the Jurassic dinosaur beds, one of which is now placed in *Cyclestherioides (Cyclestherioides janenschi)* (Kobayashi) 1954.

India has two species from the Raniganj Coal Field (Panchet Formation), Lower Triassic: *Cyclestherioides (Cyclestherioides) machkandaensis* n.sp. and *Cyclestherioides (Sphaerestheria)* sp. The Australian Upper Permian Newcastle Coal Measures has *Cyclestherioides (Cyclestherioides) lenticularis* (Mitchell).

One species of this genus occurs at two Antarctic localities: *Cyclestherioides (Cyclestherioides) alexandriae* n. sp., at Storm Peak, Upper Flow interbed, bed 10, and Blizzard Heights, Station 1, bed 9A, Lower Jurassic.

Dispersal of *Cyclestherioides* eggs between Tanzania and Antarctica appears to be one possibility in light of their Jurassic representations of this genus; or, both the African and Antarctic occurrences of *Cyclestherioides* species may have originally derived from the Lower Triassic population of India. A missing link at present in this dispersal pattern is an Upper Triassic site in either India or Africa with a *Cyclestherioides* population. Whatever dispersal tracks prevailed, the spread of *Cyclestherioides* bioprogram to South America, India, Africa, Australia, and Antarctica actually occurred.

It is evident that dispersal of eggs of species of the four genera whose fossil distribution is reviewed above, occurred to each of the five continents during the Permian–Triassic and Jurassic in the Southern Hemisphere. Multiple examples of such distributions of other conchostracan genera to four or three of the southern continents are reviewed below.

Leaia. (This genus is subdivided into two subgenera: *Leaia [=L.]*, which embraces subquadrate to subelliptical forms, and *Hemicycloleaia [=H.]*, which includes circular to ovate forms.) This ribbed conchostracan spread to all of the southern continents, with the exception of India. *Leaia (H.) gondwanella* Tasch recurred in this writer's collections from the Upper Permian *Leaia* Zone, Ohio Range, Antarctica (Tasch, 1970a, p. 185–194).

Leaia (H.) pruvosti Reed was described from the Upper Permian Rio do Rasto Formation at Poço Preto, Santa Caterina, Brazil. *Leaia (H.) leanzai* Leguizamón was found in the Tasa Cuna Formation of Cordoba, Argentina (Lower Permian, probably Sakmarian-Artinskian).

In the Republic of South Africa, Lower Beaufort beds of Natal's Upper Permian contained *Leaia (H.)* sp., and from the Upper Karoo Series, Cape Province, Upper Permian, this writer described *Leaia (H.) cradockensis* n. sp. Professor Bond's Upper Permian collection from the Madumabisa Shale, Zimbabwe, yielded *Leaia (H.) sessami* and *Leaia (H.) sengwensis.*

The greatest number of leaiid species (13 after reassignment in this memoir) occur in the Upper Permian Newcastle Coal Measures of New South Wales, Australia. (See Mitchell's types.) In the Blackwater Group, Bowen Coal Measures, Upper Permian, two leaiids species occur: *Leaia (H.)* sp. 1 and *Leaia (H.) deflectomarginis* Tasch.

Carboniferous leaiids are known from both eastern and western Australia. In the Lower Carboniferous Drummond Basin, *Leaia (H.) drummondensis* Tasch occurs in the Raymond Formation. This writer has described five species from the Canning Basin, Anderson Formation, Grant Range (all Lower Carboniferous; Tasch *in* Tasch and Jones, 1979a): *Leaia (L.) andersonae, Leaia (H.) rectangelliptica, Leaia (H.) grantrangicus, Leaia (H.) tonsa,* and *Leaia (H.) longicosta.*

From the National Collection of Morocco, Pennsylvanian (Stephanian) of the Haute Atlas, Hamalou-Agadir, this writer described *Leaia (Leaia) bertrandi* n. sp. The Termiers (1950) figured three specimens (line drawings) from Hamalou, Ida ou Zal, that have been reassigned by this writer to *Leaia (H.)* sp. 1 and 2. (See details under Morocco, this Memoir.)

There is one item of special interest in the leaiid dispersal; although India's Permian has yielded estheriinids, no *Leaia* species are known as of this writing, even though they were diligently searched for by this writer. Because India was in contact with Antarctica and Australia (see Curray and Moore, 1974; and Frontispiece herein), this question arises: Can one account for the apparent absence of ribbed conchostracans (*Leaia* sp.) in India, particularly in the Upper Permian, when dispersal to the other four continents did transpire?

There is only one known conchostracan-bearing bed in the Lower Carboniferous reported in India; i.e., the Po Series, Lipak Section, Kashmir. A single slab from this site on deposit at the Geological Survey of India (S.E.M. Laboratory, Calcutta) was examined by this writer. This slab bore a coquina of poorly preserved valves. One specimen on the slab was tentatively identified as *?Cyzicus (Euestheria)* sp. No leaiids occurred.

Rigby and Shah (1980, p. 39–41) observed that "between the Raniganj flora" (India) "and the earliest Australian Late Permian flora . . . mixing did not happen until post-Permian time." The lack of mixing of the two continental Late Permian floras parallels the apparent failure of dispersal to the Indian Permian

sites of the *Leaia* bioprogram, even though species of this ribbed conchostracan are abundant in the Upper Permian Newcastle Coal Measures. In brief, it appears that some physical impediment or barrier prevented a mixing or interchange of the respective terrestrial biotas of India and Australia during Late Permian time. When such restraints no longer existed in post-Permian time, the leaiid conchostracans were extinct everywhere in the world. (Leaiids in Permian of India, see p. 149).

The Drummond Basin, Australia, leaiids are Tournasian or Visean in age; i.e., geologically older than the Canning Basin leaiids. This appears to indicate a previous east to west dispersal to the Canning Basin. That possibility is supported by the common occurrence of hemicycloleaiids (the subgenus) in both basins. A subsequent west to east dispersal of leaiids to the Bowen Basin is also inferred. (Tasch *in* Tasch and Jones, 1979c, p. 39, items 1, 2.)

Pseudoasmussiata. (This is a new name to replace *Pseudoasmussia* n. gen. *[sic.]* Defretin, 1969, a preoccupied name; Novojilov, 1954a; for further discussion, see South America, this memoir. Defretin's genus is distinct from Novojilov's.) Species of this genus are dispersed to four southern continents. In the South American Upper Triassic of Brazil (Santa Maria Formation), Ioco Katoo's site ("Belvedere") had two species assignable to Defretin's genus: *Pseudoasmussiata katooae* n.sp. and *Pseudoasmussiata* sp. B.

The African rock column (Zaire) contained three species of this genus described by Defretin (1967): *Pseudoasmussiata ndekeensis* (Loia Series, Samba boring), Upper Wealden; *Pseudoasmussiata banduensis* (Loia Series, Bandu, Site 21), Wealden; and *Pseudoasmussiata duboisi* (Marlière) (Stanleyville Series), Kimmeridgian.

India's Lower Triassic at Mangli contained *Pseudoasmussiata indicyclestheria* n. sp., and in the Lower Triassic Panchet Formation, *Pseudoasmussiata bengalensis* n. sp. was recovered. The Lower Jurassic Kotá Formation (Tasch Station K-1) contained the species *Pseudoasmussiata andhrapradeshia* n. sp.

The Upper Basalt Flow interbed at Storm Peak, Antarctica, yielded a Lower Jurassic species, *Pseudoasmussiata defretinae* n. sp.

Because Brazil had only Upper Triassic species, the oldest Southern Hemisphere species could have been derived from the Panchet Formation of India (Lower Triassic) across a Middle Triassic transition that is not yet known. Dispersal of the Brazilian population's eggs to Antarctica during Early Jurassic time seems to be the most plausible track that can account for the Storm Peak occurrence. The Lower Jurassic species of either India or Antarctica could have been the original source of the Upper Jurassic (Kimmeridgian) species in Zaire, Africa. The latter, in turn, could account for Zaire's Cretaceous species.

Furthermore, Southern Hemisphere distribution of other members of the *Asmussia* clan include *Asmussia* and *Glyptoasmussia.* From the Lower Cretaceous Bokungu Series of Zaire, two asmussid species are known: *Asmussia dekeseensis* Defretin and *Asmussia ubangiensis* Defretin. *Glyptoasmussia* species include: *Glyptoasmussia corneti* Defretin, Stanleyville Series (Upper Jurassic) and *Glyptoasmussia luekiensis* Defretin, Haute Lueki Series (Upper Triassic or Jurassic (pre-Oxfordian), both from Zaire. This writer's Antarctic Lower Jurassic collections also yielded some members of the asmussid clan from Storm Peak (SP) and Blizzard Heights (BH): *Glyptoasmussia australospecialis* Tasch (SP, Upper Flow, bed 11) and *Glyptoasmussia* cf. *luekiensis* Defretin (SP, Upper Flow, talus). Blizzard Heights had two glyptoasmussid species: *Glyptoasmussia meridionalis* n. sp. (Station 0, bed 1 and Station 1, bed 4), and *Glyptoasmussia prominarma* n. sp. (Station 1, bed 3).

The significant point is that the above spread of the founder genus, *Asmussia* Pacht, originally described from the Lower Devonian of "Livonia" and some related genera that evolved from its species, constitutes clear evidence of Mesozoic Southern Hemisphere dispersal to each of the four continents. (Other Lower Devonian forms of *Asmussia*, as well as Upper Permian species, are known from the Russian Arctic; see Novojilov, 1958c.)

Estheriella. This genus and related forms occur on three of the southern continents. (See *"Estheriella"* under Angola, Africa.)

A member of the estheriellid clan, originally described as *Estheriella taschi* Ghosh and Shah, has been reassigned, on the basis of restudy of all samples, to a new genus, *Cornutestheriella* Tasch. This genus occurs in the Lower Triassic Panchet Formation of India (EBP bore, East Bokaro Coal Field, at depth of 250 m). (Fig. 14.)

From Africa (Tanzania), the Triassic Upper Hatambulo beds, Janensch reported *Estheriella bornhardti* Janensch (now *Estheriella (Lioestheriata) bornhardti* Janensch). Defretin described three estheriellid species from Zaire: *Estheriella (Lioestheriata) lualabensis* (Leriche), *Estheriella (Lioestheriata) evrardi*, both of which are Jurassic (Kimmeridgian), and *Estheriella (Estheriella) moutai* Leriche, Upper Triassic of Angola. The Upper Jurassic and Lower Cretaceous of South America (Brazil) has a widespread estheriellid offshoot, *Graptoestheriella brasiliensis* (Oliveira) in the states of Bahía and Sergipe (Upper Jurassic-Lower Cretaceous). Other species of this member of the estheriellid clan are known from the Lower Cretaceous of the states of Paraíba and São Paulo. In the Will Maze Collection from the Bocas Formation, Santander massif, Colombia, this writer has found a poorly preserved probable estheriellid, *Estheriella (Estheriella)* sp. (Triassic-Lower Jurassic).

There is a definite India-Africa-South America Mesozoic dispersal track of the estheriellid bioprogram which, as indicated, includes that of Cornut*estheriella* and *Graptoestheriella.* The Upper Triassic estheriellids of Angola and Tanzania and those of Zaire's Jurassic are presumably related, considering the proximity of these two countries. (See Africa map insert.) Proximity of South America and Africa can readily link the Upper Jurassic of Zaire and the Upper Jurassic-Cretaceous of Brazil.

Palaeolimnadiopsis. This genus is represented in three southern continents. Brazil has a widespread representation of this genus and related genera at multiple sites in numerous states. The geologic spread includes the Permian, Triassic, Jurassic, and Cretaceous.

The Upper Permian Rio do Rasto Formation of Poço Preto, Santa Caterina State, contained *Palaeolimnadiopsis subalata* (Reed). The Upper Triassic Santa Maria Formation of Rio Grande do Sul State yielded an undetermined palaeolimnadiopsid species. *Palaeolimnadiopsis linoi* Cardoso has been described from the Lower Cretaceous Ilhas Formation, Bahía State; an undetermined palaeolimnadiopsid is known from the Botucatú Formation, Maranhão State, in beds of the same age; and in contemporary deposits, *Palaeolimnadiopsis freybergi* Cardoso, Areado Formation, Minas Gerais State. This last-named species is also known from the Upper Jurassic-Lower Cretaceous Souza Formation, Paraíba State. From this same formation, Pernambuco State, *Palaeolimnadiopsis* cf. *reali* Teixeira (formerly *Pterograpta*) has been reported. The Aliança Formation (Mirandíba, Pernambuco) of early-Late Jurassic age, contains another genus and species of the palaeolimnadiopsid clan, *Macrolimnadiopsis barbosa* Cardoso, and another species, *Macrolimnadiopsis pauloi* Beurlen, is known from the Pastos Bons Formation, Piauí State. *?Palaeolimnadiopsis suarezi* Messalira is a questionable taxa reported from the Baura Formation (Upper Cretaceous) São Paulo.

The only other South American country that has a palaeolimnadiopsid species is Bolivia. In the past this writer identified *Palaeolimnadiopsis* (cf. *P. eichwaldi* Netshajcv (Netchaev) 1894) from a Gulf Oil core slice from the Vitiacua Limestone (Permian-Triassic).

Africa's Mesozoic (Zaire) has a few species of this genus: *Palaeolimnadiopsis lubefuensis* Defretin (Haute Lueki Series, Upper Triassic or Jurassic), *Paleolimnadiopsis ndekeensis* Defretin (Upper Wealden, Loia Series), *Palaeolimnadiopsis lombardi* Defretin (Stanleyville Series, Upper Jurassic, Kimmeridgian), and an undetermined palaeolimnadiopsid species of the last-named series. *Palaeolimnadiopsis reali* Teixeira is known from the Upper Triassic of Angola.

Australia has two sites from which palaeolimnadiopsid species are now known. *Palaeolimnadiopsis bassi* Webb comes from the Hawkesbury Sandstone, Middle Triassic, Sydney, New South Wales, and *Palaeolimnadiopsis tasmanii* Tasch from the Knocklofty Formation, Lower Triassic of Tasmania.

Relevant to palaeolimnadiopsid distribution, note was made elsewhere (Tasch, 1980, p. 12) that dispersal potential appears to have existed between the USSR, North America, and South America. The USSR presumably has the geologically oldest known occurrence of the genus *Palaeolimnadiopsis* (*P. vilujensis* Varentsov) from Yakoutie, northeast Siberia. The age of the bed, Early Carboniferous, is probable but uncertain (Novojilov, 1958c, p. 99). Whenever it occurred in Carboniferous time, the Siberian species could be related to *Palaeolimnadiopsis carpenteri* Raymond, from Kansas Leonardian.

The Lower Permian of the United States could account for the Upper Permian Rio do Rasto species of *Palaeolimnadiopsis* from Brazil. All South American Mesozoic species of this genus could be linked to the Upper Permian population of Brazil.

In Africa, the oldest Mesozoic palaeolimnadiopsid came from the Angola Upper Triassic. Younger Mesozoic species (Zaire) could have evolved by serial egg dispersal from the Angola site, considering the proximity of both countries (see Fig. 1).

Because the geologically oldest palaeolimnadiopsid species in Australia comes from the Lower Triassic of Tasmania, it is the likely source of the Middle Triassic Australian mainland species noted above. What dispersal track can account for the occurrence of the Tasmanian species? A USSR–Australia track is one possibility; another is a southern dispersal track between the USSR and Antarctic, which is discussed below.

IDENTICAL AND/OR RELATED FOSSIL CONCHOSTRACAN SPECIES IN THE SOUTHERN CONTINENTS

Antarctica–Africa. Marlière's species, *Cyzicus (Lioestheria) malangensis,* is known from Angola (Cassanje III, region of Marimba, Upper Triassic) and Zaire (Haute Lueki Series equivalent to Cassanje III, Congo Basin, River Lueki and four other rivers). The Zaire specimens are dated as Upper Triassic or Jurassic. The same species was found in the Antarctic collection (Lower Jurassic) of this writer, from Blizzard Heights (Station 1, bed 7), and is represented by several specimens. Valve characteristics and parameters of the BH specimens overlap those described by Defretin for the Zaire representatives of the species.

The following three other species of the asmussid clan from the Congo Basin, Africa, have close equivalents or relatives in the Tasch Antarctic Collection.

1. *Glyptoasmussia luekiensis* Defretin from the Haute Lueki Series, Zaire, Cassanje III, Luhuhu River, and along some other rivers in the region (Upper Triassic or Jurassic). This species is also found in the Storm Peak, Antarctic samples; i.e., *Glyptoasmussia* cf. *luekiensis* Defretin, Upper Flow, beds 2 and 7, Lower Jurassic.

2. *Glyptoasmussia corneti* (Marlière) Defretin, Stanleyville Series, Upper Jurassic, overlaps the Blizzard Heights species (Station 1, beds 2 through 4, and 7), *Glyptoasmussia prominarma* n. sp., in five valve parameters while differing in configuration.

3. *Pseudoasmussiata defretini* n. sp. in Storm Peak, Upper Flow, bed 4 (Lower Jurassic) has near relatives in the Congo Basin Mesozoic (Upper Jurassic–Cretaceous). However, the three African species differ from the Antarctic species in configuration and many parameters. Evolution between the time lapse (Middle-to-Late Jurassic) that separates the occurrence in Antarctica from those in Africa could account for the observed variation.

Nonmarine dispersal of specific bioprograms, via conchostracan eggs, between the Congo Basin, Africa, and what is now the Blizzard Heights/Storm Peak region of Antarctica, may explain the above distribution. There is no evidence that conchostracan eggs can survive seawater immersion and transport. However, this writer's experiments have developed definite evidence that all the living conchostracans tested, immersed in 1 to 20% salinities, succumbed in 24 h. The longest surviving individuals were in the 1% solution (Tasch, 1963b, p. 1242). Eggs taken from living conchostracans in other experiments failed to hatch in solutions of any salinity, but did hatch in nonsaline water.

Another consideration is that in treating the same bioprogram at widespread and/or far removed sites, an originating time and place for the given genetic plan is a necessary requirement. It is such links that are denoted when speaking of dispersal tracks. The latter are invariably nonmarine and/or slightly brackish.

Australia–Africa. The Upper Triassic Phyllopod Beds of northern Angola contained fossils of *Palaeolimnadia (Palaeolimnadia)* cf. *wianamattensis* (Mitchell) (Cassanje I Series), originally described from the Australian rock column (Middle to Upper Triassic, Wianamatta Group). This species also occurs in the Upper Permian and, in the Canning Basin, Lower Triassic, as well as the Bowen Basin, Queensland Lower Triassic (Blackwater, Rewan Group). (See Tables 1 and 2, Appendix.)

A second Angola species, *Estheriina (Nudusia)* cf. *rewanensis* Tasch, also from the Phyllopod Beds, is a close equivalent of the Lower Triassic, *Estheriina (N.) rewanensis* Tasch, 1979c from Duaringa, Bowen Basin. This writer has also found this species in subsurface cuttings from the Blina Shale (Canning Basin, Lower Triassic).

There is one possibility that might account for the Bowen Basin estheriinids of the species cited above; via Antarctica, a South America-Australia (Bowen Basin) dispersal track during early Triassic time. Because most of Antarctica is covered by a continental ice sheet, a large part of its inaccessible rock column could contain fossil data on additional dispersal tracks to supplement those for which we have present knowledge. A more specific indicator for an Australian–Antarctica Triassic dispersal track will be noted in a subsequent section.

Australia–Antarctica. Mitchell (1927) described *Palaeolimnadia (Grandilimnadia) glenleensis* (M.) from the Australian Wianamatta Group (Middle to Upper Triassic. See Mitchell's types for description.) Sharing umbonal position and size, and valve configuration, as well as several valve parameters with Mitchell's species, is *Palaeolimnadia (Grandilimnadia)* cf. *glenleensis* (Mitchell) in the Tasch Antarctic Collections from Storm Peak (Lower Jurassic, Upper Flow, beds 2, 7, and 8).

Close proximity of Australia–Antarctica into Early Tertiary is well established by varied data, and such closeness can readily accommodate the Mesozoic linkage implied by the joint occurrence of the indicated species.

OTHER INTERRELATIONSHIPS BETWEEN SOUTHERN CONTINENTS

India–Antarctica. A unity in morphological pattern characterizes *Estheriina (Nudusia) indijurassica* Tasch (Kotá Formation, Lower Jurassic of India) and *Estheriina (Nudusia) stormpeakensis* Tasch (Storm Peak, Antarctica, Upper Flow, beds 1, 6, and 10). The pattern includes an arched dorsal margin, an umbo rising above the dorsal margin, a smooth, tapered umbo, and closeness of several ratios (H/L, a/H, b/H, and c/L). Nevertheless, the two species differ in configuration (ovate vs. subovate-elliptical) and in several ratios. It cannot be determined if the India population of *"indijurassica"* also had a larger dimorph, as did *"stormpeakensis."* The specimen available fits several of the parameters and valve characteristics, as noted above, of the smaller dimorph of the Antarctic equivalent. Furthermore, the umbo is inset in the India species but subterminal in the Antarctic species.

Obviously, these two species are distinct, while related. Accordingly, these data support the inference of a divergence of the two populations from a common source in pre-Early Jurassic time. Of significance is the restricted occurrence of Jurassic estheriinids, which were found only in India and Antarctica where the respective deposits had many estheriinid species.

India-Australia. If we follow the Curray and Moore (1974) predrift placement of India, that positions it proximate to both Antarctica and Australia, or the Veevers, and others (1971) placement of India proximate to Australia only, then it is logical to inquire what fossil conchostracan assemblages or individual species indicate dispersion between them?

Because west Australia (Canning Basin and elsewhere) has a Triassic conchostracan fauna, as does east Australia (Bowen and Sydney basins), as well as the Raniganj and East Bokaro coal fields of India, these will be compared.

The Andal area of west Bengal, India, contained *Palaeolimnadia* species as well as *Cornia panchetella* n. sp. in a subsurface bore (Panchet Formation, Lower Triassic). Similarly, as reviewed in a previous section, *Palaeolimnadia (P.) wianamattensis* (Mitchell) is known from the Lower Triassic of western Australia as well as eastern Australia (Arcadia Group), and *Cornia coghlani* (Etheridge, Jr.) occurs in the Lower Triassic (Sydney Basin). In the Middle-Upper Triassic Wianamatta Group, both of these species occurred together. The Australian rock column has yielded palaeolimnadid species from the Permian (Croudace Bay Formation, Tartarian) as well as Middle-Upper Triassic. The corniid fauna from India that were originally collected by this writer from a surface exposure at the village of Mangli (central India) included numerous specimens, but subsequent data from the subsurface of the East Bokaro Coal Field (Panchet Formation, Lower Triassic) produced abundant representatives of the new corniid species (*Cornia panchetella* n. sp.). No palaeolimnadids are known from the India Permian.

Clearly, the *Palaeolimnadia* and *Cornia* bioprograms were operative in India and Australia during Lower Triassic time. Where did India, lacking any Permian representatives of these genera, derive its Mesozoic species? One possibility is the dispersal of eggs from either the north or the south, the Russian Upper Permian, Kouznetsk Basin *(Cornia),* or the Lower Triassic of the Australian Basins, Sydney Basin *(Cornia),* and/or Bowen Basin *(Palaeolimnadia).*

South America-Africa-India. The joint occurrence of *Gabonestheria* and *Cornia* in the Upper Permian Rio do Rasto Formation of São Pascoal (Brazil) and in the older Permian of Gabon, Africa, as well as in the Upper Triassic of northern Angola (Cassanje I Series), and the Lower Triassic of India (Panchet Formation, East Bokaro Coal Field), points to a three-continent Paleozoic–early Mesozoic dispersal. As noted in a previous section, *Cornia* alone had a much wider distribution.

India-Antarctica-Australia. The latitudinal position of India and its geographic placement relative to Antarctica and Australia, as well as the timing of its northward drift, are important considerations in interpretation of the fossil conchostracan data of this memoir. This is especially so in the context of the findings of geophysical investigators and those of vertebrate paleontologists.

Physical contact of "southeastern peninsular India and Sri Lanka with a part of east Antarctica" has been postulated (Fedorov and others, 1982, p. 73): "The Mahanadi Graben (India) filled by the Permian coal measures of Gondwanic type is thought to have continued into the rift zones of Avery Ice Shelf and Lambert Glacier (East Antarctica) where similar coal measures with glossopterid flora are wedged among the crystalline basement blocks."

The fossil conchostracan data of this memoir (above, and see *Estheriina, Palaeolimnadia,* and *Cornia* discussions in this section as well as the India section) are bolstered by fossil fish evidence:

Seaways are great barriers to free migrations of freshwater fishes. It is difficult to explain the occurrence of freshwater pholidophorids unless the continents of India, Australia and Antarctica were in close proximity for migration during Early Jurassic. (Jain, 1980, p. 120).

Other Jurassic vertebrates (sauropod dinosaurs) in the Kotá Formation of India and elsewhere, led Colbert (1981, p. 279) to conclude:

. . . at this date peninsular India must have been in close contact with some other Gondwana areas.

Colbert envisioned a sauropod dinosaur migration between the Gondwana continents throughout Jurassic time (Colbert, 1981, p. 278, Fig. 3).

Gondwana was still largely intact at the end of Jurassic history. (Colbert, 1981, p. 280).

Thus, data from three independent sources; structural (Archean–Permian), the Paleozoic-Mesozoic invertebrate data (this memoir), and Jurassic vertebrates, converge in agreement on the proximity of India and Antarctica (Archean, Permian, Jurassic), India and Australia (Triassic), and, in fact, all of Gondwanaland (Jurassic) (Colbert, 1981).

As for the northward drift of subcontinent India, this is now thought to have begun 70 Ma (Late Cretaceous), whereas collision with Eurasia did not occur until some 20–30 Ma (Miocene). The velocity of the northward drift of India was an estimated 10 cm/yr for the first 30 m.y. and about one-half that rate for the next 40 m.y. (Molnar and Tapponier, 1977, p. 35). These data point to the relative slowness of India's northward drift and its late beginning, starting from lat 40°S in Cretaceous time (Molnar and Tappoiner, 1977, Fig., p. 35; cf. Colbert, 1973, p. 407), and sustain the other India evidence reviewed above.

South America–Africa. A review of all the available data on fossil conchostracans of Africa, after some reassignments, reveals a surprising fact—there is only a single endemic genus, *Afrolimnadia* (Cave Sandstone, Siberia, Republic of South Africa, Upper Triassic). By contrast, South America has five such genera known only from Brazil: *Aculestheria* (Candeias Formation, Upper Jurassic-Lower Cretaceous), *Graptoestheriella* (Botucatú and Souza Formations, Lower Cretaceous, and Japoata Formation, Upper Jurassic), *Acantholeaia* (Estrada Nova Group, Upper Permian), *Unicarinatus* (Iratí Formation, Lower Permian?), and *Macrolimnadiopsis* (Aliança and Pastos Bons Formations, Upper Jurassic).

The rest of the African genera share their respective bioprograms with South America. Zaire: *Palaeolimnadiopsis, Pseudoasmussiata, Echinestheria, Estheriella.* Angola: *Palaeolimnadiopsis, Estheriina, Cornia, Echinestheria, Gabonestheria,* and *Palaeolimnadia.* (The same applies to the Permian leaiid bioprogram; Zimbabwe, Republic of South Africa, Brazil, and Argentina, as remarked earlier.) Ubiquitous *Cyzicus* occurs all over South America and the other continents as well.

How can the sparse endemism in Africa be explained? Because endemism requires isolation through the development of fractionated gene pools, the implication follows that such opportunities were unavailable as far as conchostracans were concerned. The result was the mix of genera known in South America and other continents as well.

Of the several endemic genera in Brazil, two are Paleozoic and the rest, Mesozoic, ranging from Upper Jurassic to Lower Cretaceous. Their appearance in this time frame may be related to the postulated drift of South America relative to Africa from Early Jurassic in the south, to Early Cretaceous northward (off Gabon-Cameroon; Emery and others, 1975, p. 2257).

The drifting apart of South America from Africa was not complete until Late Cretaceous and occurred over a 30 m.y. span. During this protracted time, by drift-induced cumulative stress, a few former continuous areas in South America could have been sliced into several isolated segments which subsequently contained small freshwater situations. That would allow corresponding isolation of at least some conchostracan eggs from the parent population and hence be ultimately conducive to the appearance of a few endemic species. That is one possibility. Shrinkage of larger water bodies leading to separated ponds and pools is another possibility.

Successive conchostracan generations traceable back to the original wind-transported eggs, could undergo genetic changes not shared with their exotic ancestral species.

Since the drift of South America began in Early Jurassic time, it is apparent that, during all of the Paleozoic and most of the Mesozoic, it was in contact with Africa to a greater or lesser extent. Such contiguity facilitated the persistence, through the Early Cretaceous, of nonmarine dispersal potential. This readily explains shared pre-Tertiary bioprograms of conchostracan genera of the two continents noted above. (Cf. Krommelbein, 1966, on shared freshwater Cretaceous ostracodes.)

USSR–China–Antarctica–Australia. A two-way migration of terrestrial biota during the Permian and Triassic linked Gondwana and part of Asia (Du Toit, 1937, p. 127–128; Tasch, 1970b, p. 589). Analysis of the fossil insect fauna and its origin led Riek (1970, p. 598) to infer that the late Paleozoic influx (into Australia)

may have taken place over a long archipelago extending from southeast Asia through an island chain in the vicinity of New Caledonia and New Zealand.

Du Toit (1937) envisioned Madagascar and India as the "stepping stones" linking China and Australia. There are other interpretations. The interpretation that fits the data of this memoir postulates that peninsular India and "perhaps" China and Indochina remain joined to Gondwanaland up to the end of Cretaceous time (Colbert, 1973). Other evidence indicates a Devonian–Carboniferous dispersal link between mainland China and Australia (see below).

Placed in the above context, certain Australia, Antarctica, China, and other Asian fossil conchostracan and/or fossil insect occurrences during portions of Paleozoic–Mesozoic time can be better understood. The Carboniferous *Leaia*-bearing beds of west Australia (Canning Basin) and east Australia (Drummond Basin) are faunally related to equivalents in the Middle Devonian, Gitou Formation, People's Republic of China (cf. Shen, 1978).

Originally, this writer inferred that the link between China and Australia was directly to the Canning Basin (Tasch *in* Tasch and Jones, 1979a, p. 12). Now, however, a further study of all the data has led to the inference of direct dispersal from China to the Drummond Basin (east Australia), and from there a gradual dispersion of Leaia eggs occurred, via "stepping stones" to the Canning Basin. Because the Chinese Middle Devonian as well as the Lower Carboniferous of the Canning Basin both have *Leaia* and *Rostroleaia* species, a second later dispersion from China could have brought eggs of *Rostroleaia* to the Canning Basin as well as more eggs of *Leaia* species. The Grant Formation of the Canning Basin yielded four *Leaia (Hemicycloleaia)* species and one *Leaia (Leaia)* species (Tasch *in* Tasch and Jones, 1979a, Figure 2).

Hemicycloleaia was reassigned as a subgenus of *Leaia* by this writer. This subgenus occurs in abundance in the lower Stephanian of the Donetz Basin (USSR) and the Westphalian of the Karagand Coal Basin, as well as in the Chinese Middle Devonian and Permian. It also is known from the Russian Permian and the Australian Carboniferous and Permian. (See Australia section, Newcastle Coal Measures, this memoir.) (Professor W. T. Chang [1980, personal commun.] informed the writer of a discovery of more Devonian conchostracan-bearing sites and the first Chinese Upper Permian leaiid collections from three provinces: Hunan, Kwangsi, and Kansu).

At the time (1979a) this writer was surprised to find a conchostracan genus, *Ellipsograpta,* described originally from the

Chinese Cretaceous (Nenkiang Shale), represented in the Grant Formation of the Australian Carboniferous. However, subsequent discoveries by Chinese colleagues in the Devonian and Permian made a Chinese connection quite plausible.

A probable homopteran insect from the Permian Polarstar Formation, Sentinel Range, [Ellsworth Mountains], Antarctica, had relatives in Australia and the USSR (Tasch and Riek, 1969, p. 1529–1530). The Russian connection extends to the Antarctic Peninsula, Mt. Hope, from which site Zeuner described a Jurassic beetle. It also extends to equivalent forms in the Triassic of the Kouznetsk Basin as well as the Upper Triassic of New South Wales, Australia. Another Antarctic site (Carapace Nunatak, southern Victoria Land) had mayfly nymphs with analogues in the Jurassic of Asia and Siberia (USSR). (For further details, see Carpenter, 1969, p. 418–425; Tasch, 1970b, p. 589–590.)

Fossil conchostracan Paleozoic ties to the Russian rock sequence were discussed above for *Leaia (Hemicycloleaia)* species. Here, note can be made of an east Siberian conchostracan, *Cyzicus (Euestheria) ipsviciensis* (Mitchell) 1927, found in the lower-Middle Jurassic (Novojilov and Kapel'ka, 1968, p. 117–129, Pl. B, Fig. 3 and text-Fig. 1); site F 15, depth 257 m, Ienissei, 60 km north of Krasnoviarsk, USSR. This species was originally described from the Upper Triassic Wianamatta Series of Denmark Hill, Queensland, Australia. Kobayashi (1975, p. 65) found this report "quite astonishing," but it is readily comprehensible when one considers the above-cited data as well as other data of this memoir, despite the time lapse. Yet undiscovered links are indicated.

The evidence is not confined to the example cited above, nor is it restricted to an example with a time gap. There are Russian specimens from the Permian (Tartarian) that were assigned to several of Mitchell's species from the Newcastle Coal Measures (Novojilov, 1958c). Included are: *Glyptoasmussia belmontensis* (Mitchell) (=*Cyzicus (Euestheria) belmontensis* (Mitchell), from the deposits of the Laptev Sea, as well as from the synchronous beds of the Korenevskaia Series in the Dniepr-Donetz Basin (Novojilov, 1958b, p. 38); *Pseudestheria novocastrensis* (Mitchell) [=*Cyzicus (Euestheria) novocastrensis* (Mitchell)]

from the littoral deposits of the Laptev Sea and the Lower Toungouskaia (Novojilov, 1958b, p. 33–34); *Pseudestheria obliqua* (Mitchell) [=*Cyzicus (Euestheria) obliqua* (Mitchell)], from a river south of the Gulf of Khatanga, Cape Ilia, Misailap Series (Novojilov, 1958b, p. 33).

Palynomorph analysis of this writer's Antarctic samples from Carapace Nunatak, Storm Peak, and Coal Sack Bluff (Tasch and Lammons, 1978, p. 455–460) led to some interesting findings: "The assemblage reported by Filatoff, 1975, from the Jurassic Evergreen Formation of the Perth Basin included 67 percent of the species found in the Transantarctic Mountains (=TAM). Similarly, Sub-Assemblage A_2 (Carapace Nunatak, Station 2, bed 1) included 62 percent of the same TAM composite assemblage" (idem p. 457). Perth Basin is in Western Australia and so it is equally relevant that correspondences were found, though in "lesser degrees . . . between the Antarctic samples and assemblages from the Jurassic Northeast Surat Basin" (idem, p. 455). (Cf. Kyle and Fasola, 1978, p. 313–318, for Beardmore Glacier, Antarctica, Triassic correspondences, to equivalent palynomorph suites in the Triassic of Queensland, Australia.)

The above correspondence in palynomorph assemblages of the Jurassic of Antarctica and Australia is readily explained by Heirtzler's report (*in* Frakes and Crowell, 1971, p. 731) that the oldest age of the ocean floor between the two continents is 40 Ma, thus denoting a generally accepted long-term proximity of the two continents into Tertiary time.

Absence of conchostracan-bearing beds in the Australian Jurassic contrasts with its productive Jurassic microflora. Because Antarctican and Australian basins during Jurassic time had high percentages of common palynomorph assemblages, it is plausible to assume that had conchostracan eggs of that age been available in Australian basins, they would also have been dispersed by winds. That would have established a Jurassic link with Antarctica. In this connection present data indicate that no airborne conchostracan eggs were dispersed from Antarctica to Australia during Jurassic time. This probably denotes restricted directions for wind currents at the time.

Summary

1. Evidence assembled and reviewed in this memoir establishes the recurrent existence in the geologic past of nonmarine dispersal opportunities between continents of the Southern Hemisphere. Traceable dispersal tracks deciphered from fossil conchostracan distributions of key taxa link Antarctica, South America, Africa, India, and Australia during some of Paleozoic and Mesozoic time. Independent of any other evidence, these data require proximity of continents.

2. Other geological, geophysical, oceanographic, fossil vertebrate, fossil insect, and palynological evidence reenforce the conclusions reached in this memoir. These other studies, exclusive of Tasch and Lammons (1978) at the time undertaken, were unrelated to each other or to this research report.

3. Fossil conchostracan data from parts of the Paleozoic/-Mesozoic of Antarctic and/or Australia point to nonmarine dispersal tracks from the USSR and the Peoples Republic of China to these southern continents.

4. During the Mesozoic, parts of India, Australia, and Antarctica shared conchostracan and other bioprograms. That requires a continental reassembly that places India in contact with both of these southern continents. At present, proximity is generally shown with one or another continent in published reconstructions of Gondwanaland.

5. The evidence of shared Jurassic conchostracans between Africa and Antarctica requires a more precise reassembly of these two land masses.

6. If the nonmarine distribution cited in items 3 to 5 above are ignored or bypassed in reassemblies of the pre-drift Gondwana continents, some erroneous continental placements are bound to ensue.

7. In Antarctica, Jurassic fossil conchostracan data documents the spread of similar bioprograms between Storm Peak and Blizzard Heights and its environs; and between Storm Pcak, Carapace Nunatak (southern Victoria Land), Agate Peak and Gair Mesa (study in progress; northern Victoria Land). Large distances separate the several fossiliferous sites in northern and southern Victoria Land. The likelihood of some intervening fossil conchostracan sites between these four localities is probable.

Note added in proof. Correlation Coefficients for variables and ratios, clustered collections of *Cyzicus (Lioestheria) disgregaris* from Gair Mesa stations with representatives of this species at Storm Peak, Carapace Nunatak and Agate Peak stations. A check by Comparable Population statistics (90 percent confidence level) sustained this finding (cf. p. 71).

Note added in proof. A preliminary report (personal communication, 6/12/87) by Shekkar Chandra Ghosh (Geological Survey of India) included sketches of fossils recently found in Central India. These unequivocally represented ribbed forms of the genus *Leaia.* Until now, leaiids were known from South America, Africa, Australia, and Antarctica. Their absence from India was a missing link in the Gondwanaland dispersal and puzzled the writer who had reasoned that they should be there. Tasch and S. C. Ghosh searched for them during the 1971–1972 season in the Raniganj Basin but found only nonribbed forms described herein. This new report, if confirmed by detailed study, will solve the puzzle and further support this Memoir's thesis.

Plate 1

AFRICA

All Specimens from British Museum (Natural History) (=BM)

Figures 1, 2. *Cyzicus (Lioestheria) draperi* (Jones). **1:** I-3233, syntype, left valve, L-17.0 mm. Arrow points to smaller valve partly covered by larger valve, but showing growth bands on umbo that are eroded from larger specimen. **2:** Detail of punctate ornamentation of Figure 1, enlarged. Cave Sandstone, east side of Platberg, Drakensberg, Natal, Republic of South Africa (R.S.A.) Upper Triassic.

Figure 3. *Cyzicus (Euestheria) anomala* (Jones). BM-6130 (formerly 302A-1), holotype, *non* Syntype (Morris, 1980), left valve and eroded upper interior of right valve. Ornamentation close-packed granules, L-4.8 mm. Basal unit Uitenhage Series, Swellendam District, R.S.A. Lower Cretaceous.

Figures 4, 5. *Cyzicus (Euestheria) greyi* (Jones). **4:** BM I-2436a, holotype, right valve, arrow points to faint growth bands on umbo, L-3.2 mm. **5:** BM I-2436b, paratype, right valve shows faint growth bands on umbo, L-3.2 mm. Near Cradock, Cape Province, R.S.A. Upper Permian.

Figures 6, 7. *Leaia (Hemicycloleaia) cradockensis* n. sp. **6:** Paratype, BM-I 2436b, left valve, enlarged (incomplete, covered), length undetermined. Note break of valve below posterior rib indicating that the posterior termination of rib may have broken off with a portion of valve. **7:** BM-I-2436a, holotype, left valve on same slab as multiple specimens of *Cyzicus (Euestheria) greyi* (Jones), L-2.8 mm. Same as *"greyi"* above.

Plate 2

AFRICA

Morocco, Republic of South Africa, Zimbabwe

Figures 1, 2. *Leaia (Leaia) bertrandi* n. sp. National Collection of Morocco (NCM), hs, g λ **1:** Slab 1, side B, specimen "a," holotype, mold of right valve, L-5.5 mm. Haut-Atlas, Hamalou-Agadir, L-4, Morocco (Stephanian). **2:** Paratype, slab 2, side B, mold of right valve, enlarged.

Figures 3, 4. *Cyzicus (Euestheria) simoni* (Pruvost). NCM, hw-93-q 14, dimorphs. **3:** short form, left valve, Djerada, Morocco, L-3.8 mm. **4:** long form, right valve, L-4.6 mm (Westphalian).

Figure 5. *Leaia (Hemicycloleaia) sessami* Bond. Geological Collection, National Museum of Zimbabwe (NMZ) 7196(A), rubber imprint of holotype, right valve, L-7.0 or more mm. Near Madziwadzide's kraal, Madumabisa shale, Lower Beaufort, Zimbabwe, Upper Permian.

Figure 6. *Leaia (Hemicycloleaia) cradockensis* n. sp. British Museum (Natural History) slab I-2436A (see arrow which indicates the location of holotype on slab). See Pl. 1, Fig. 7, for details.)

Figure 7. *Leaia (Hemicycloleaia) sengwensis* Bond. NMZ 7148, rubber imprint of holotype, representing an external mold of left valve, L-4.0 mm. East of Sengwe River and north of Nenyunka's kraal, Upper Madumabisa Mudstone (K^{5c} Zimbabwe). Upper Permian.

Plate 3

AFRICA

Republic of South Africa, Lesotho, Algeria

Figures 1, 2. *Afrolimnadia siberiensis* n. g., n. sp. **1:** TC 30027, right valve, holotype, L-2.8 mm; note large, anteriorly directed umbo and variable spacing of growth bands. **2:** TC 30028, paratype, left valve, L-12.7 mm. Siberia, R.S.A., Upper Triassic.

Figures 3, 4, 6. *Cyzicus (Euestheria) lefranci* n. sp. **3:** TC 4055, right valve, holotype, L-4.0 mm. **4:** 40061, internal mold of left valve, paratype I, L-3.8 mm. **6:** TC 4007, paratype 2, right valve, with well-preserved euestheriid ornamentation, L-3.2 mm. Äin-el-Hadjadj, Plateau du Tademait, Algerian Sahara. (Senonian) Cretaceous. (See Appendix for details.)

Figure 5. *Cyzicus (Lioestheria?)* sp. undet. (Photo taken by the curator, Albany Museum, Grahamstown, R.S.A.) No. 4165, portion of left valve and underlying internal mold, L-7.5 mm. Witteberg Series, Cape System, R.S.A., Lower Devonian.

Figure 7. *Cyzicus (Euestheria) thabaningensis* n. sp. TC 30337c, left valve, holotype, L-2.1 mm. Cave Sandstone, Thabaning, Lesotho. Upper Triassic.

Plate 4

AUSTRALIA, AFRICA

Figure 1. *Cyzicus (Euestheria) talenti* n. sp. TC 10901, syntype, external mold of right valve, L-4.1 mm. Fairy Beds, Buchan Caves Limestone, Victoria, Australia.

Figure 2. *Palaeolimnadia (Grandilimnadia) glenleensis* (Mitchell). Macquarie University New South Wales (NSW). M.U. 3682, internal mold of left valve, L-2.2 mm. Hornsby diatreme, Hawkesbury Sandstone, New South Wales, Australia.

Angola

Figures 3, 6. *Cornia angolata* n. sp. **3:** TC 90004, holotype, Oesterlen site XP/63/2, internal mold of right valve, L-2.5 mm. Upper Cassanje Series, Karoo System, Northern Angola. Upper Triassic. (See Appendix for details.) **6:** TC 90742, paratype, L-2.5 mm, internal mold of right valve. Same site as Figure 3.

Figure 4. *Palaeolimnadia (Palaeolimnadia)* cf. *wianamattensis* (Mitchell), TC 90001, internal mold of left valve. Oesterlen site XP/63/2. L-1.3 mm. (See 3 above for series, system, age.)

Figures 5, 7. *Palaeolimnadia (Grandilimnadia) africania* n. sp. **5:** TC 90900, holotype, left valve, L-1.9 mm, Oesterlen site, XP/72/6b. (see 3 above for series, system, age.) **7:** TC 90729, paratype, left valve, L-1.8 mm. Same site as holotype.

Figure 8. *Palaeolimnadia (Grandilimnadia) oesterleni* n. sp. TC 90010, holotype, left valve, L-2.2 mm. Oesterlen site, XP/72/6. (See 3 above for series, system, age.)

Figure 9. *Estheriina (Nudusia)* cf. *rewanensis* Tasch 1979. TC 90008, holotype, right valve, note distinct definition of umbo in contrast to rest of valve. Oesterlen site XP/72/6b. (See 3 above for series, system, age.)

Figure 10. *Gabonestheria gabonensis* (Marlière). TC 90009, internal mold of left valve; note large umbonal area terminating in a prominent node, L-2.4 mm. Oesterlen site XP/72/7. (See 3 above for series, system, age.)

Plate 5

AFRICA

Morocco, Mozambique, Zambia

Figures 1, 2, 3. Morocco ribbed forms. Ink sketches by Termier and Termier, 1950. Reassigned herein from *Leaia* cf. *tricarinata* Meek and Worthen (Termier and Termier, 1950). **1:** *Leaia (Hemicycloleaia)* sp. (Termier and Termier, 1950, Fig. 42), L-9.0 mm. **2:** *?Estheriella* sp. (Termier and Termier, 1950, Fig. 43), L-8.0 mm. **3:** *?Leaia* (Termier and Termier, 1950, Fig. 44), L-5.5 mm. Nord de Hamahou, Ida ou Zal, Morocco. Westphalian.

Figure 4. *Cyzicus (Lioestheria) lebombensis* (Rennie), (sketch by Rennie, 1937). Slab No. 11, holotype. Sedimentary interbeds of the Rhyolite Lebombo Volcanic Formation, Namahacha, west of Lourenço Marques, Mozambique, L-6.9 mm. Jurassic (Lias).

Figure 5. *?Asmussia borgesi* (Teixeira). Sketch by Teixeira, 1943 (photographs are less clear than the sketch); left valve, plant-bearing. Tête, Mozambique. L-1.8 mm. Tentatively assigned to *Cyzicus (Lioestheria)*.

Figure 6. *Cyzicus (Lioestheria) welleri* (Bond). (Sketches by Bond, 1964.) Right valve, irregular punctation. Bedded, micaceous siltstone above the Escarpment Grit, Luano Valley, Zambia. Upper Permian or Lower Triassic.

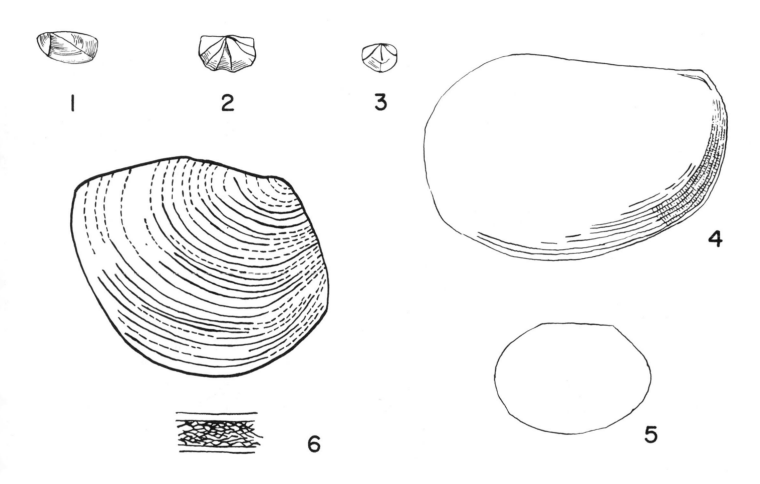

Plate 6

ANTARCTICA

Brimstone Peak, Lashly Mountains, Portal Mountain, Mauger Nunatak

Figure 1. *Cyzicus (Lioestheria) rickeri* n. sp. Syntype, British Museum (Natural History) No. L-3452, internal mold of right valve. Other components of the syntype are fragments; note hachure-like markings on growth bands. Locality M.S. 116, Brimstone Peak. Jurassic. L-3.2 mm.

Figures 2, 3. *Cyzicus (Euestheria) lashlyensis* n. sp. Syntype, New Zealand Geological Survey. 2: No. GS 7399/8, internal mold of left valve. Note contiguous granules on growth bands L-3.2 mm. 3: Syntype, GS 7399/4. Crushed and eroded valve fragment with ornamentation as in 2 above, enlarged. Sites: Transantarctic Expedition Localities 388 and 432, Lashly Mountains, Middle–Upper Devonian, enlarged.

Figures 4, 5. *Cyzicus (Euestheria) ritchiei* n. sp. 4: Holotype, Australian Museum (Sydney), F 54956f, crushed portion of left valve underlain by an internal mold; euestheriid ornamentation clearly seen on growth bands. L-3.8 mm. 5: Paratype, F 54956c, crushed portion of left valve underlain by an internal mold. Portal Mountain, southern Victoria Land, Upper Devonian, L-3.6 mm.

Figures 6, 7. *Cyzicus (Lioestheria) longacardinis* n. sp. 6: Syntype, TC 8033a, internal mold of left valve, L-4.2 mm. 7: detail of 6 showing characteristic hachure-type markings on growth bands. Mauger Nunatak, Queen Maud Mountain, Antarctica, D.E. Site 62-10-BC, Jurassic.

Figures 8, 9. *Palaeolimnadia (Grandilimnadia)* sp. 1. 8: Internal mold of left valve, TC 8055a; the specimen is cracked down the middle and has been glued together. Note large umbo and five growth bands; L-1.9 mm. 9: enlargement of 8. Mauger Nunatak, Antarctica, lat 85°44′S, long 176°40′E. D.E. Site 62.4, Jurassic. (See Appendix and text for section data.)

Plate 7

ANTARCTICA

Mauger Nunatak

Figures 1, 2, 3, 4. *Cyzicus (Lioestheria) maugerensis* n. sp. **1:** Syntype, TC 8316a, internal mold of left valve overlain by patches of inner layer of original valve; note cancellate appearance due to close spacing of growth bands bearing hachure-like markings, L-3.3 mm. **3:** detail of 1, showing the pseudo-cancellate ornamentation, enlarged. **2:** Syntype, TC 8310b, internal mold of left valve overlain by fragments of inner layer of original valve, L-2.9 mm. **4:** Syntype, TC 8313, detail, showing common ornamentation of the species, enlarged. D. E. Sites 62-10-1A, 62-10-B, 62-10-BC, and 62-10-2B, Mauger Nunatak, Jurassic.

Figure 5. *Cyzicus (Lioestheria) longacardinis* n. sp. Syntype, TC 8030b, internal mold of left valve eroded along dorsal margin, under remnant of inner layer of original valve. L-4.2 mm. Mauger Nunatak, D. E. Site 62-10-BC, Jurassic.

Figures 6, 7, 8. *Cyzicus (Euestheria) juravariabalis* n. sp. **7:** Syntype, TC 8027 and fragment of inner layer of left valve underlain by internal mold. L-2.1 mm. **8:** Syntype, TC 8027c, same as 7; L-3.4 mm. **6:** TC 8311, fragment showing euestheriid ornamentation of contiguous granules, enlarged. D. E. Site 62-10-BC, Mauger Nunatak, Jurassic.

Plate 8

ANTARCTICA

Blizzard Heights

Figures 1, 3. *Cyzicus (Lioestheria) malangensis* (Marlière). **1:** TC 11109A, internal mold of right valve, L-5.5 mm, Blizzard Heights (BH), Station 1, bed 6. **3:** Detail of 1; note hachure-type markings on growth bands, enlarged.

Figures 2, 4, 5, 6. *Cyzicus (Euestheria) formavariabalis* n. sp. **2:** Syntype, TC 11331a, internal mold of subovate left valve. L-6.2 mm, BH, Station 1, bed 5. **4:** Detail of 2 showing euestheriid ornamentation on growth bands, enlarged. **5:** Syntype, TC 11331c, remnant of shell of original right valve overlying internal mold. L-5.3 mm, BH, Station 1, bed 4. **6:** Syntype, TC 11331d, internal mold of left valve with slight remnant of inner layer of original shell. L-5.0 mm, BH, Station 1, bed 4.

Figure 7. *Glyptoasmussia* sp. 1. TC 11502, internal mold of left valve. L-3.4 mm, D. E. Site 23 (equivalent to Tasch BH, Station 0).

Figure 8. *Cornia* sp. 1. TC 10014A, internal mold of left valve. L-3.1 mm. BH, Station 0, bed 3. (Arrow indicates terminus of tapered, coiled-under, spinous terminus of an umbonal node.)

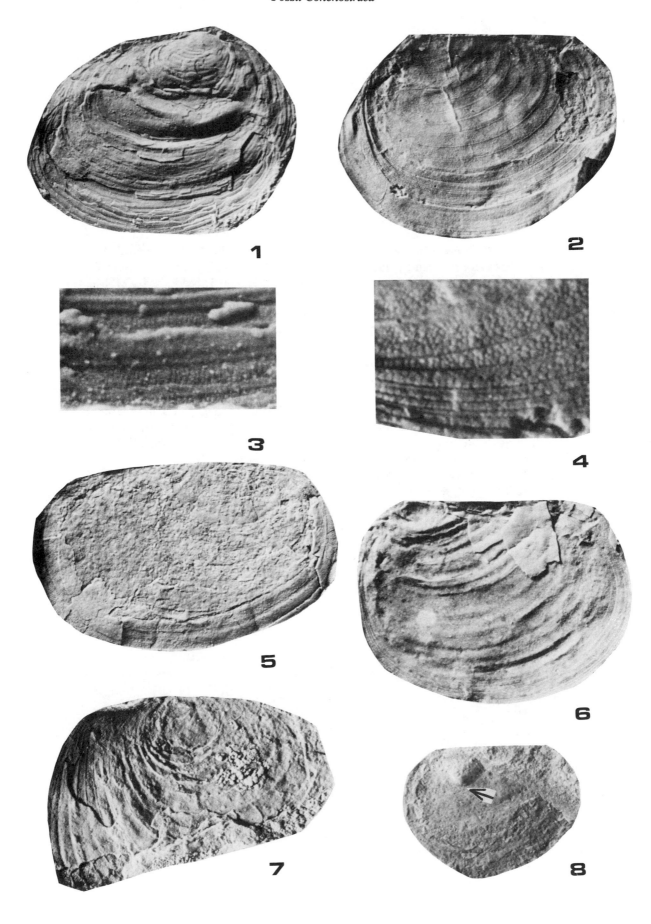

Plate 9

ANTARCTICA

Blizzard Heights

Figures 1, 2, 3. *Cyzicus (Lioestheria) malangensis* (Marlière). **1:** TC 11214A, internal mold of right valve overlain by fragment of shell. L-4.6 mm. BH, Station 1, bed 7. **2:** TC 11233B, internal mold of left valve, L-4.5 mm, BH Station 1, bed 7. **3:** TC 11184c, internal mold of left valve, L-5.5 mm. BH, Station 1, bed 7.

Figure 4. *Cyclestherioides (Cyclestherioides) alexandriae,* n. sp. For type, see Storm Peak entries (Pl. 14, Figs. 7, 8). TC 11225c. L-5.1 mm. BH, Station 0, bed 6.

Figures 5, 7. *Cyzicus (Lioestheria) antarctis,* n. sp. **5:** TC 11117A, paratype, internal mold of right valve and remnants of inner layer of shell. L-5.7 mm. **7:** Paratype, TC 11117C, left valve (same slice as 5). L-6.5 mm. BH, Station 1, bed 4. (For holotype, see Pl. 12, Fig. 3.)

Figure 6. *Estheriina (Estheriina)* sp. 1. TC11319-05A, right valve and part of inner layer of original valve with underlying internal mold (visible umbo and growth bands on it). L-3.5 mm. BH Station 1, bed 4. (Note distinctness of umbonal area compared to rest of valve.)

Figure 8. *Cyzicus (Lioestheria)* cf. *brevis* (Shen). TC 11074, right valve, internal mold and overlain fragments of original valve. L-5.7 mm. BH, Station 0, bed 3.

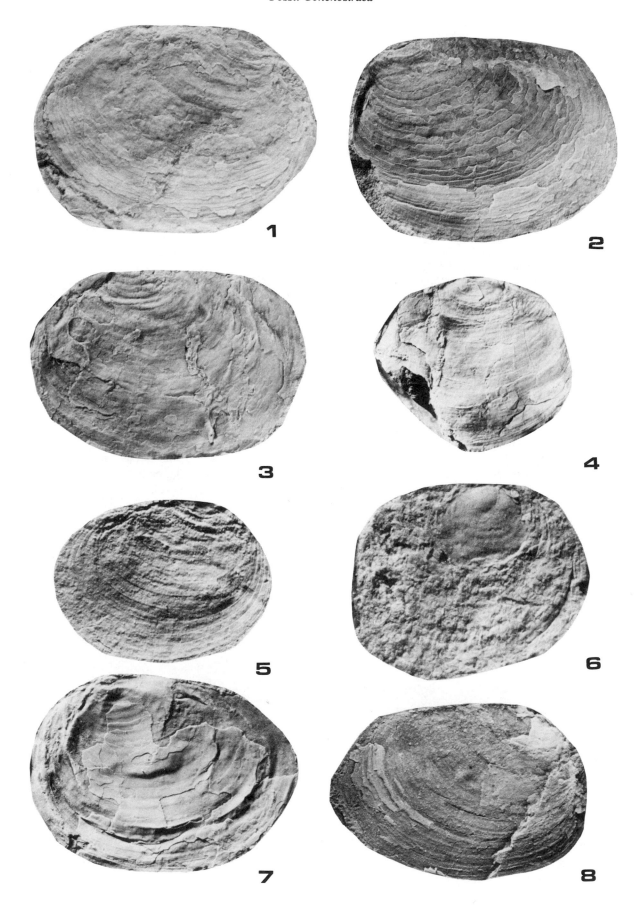

Plate 10

ANTARCTICA

Blizzard Heights

Figure 1. *Cyzicus (Euestheria) formavariabalis,* n. sp. TC 11149, internal mold of right valve, enlarged. BH, Station 1, bed 2b.

Figure 2. *Estheriina (Nudusia)* sp. 2. TC 11400A, external mold of left valve; note smooth umbo. L-5.2 mm. BH environs. D. E. Site 23.59b (equivalent to Tasch Station 0, bed 3).

Figure 3. *Estheriina (Nudusia) stormpeakensis* n. sp. TC 11066-10. L-5.5 mm. BH, Station 0, bed 3. (Cf. SP, TC 9300 under "Storm Peak" section, this memoir.)

Figures 4, 6. *Cyzicus (Euestheria) beardmorensis* n. sp. **4:** Holotype, slab 1a, internal mold of left valve with remnant of inner portion of the original valve overlying. L-5.1 mm. **6:** Paratype, slab 1b, internal mold of right valve, L-4.9 mm. D. E. Site 23.59AA (equivalent to Tasch Station 0, bed 2). (See text-Fig. 4 for D. E. beds.)

Figure 5. *Cyzicus (Lioestheria) longulus* n. sp. Holotype, TC 11097, remnants of inner layer of subelliptical right valve and its internal mold. L-5.5 mm. BH, Station 0, bed 3.

Figure 7. *Glyptoasmussia meridionalis* n. sp. Holotype, TC 11245A, left valve, fragment of inner layer of valve and underlying internal mold. L-5.5 mm. BH, Station 0, bed 3.

Figure 8. *Glyptoasmussia prominarma* n. sp. Holotype, TC 11127.01c, left valve impressed into substrate; one carbonized layer beneath patches of a white enameloid layer is exposed. L-3.5 mm. BH, Station 1, bed 6.

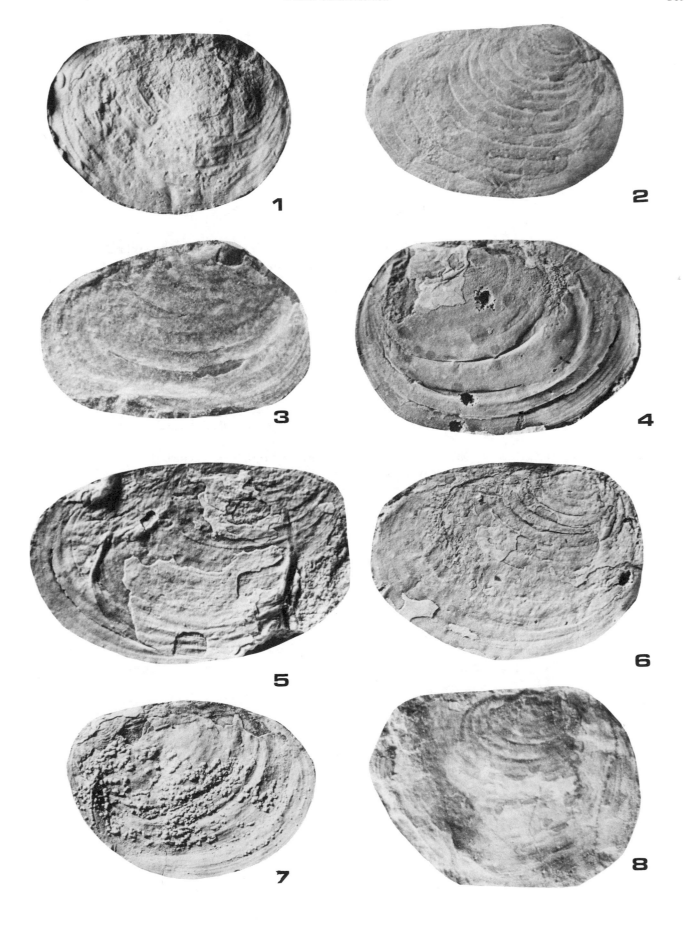

P. Tasch

Plate 11

ANTARCTICA

Blizzard Heights and Peak

Figures 1, 8. Cyzicus (Euestheria) ellioti n. sp. **1:** Holotype, TC 11010, internal mold of right valve. L-6.0 mm. **8:** Paratype 1, TC 11039, right valve remnant and underlying mold. L-5.4 mm. D. E. Site 23.59AA (equivalent to BH, Tasch Station 0, bed 2).

Figures 2, 3, 4. *Cyzicus (Lioestheria) malangensis* (Marlière). **2:** TC 11328A, part of left valve and its underlying internal mold. L-4.5 mm. **4:** TC 11328B, internal mold of right valve. L-5.1 mm. **3:** Detail of TC 11328B, showing ornamentation, enlarged. BH, Station 1, bed 2A.

Figure 5. *Estheriina (Nudusia?) brevimargina* n. sp. Holotype, TC 11405A, inner layer of flattened left valve. L-3.8 mm. Blizzard Peak, D. E. Site 33.56A (See text-Fig. 4 for locality map.)

Figure 6. *Cornia* sp. 2. TC 11420, inner part of valve and underlying mold. Arrow points to minute tapered spine. L-2.7(?) mm. D. E. Site 23.18c (equivalent to Tasch Station 0, bed 3).

Figure 7. *Cyzicus (Euestheria) minuta* (von Zieten) emend. Defretin-LeFranc. TC 11293, internal mold of left valve. L-3.3. BH, Station 1, bed 6.

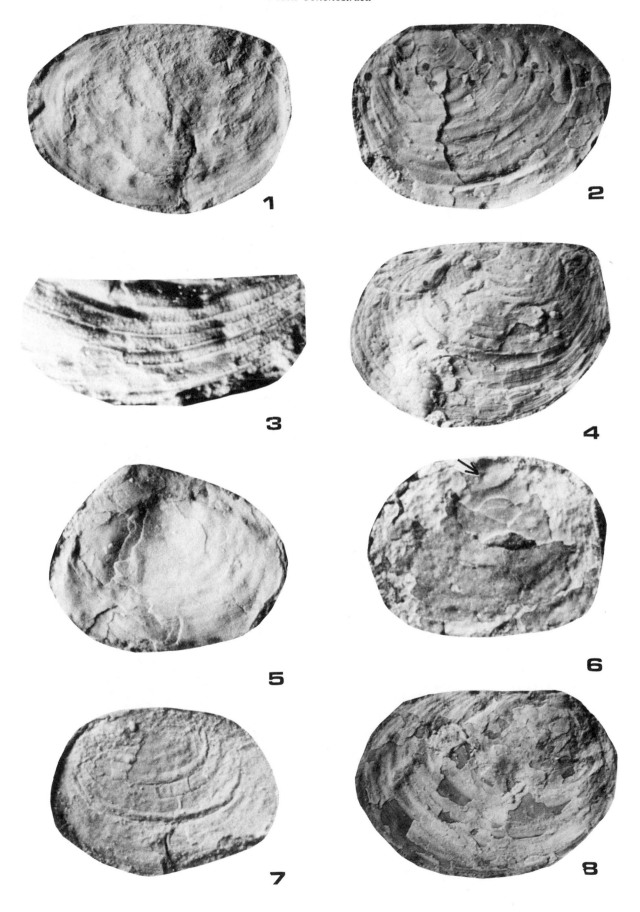

Plate 12

ANTARCTICA

Blizzard Heights

Figures 1, 2. *Cyclestherioides (Cyclestherioides) alexandriae* n. sp. **1:** TC 11153B, internal mold of left valve. Note short dorsal margin and submedial umbo. L-4.7 mm. **2:** TC 11153A, internal mold of left valve. L-3.6 mm. (See Storm Peak section for species diagnosis.) BH, Station 1, bed 2A.

Figure 3. *Cyzicus (Lioestheria) antarctis* n. sp. Holotype, TC 11313, internal mold of right valve, L-5.5 mm. BH, Station 1, bed 4.

Figure 4. *Cyclestherioides (Sphaerestheria)* cf. *aldensis* Novojilov. TC 11280b, external mold of right valve. L-2.7 mm. BH, Station 1, bed 5.

Figures 5, 7. *Cyzicus (Euestheria) formavariabalis* n. sp. **5:** TC 11189B, left valve, internal mold and partial overlay by inner valve layer. L-5.5 mm. **7:** TC 11189A, carbonized internal mold of right valve, partly overlain. L-5.4 mm. BH, Station 1, bed 2b.

Figures 6, 8. *Glyptoasmussia meridionalis* n. sp. **6:** TC 11245G, paratype 1, internal mold of left valve. L-5.4 mm. **8:** TC 11245B, paratype 2, right valve, remnant of inner part of original valve and underlying mold. L-5.6 mm. (All types from BH, Station 0, bed 4.)

Plate 13

ANTARCTICA

Storm Peak

Figures 1, 2, 3, 4. *Estheriina (Nudusia) stormpeakensis* n. sp. **1:** Syntype, TC 9300, internal mold of left valve, L-1.9 mm. **2:** Syntype, dimorph 1, TC 9015, internal mold of right valve with partially eroded umbo. L-5.0 mm. Storm Peak (SP), Upper Flow, bed 6, lower half (for 1, 2 above). **3:** Syntype, TC 10121c, external mold of right valve. L-1.7 mm. SP, Upper Flow, bed 1, lower. **4:** Syntype, dimorph 2, TC 9119c, internal mold of left valve. L-4.0 mm. SP, Upper Flow, bed 10.

Figures 5, 6. *Cyzicus (Euestheria) ichthystromatos* n. sp. **5:** Syntype, TC 9200, juvenile, left valve with internal layer of original valve. L-3.3 mm. D. E. Site 27.85 (equivalent to SP, Tasch Station 0, bed 4) Lower Fish Bed. **6:** Syntype, TC 9143B, inner portion of right valve, L-5.5 mm. SP, Lower Flow, Tasch Station 2, bed 4 (a correlate of SP, Station 0, bed 4). (See Appendix for section.)

Figure 7. *Glyptoasmussia australospecialis* n. sp. Holotype, TC 9205, internal mold of right valve. Note eccentrically ovate shape and brief dorsal margin. L-3.0 mm. SP, Upper Flow, bed 11.

Figure 8. *Palaeolimnadia (Grandilimnadia)* cf. *glenleensis* (Mitchell). TC 9857a, external mold of right valve. L-2.6 mm. SP, Upper Flow, bed 7.

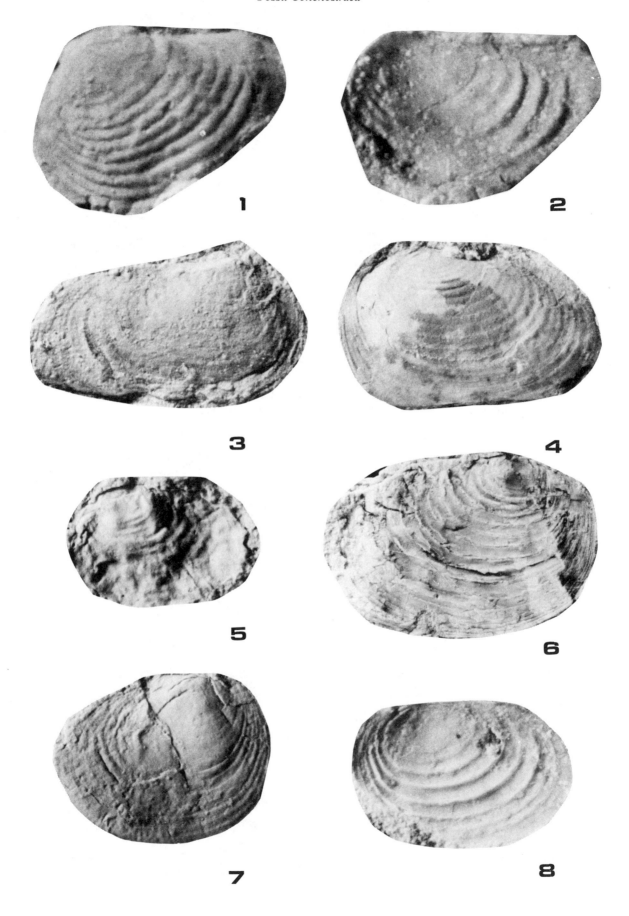

Plate 14

ANTARCTICA

Storm Peak

Figures 1, 2, 3, 4. *Cyzicus (Euestheria) crustapatulus* n. sp. **1:** Syntype, TC 10016b, dimorph-D2, internal mold of right valve with overlay of remnant of inner portion of original valve. L-4.2 mm. **2:** Syntype, 10016a, internal mold as in 1. L-6.3 mm. **3:** Syntype, TC 10016d, detail of ornamentation. Greatly enlarged. **4:** Syntype, TC 10016f, internal mold as in 1. L-5.4 mm. SP, Upper Flow, bed 5.

Figures 5, 6. *Palaeolimnadia (Grandilimnadia) minutula* n. sp. **5:** Paratype, TC 10007, internal mold of left valve. L-0.97 mm. **6:** Holotype, TC 10004, external mold of right valve, inner portion of valve preserved. L-1.4 mm. SP, Upper Flow, bed 6, lower half.

Figures 7, 8. *Cyclestherioides (Cyclestherioides) alexandriae* n. sp. **7:** Holotype, TC 9095-01, flattened right valve, inner layer of valve. L-4.2 mm. **8:** Paratype, TC 9151, crushed inner layer of left valve L-4.7 mm. SP, Upper Flow, bed 10.

Plate 15

ANTARCTICA

Storm Peak

Figure 1. *Glyptoasmussia* cf. luekiensis Defretin-Le Franc. TC 9090, paired internal molds of right (a′) and left (a) valves. L-4.0 mm (a′). SP, Upper Flow, loose piece.

Figures 2, 4. *Cyzicus (Euestheria) crustapatulus* n. sp. **2:** Syntype, TC 10122b, dimorph 1-D-1, internal mold of left valve. L-5.5 mm. **4:** Syntype, TC 9077A, internal mold of left valve. L-5.2 mm. SP. Upper Flow debris (both types traced to bed 5).

Figure 3. *Estheriina (Estheriina)* sp. 3. TC 9185B, internal mold of left valve. L-3.8 mm. SP, Upper Flow, bed 5. Arrow points to pinnacle-like terminal segments of raised umbo.

Figure 5. *Estheriina (Nudusia)* sp. 3. TC 9097b, internal mold of right valve; note bubble-shaped umbo; length of umbo, 1_u-0.7 mm. SP, Upper Flow, loose piece.

Figure 6. *Cyzicus (Euestheria) rhadinis* n. sp. Holotype, TC 9187c, internal molds of right and left valves; note tapering of valve posteriorly and subterminal umbo. L-5.4 mm. SP, Station 1, bed 4.

Figures 7, 8. *Cyzicus (Euestheria) transantarctensis* n. sp. **7:** Holotype, TC 9081C, flattened right valve. L-3.5 mm. SP, Upper Flow, bed 2. **8:** Paratype, TC 9112B, left valve and underlying mold. L-3.5 mm. Both types from SP, Upper Flow, bed 2.

Plate 16

ANTARCTICA

Storm Peak and Blizzard Heights

Figures 1, 2, 4. *Cyzicus (Lioestheria) disgregaris* n. sp. **1:** Syntype, TC 10006d, internal mold of left valve. L-6.4 mm. **2:** Syntype, TC 9072, a dimorph-1, internal mold of left valve. L-5.0 mm. **4:** Syntype, dimorph-2, TC 9045b, L-4.8 mm. (All syntypes from SP, Upper Flow, bed 6. (See also Pls. 17, 18, 19, 20.)

Figure 3. (Blizzard Heights.) *Cyclestherioides (Cychestherioides) alexandriae* n. sp. TC 11156, right valve and underlying mold. L-4.0 mm. BH, Station 1, bed 2. (Cf. SP, Upper Flow, bed 10.)

Figures 5, 6, 8. *Cyzicus (Euestheria) castaneus* n. sp. **5:** Syntype, TC 9055a, dimorph D-1, internal mold of left valve, L-6.8 mm. SP, Station 1, bed 6. **6:** enlarged detail of 5. Contiguous granule ornamentation and double growth bands. **8:** Syntype, TC 9055c, dimorph D-2, internal mold of right valve. L-4.9 mm. SP, Station 1, bed 6.

Figure 7. *Cyzicus (Euestheria)* sp. 1. TC 10099b, inner portion of right valve impressed in substrate. L-3.8 mm. SP, Upper Flow, bed 1, lower.

Plate 17

ANTARCTICA

Storm Peak

Figure 1. *Asmussia* sp. 1. TC 9183A, right valve, remnant of inner portion of shell overlying mold; note subquadrate configuration. L-5.1 mm. SP, Upper Flow, bed 9.

Figure 2. *Estheriina (Nudusia) stormpeakensis* n. sp. Syntype, TC 9196f, right valve and underlying mold. L-3.2 mm. SP, Upper Flow, float.

Figures 3, 6. *Cyzicus (Lioestheria) disgregaris* n. sp. **3:** TC 9894, portion of left valve with underlying mold. L-3.3 mm. **6:** detail of 3 with lioestheriid ornamentation, enlarged. SP, Upper Flow, bed 11.

Figures 4, 5. *Cyzicus (Euestheria) crustapatulus* n. sp. Juveniles. **4:** TC 9053A, internal mold of left valve. L-5.5. **5:** TC 9075A, portion of right valve. L-6.7 mm. SP, Upper Flow, bed 5.

Figure 7. *Pseudoasmussiata defretinae* n. sp. TC 10114A, inner layer of right valve and underlying mold. L-5.0 mm. SP, Upper Flow, bed 4.

Figure 8. *Estheriina (Estheriina)* sp. 2. TC 11431, external mold of right valve; note the smooth, terminal umbonal segment. L-4.8 mm. SP, Lower Flow, Station 0, bed 2.

Plate 18

ANTARCTICA

Agate Peak, Blizzard Heights, Storm Peak, and Mauger Nunatak

Figure 1. *Cyzicus (Lioestheria) disgregaris* n. sp. D. E. No. 81-1-2B, specimen "c." Syntype, subovate dimorph, eroded left valve partly overlying a right valve. L-5.7 mm (left value). Agate Peak, northern Victoria Land. (See also Pls. 17, 19, and 20.)

Figure 2. *Cyzicus (Euestheria) juracircularis* n. sp. Holotype, TC 11270-01, carbonized left valve. L-3.7 mm. BH, Station 1, bed 6.

Figure 3. *Palaeolomnadia (Grandilimnadia)* cf. *minutula* n. sp. TC 10015, internal mold of left valve. L-0.50 mm. Storm Peak, Upper Flow, bed 11, lower.

Figure 4. Conchostracan appendages; segmented rami of more than one antenna bunched together in a brief space. Arrow points to impressions of setae. TC 8400. Mauger Nunatak, D. E. Site 62-10-1B. Distance from left to right of all appendages at widest spread, 2.7 mm.

Figures 5, 7. *Glyptoasmussia meridionalis* n. sp. **5:** TC 11312, internal mold of right valve. BH, Station 1, bed 4. L-5.6 mm. **7:** TC 9065b, internal mold of right valve. L-6.2 mm. Storm Peak, Upper Flow, float. (See Blizzard Heights, Pl. 10, Fig. 7.)

Figure 6. *Cyclestherioides (Cyclestherioides) alexandriae* n. sp. TC 9132c, paratype 1, inner layer of right valve. L-4.6 mm. Storm Peak, Upper Flow, bed 10.

Figure 8. *Cyzicus (Euestheria) ichthystromatos* n. sp. TC 9143A, syntype, inner portion of right valve. L-5.5 mm. Storm Peak, Station 2, bed 2.

Plate 19

ANTARCTICA

Blizzard Heights, Storm Peak, Agate Peak, and Carapace Nunatak

Figure 1. *Palaeolimnadia (Grandilimnadia)* cf. *glenleensis* (Mitchell). TC 9033, internal mold of right valve. L-1.8 mm. Storm Peak, Upper Flow, bed 6, lower half.

Figure 2. *Estheriina (Nudusia)* sp. 1. TC 11134, mold of left valve. L-2.3 mm. Blizzard Heights, Station 1, bed 8, lower half.

Figure 3. *Cyzicus (Lioestheria) disgregaris.* D. E. Site 81-1-2B, specimen "b," laterally ovate, eroded right valve. Note lioestheriid ornamentation (hachure markings) on anterior face of the valve; valve somewhat distorted by infolding of ventral growth bands. L-5.7 mm. Agate Peak, northern Victoria Land.

Figures 4, 5, 6, 7, 8. *Cyzicus (Lioestheria) disgregaris.* **4:** TC 80403e, subcircular valve, female; internal mold of right valve. L-5.0 mm. Carapace Nunatak, Tasch Station 2, bed 1. **6:** detail of 4 showing characteristic ornamentation. Enlarged. **5:** TC 80201B, ovate valve, male; internal mold of right valve. L-6.0 mm. Carapace Nunatak, Tasch Station 1, bed 3. **7:** detail of 5 showing ornamentation and double bands (cf. 6). **8:** TC 80278b, subcircular valve, female, internal mold of right valve. L-4.7 mm. Carapace Nunatak, Tasch Station 1, bed 1. (See also Pls. 17, 18, and 20.)

Plate 20

ANTARCTICA

Carapace Nunatak

Figures 1, 2, 3, 4. *Cyzicus (Lioestheria) disgregaris* n. sp. **1:** TC 80375, laterally ovate valve, male; internal mold of left valve. L-4.7 mm. Tasch Station 2, bed 2. **2:** TC 80374b, ovate valve, male; internal mold of right valve. L-4.6 mm. Tasch Station 2, bed 2. **3:** TC 80310B, laterally ovate valve, male; internal mold of right valve, L-4.9 mm. Tasch Station 0, bed 3. **4:** TC 80017A, b, laterally ovate valve, male; internal mold of right valve. L-5.0 mm. Schopf Collection, southeast spur, ~3.0 m above the siltstone.

Figures 5, 6. 5: TC 80025c, ovate valve, male; internal mold of left valve, L-5.0 mm. **6:** TC 80025b, subcircular valve, female; internal mold of left valve. L-4.3 mm. Schopf Collection, north spur. (See also preceding four plates.)

Plate 21

ANTARCTICA

Mauger Nunatak

Figure 1. *Palaeolimnadia (Palaeolimnadia)* sp. TC 8308, fragment of crushed left valve with very large umbo and three growth bands. Enlarged. D. E. Site 62-10-1A'.

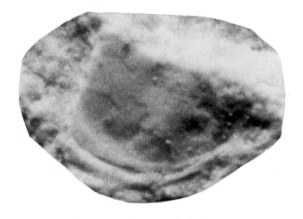

P. Tasch

Plate 22

AUSTRALIA (PERMIAN)

Newcastle Coal Measures

Figure 1. *Cycloleaia discoidea* (Mitchell). TC 10521b, external mold of right valve. d_2-3.8 mm. Station 5A, bed 2g. New South Wales (=N.S.W.).

Figure 2. *Leaia (Hemicycloleaia) belmontensis* (Mitchell). TC 10552, internal mold of left valve. L-4.6 mm. Station 7a, bed 3, N.S.W. [= *L. (H.) mitchelli*].

Figure 3. *Leaia (Hemicycloleaia) collinsi* (Mitchell). TC 10617b, internal mold of right valve. L-3.9 mm. Station 5, bed 5. N.S.W.

Figure 4. *Leaia (Hemicycloleaia)* cf. *compta* (Mitchell). TC 10635b, external mold of left valve. d_1-5.0 mm. Station 8, unit 8d, N.S.W.

Figure 5. *Leaia (Hemicycloleaia) magnumelliptica* n. sp. TC 10618a, holotype, external mold of left valve. L-9.2 mm. Station 1, bed 3. N.S.W. Note wide separation of the two ribs on the umbo (=d_1).

Figure 6. *Leaia (Hemicycloleaia)* cf. *etheridgei* (Kobayashi). TC 10105, internal mold of left valve. L-3.15 mm. Station 13, bed 2g. N.S.W.

Plate 23

AUSTRALIA (PERMIAN)

Newcastle Coal Measures

Figure 1. *Leaia (Hemicycloleaia) mitchelli* (Etheridge Jr.). TC 10623a, external mold of left valve. L-5.2 mm. Station 1, bed 3. TC 10619, external mold of right valve. L-6.0 mm. Station 13, bed 2g. N.S.W.

Figure 2. *Leaia (Hemicycloleaia) immitchelli* n. sp. TC 10547, internal mold of left valve. L-5.5 mm. Station 7a, bed 1. N.S.W.

Figure 3. *Cyzicus (Euestheria) lata* (Mitchell). TC 10591-01, internal mold of right valve. L-4.5 mm. Station 14, 4.75 m below Dudley seam. N.S.W.

Figure 4. *Leaia (Hemicycloleaia) elliptica* (Mitchell). TC 10564, external mold of right valve. L-6.6 mm. Station 7, bed 2. N.S.W.

Figure 5. *Cyzicus (Euestheria) novocastrensis* (Mitchell). TC-10594, internal mold of right valve. L-5.5 mm. Station 14, 4.75 m below Dudley seam. N.S.W.

Figure 6. *Leaia (Leaia) oblonga* (Mitchell). TC 10625a, external mold of right valve. L-3.6 mm. Station 5, bed 5. N.S.W.

Figure 7. *Cyzicus (Euestheria)* sp. undet. TC 10113, external mold of left valve. L-3.1 mm. Station 13, bed 2g. N.S.W.

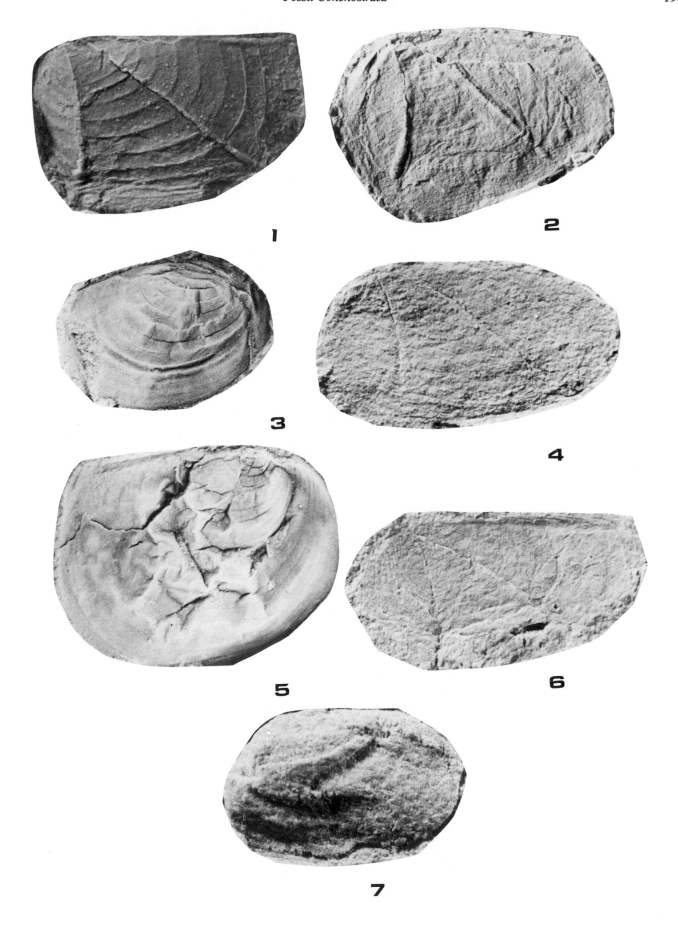

Plate 24

AUSTRALIA (PERMIAN)

Newcastle Coal Measures

Figure 1. *Leaia (Leaia) oblongoidea* n. sp. TC 10608, paratype, internal mold of left valve. L-6.6 mm. Station 4, bed 3. N.S.W. (For holotype Australian Museum 25420, see Mitchell, 1925, Pl. 43, Fig. 18 and Pl. 40, Fig. 8 herein).

Figure 2. *Leaia (Hemicycloleaia) ovata* (Mitchell). TC 10583, external mold of right valve. L-3.8 mm. Station 4, bed 3. N.S.W.

Figure 3. *Leaia (Hemicycloleaia) kahibahensis* n. sp. TC 10101, internal mold of left valve. L-7.0 mm. Station 13, bed 2g. N.S.W.

Figure 4. *Leaia (Hemicycloleaia)* cf. *paraleidyi* (Mitchell). TC 10533, external mold of right valve. L-4.6 mm. Station 6a, bed 5. N.S.W.

Figure 5. *Palaeolimnadia (Palaeolimnadia) glabra* (Mitchell). TC 10110, internal mold of right valve. L-3.0 mm. Station 1, bed 2. N.S.W.

Figure 6. *Palaeolimnadia (Palaeolimnadia) wianamattensis* (Mitchell). TC 10603A, internal mold of right valve. L-3.5 mm. Station 4, bed 1. N.S.W.

Figure 7. *Leaia (Hemicycloleaia)* cf. *sulcata* (Mitchell). TC 10829, external mold of part of right valve. d_1-0.8 mm. Station 13, bed 2e. N.S.W.

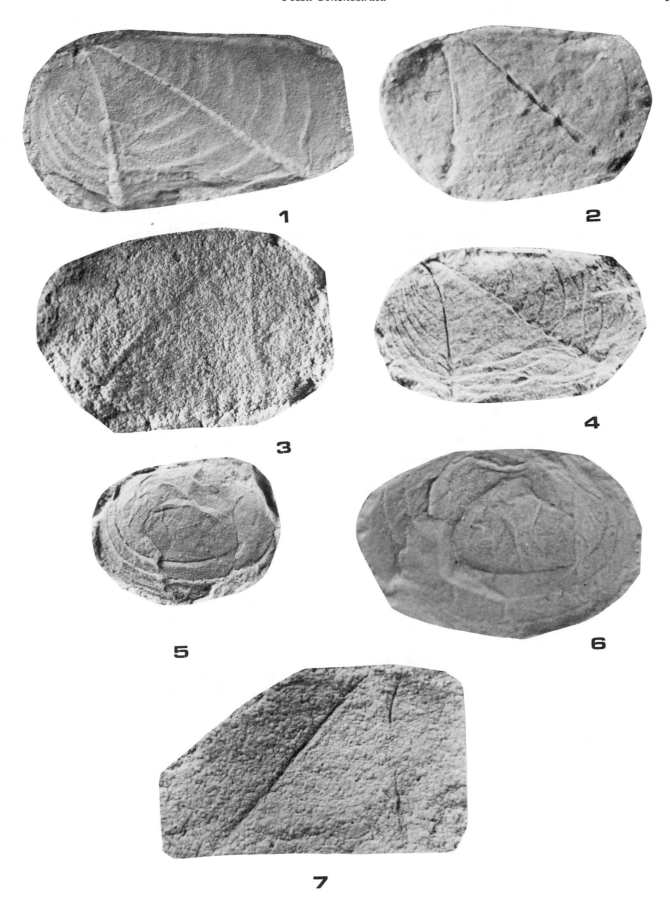

1

2

3

4

5

6

7

Plate 25

AUSTRALIA (TRIASSIC, CRETACEOUS)

Figure 1. Palaeolimnadia (Grandilimnadia) arcadiensis n. sp. TC 10592A, holotype, internal mold of left valve. L-2.3 mm. Note large umbonal area. Arcadia Formation, Narrabeen Group, Duckworth Creek near Bluff, Queensland. Lower Triassic.

Figure 2. *Palaeolimnadia (Grandilimnadia) glenleensis* (Mitchell). TC 10591, internal mold of left valve. L-2.2 mm. Same locality and age as 1 above. Note comparatively large subterminal umbo.

Figures 3, 4. *Cyzicus (Euestheria) triassibrevis* n. sp. **3:** TC 10581b, holotype, external mold of right valve. L-3.5 mm. **4:** TC 10582d, subcircular dimorph, external mold of left valve; note greater height of posterior sector. L-3.2 mm. Garie Beach, lat 34°8′S, long 151°73′E, 10 km southwest of Port Hacking Point, Narrabeen Group. Lower Triassic.

Figure 5. *Cyzicus (Lioestheria) fictacoghlani* n. sp. Geological Survey of New South Wales, Mining Museum MMF-3113, internal mold of right valve. L-3.2 mm. Cremorne Bore No. 2, depth 674.4 m, Narrabeen Group. Lower Triassic.

Figures 6, 7. *Cyzicus (Lioestheria) branchocarus* Talent. **6:** National Museum, Victoria (NMV) - P-34290, paratype, paired valves. L-3.7 mm. **7:** NMV-34291, holotype, right valve. L-5.4 mm. Korumburra Group, ~346 km east of Koonwarra, Victoria. Lower Cretaceous (Valangian–Aptian).

Figure 8. *Palaeolimnadiopsis bassi* Webb. Australia Museum Collection F-36275, holotype, rubber imprint prepared by Australian Museum L-17.5 mm. Hawkesbury Sandstone, Brookvale, Sydney. Middle Triassic.

Plate 26

INDIA

Raniganj and East Bokaro Coal Basins

Figures 1, 2. *Cornutestheriella taschi* (Ghosh and Shah), Tasch, n.g. **1:** Geological Survey of India (GSI), GSI-EBP-4, Tasch Reference No. 46000. Internal mold of right valve showing marginal, apron-like, incomplete ribs, which, higher up on the valve, are represented by nodes, and an umbonal, curved segmented spine (arrow) that, in ultra-structure, is fibrous. L-3.2 mm. **2:** Detail of nodes in 1, SEM, greatly enlarged. East Bokaro Coal Basin.

Figures 3, 4, 5, 6, 7, 8. *Cornia panchetella* n. sp. Syntypes. **7:** TC 40197A and B, dimorphs occurring together; left valve with umbonal spine, designated male; right, incomplete valve with umbonal node, designated female; representing respective internal molds. L-1.9 mm (female). Raniganj Coal Field, west Bengal, GSI, RNM-4 borehole, 250 m above Panchet/Raniganj contact. **8:** enlargement of female valve in 7. **3, 4:** Another valve from same horizon as above, enlarged. **3:** SEM photograph of female valve. **4:** SEM photograph showing that the pores of the valve are continued onto the nodose rise, indicating that the rise occurred *during* formation of the larval valve. **5, 6:** Another valve, same horizon as above, enlarged. **5:** SEM photograph of a broken prominent spine on a male valve. **6:** SEM photograph showing the projection of the tapered spine and its position markedly dorsad on the umbo. The above *Cornia* species also occurs at Mangli (M-1) in beds 2, 7, and 8, and in other boreholes (west Bengal) in the Raniganj Basin, RNM-4, 106 m; and CMPDI at 227 m above the Panchet/Raniganj contact. See text—Figure 15 for subsurface data.

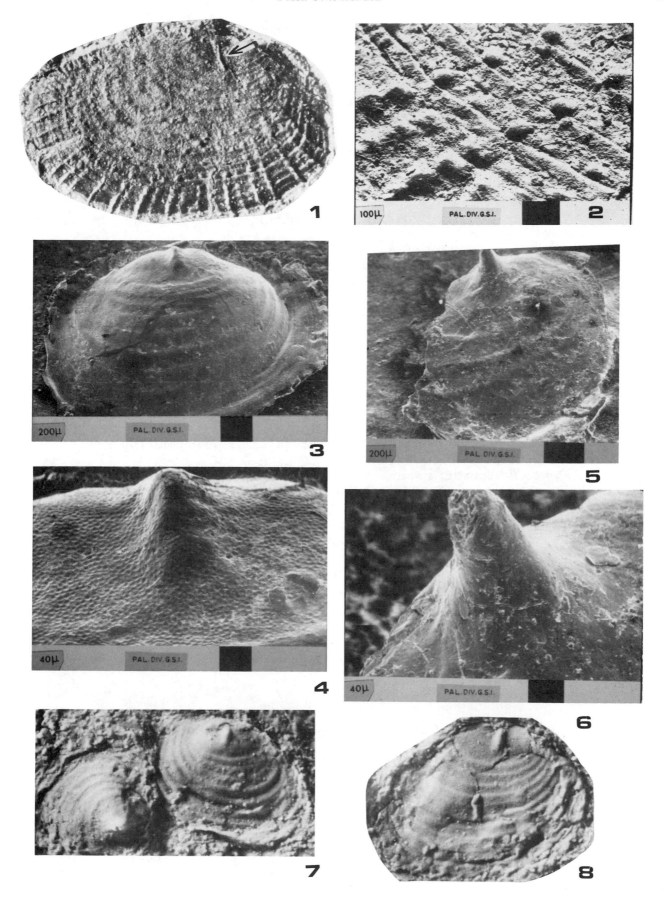

Plate 27

INDIA

Panchet and Kotá Formations

Figure 1. *Estheriina (Nudusia) adilabadensis* n. sp. TC 40001, holotype, internal mold of left valve. L-1.7 mm. Tasch Site K-1, bed 8A, Kotá Formation, Andhra Pradesh State. Lower Jurassic.

Figure 2. *Cyzicus (Euestheria) mangaliensis* (Jones). TC 40215, topotype, internal mold of left valve. L-7.6 mm. Tasch Site M-1, bed 5, lower. Panchet Formation correlate, Mangli Village. Lower Triassic.

Figure 3. *Cyzicus (Lioestheria) bokaroensis* n. sp. TC 40117, Geological Survey of India Site No. EBP-4, holotype, external mold of left valve. L-3.0 mm. East Bokaro Coal Field, Panchet Formation, Bihar. Lower Triassic.

Figures 4, 6. *Estheriina (Nudusia) indijurassica* n. sp. **4:** TC 40055, holotype, internal mold of left valve. L-2.1 mm. Tasch Site K-2, bed 2, lower. Kotá Formation, state and age as in 1 above; lighted from upper left. **6:** same as 4, with light off the umbo to accent its shape and provide a better definition of the antero-dorsal growth bands relative to umbo.

Figure 5. *Pseudoasmussiata indicyclestheria* n. sp. TC 40208, holotype, internal mold of right valve. L-2.3 mm. Tasch Site M-1, float. Panchet Formation correlate, Mangli Village. (Same bedding plane as *Cornia panchetella.) Lower Triassic.*

Figure 7. *Pseudoasmussisata bengaliensis* n. sp. TC 40024, syntype (female?), internal mold of right valve. L-4.0 mm. Tasch Site RN-1-B, bed 2. Panchet Formation, Raniganj Basin. Lower Triassic.

Figure 8. *Pseudoasmussiata andhrapradeshia* n. sp. TC 40002, holotype, external mold of right valve. L-2.1 mm. Tasch Site K-1, bed 6a, upper. Kotá Formation, state and age as in 1 above.

Plate 28

INDIA

Panchet and Kotá Formations

Figure 1. *Cyzicus (Euestheria) dualis* n. sp. TC 40092a, syntype (male?), complete external mold of right valve. L-3.3 mm. Site RN-2(?). Raniganj Coal Field, Panchet Formation. Lower Triassic.

Figure 2. *Estheriina (Estheriina) pranhitaensis* n. sp. TC 48004b, holotype, external mold of crumpled left valve. L-3.5 mm. Kotá Formation, Tasch Site K-7, bed 6, southwest of Kotá Village, Maharashtra State. Lower Jurassic. (Naupliid valve appears superimposed on a broad apron of the rest of valve.)

Figure 3. *Estheriina (Nudusia) adilabadensis* n. sp. TC 40061, paratype, external mold of left valve. L-2.8 mm. Tasch Site K-1, bed 8A. Kotá Formation, Adilabad District, Andhra Pradesh State. Lower Jurassic.

Figure 4. *Cyclestherioides (Cyclestheriodes) machkandaensis* n. sp. TC 41002, holotype, eroded internal mold of left valve. L-2.0 mm. Site RM-1, bed 1b, Raniganj Coal Field, Panchet Formation. Lower Triassic.

Figure 5. *Estheriina (Nudusia) indijurassica* n. sp. TC 48002, paratype, naupliid, internal mold of left valve. L-1.5 mm. Tasch Site K-2, bed 2. Kotá Formation, Andhra Pradesh State. Lower Jurassic.

Figures 6, 7, 8. *Cyzicus (Lioestheria) crustabundis* n. sp. **6:** TC 40189a, syntype, internal mold of right valve. L-4.3 mm. Tasch Site K-4, bed 11, Kotá Formation, Andhra Pradesh State. Lower Jurassic. **7:** TC 40189b, syntype, internal mold of right valve. posterior margin partially eroded. L-4.6 mm. Site, K-4, bed 11. **8:** TC 40207a, an apparent dimorph, L-3.3 mm. Site K-4, bed?

Plate 29

INDIA

Panchet and Kotá Formations

Figure 1. *Cyzicus (Euestheria) mangaliensis* (Jones). British Museum In-28223, syntype, internal mold of left valve. L-6.8 mm. Panchet Formation correlate, Mangli Village. Lower Triassic. (Cf: Pl. 27, Fig. 2 for topotype showing a more complete dorsal margin and umbo.)

Figure 2. *Cyzicus (Euestheria) raniganjis* n. sp. TC 40094, holotype, mold of left valve. L-3.3 mm. Raniganj Coal Field, Panchet Formation, Station RN-2, bed 4, Numa River, Lower Triassic.

Figure 3. *?Cyzicus (Lioestheria?) kotahensis* (Jones). British Museum In-28268, syntype, mold of a presumably right valve. L-3.5? mm. Very poorly preserved. (Note ostracod mold in upper right section and growth band remnant in upper left and lower right on the photograph taken of the most complete of Jones' very poor type material.) Pranhita River, southwest of Kotá Village, District of Chandrapur, Maharashtra State.

Figure 4. *Palaeolimnadia (Grandilimnadia?)* sp. TC 48001, mold of left valve. L-2.5?mm. Tasch Site K-2, bed 2, Kota Formation, east-southeast of Kadamba Village (Sirpur Taluka), Adilabad District, Andhra Pradesh. Lower Jurassic.

Figure 5. *Cyzicus (Euestheria) basbatiliensis* n. sp. TC 40011b, syntype, internal mold of right valve. L-3.9? mm. Raniganj Coal Field, Panchet Formation. Site RB-2, bed 5. Lower Triassic.

Figure 6. *Cyclestherioides (Sphaerestheria)* sp. TC 40077a, internal mold of right valve, eroded along antero-ventral margin and around and below umbonal area. L-2.5 mm. Site RN-1-A, bed 4. Raniganj Coal Field, Panchet Formation, Lower Triassic.

Figure 7. *Estheriina (Nudusia) bullata* n. sp. TC 40155a, holotype, external mold of left valve. L-2.4 mm. Tasch K-2, bed 3, upper "c," Kotá Formation, Andhra Pradesh. Lower Jurassic.

Figure 8. *Cyzicus (Lioestheria) miculis* n. sp. TC 40999, syntype, internal mold of right valve. L-3.3 mm. RN-1-B(?), Panchet Formation, Raniganj Basin. (Probably a female valve.)

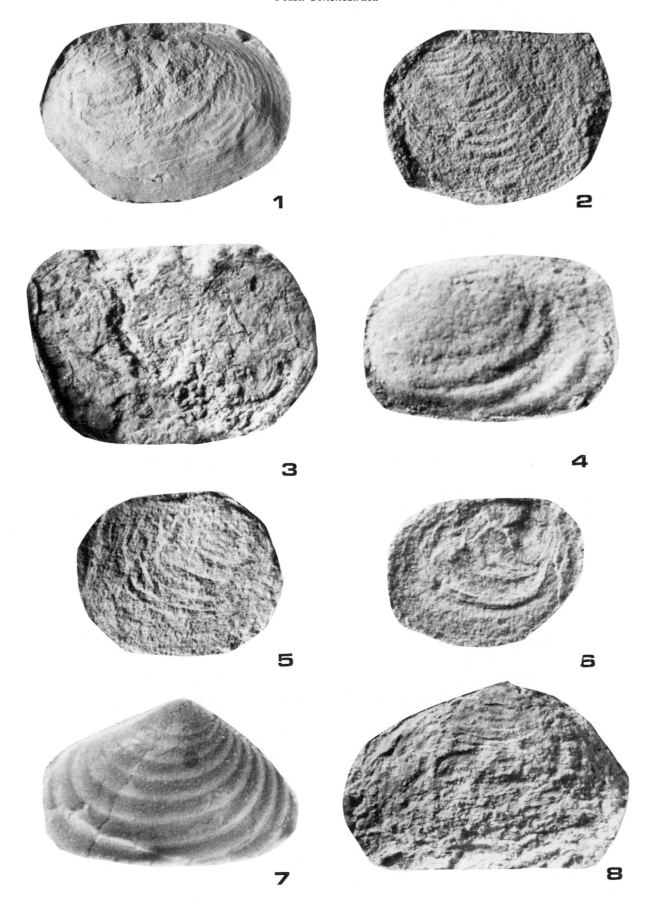

Plate 30

INDIA

Panchet and Mahadeva Formations

Figure 1. *Cornia panchetella* n. sp. TC 40188-22, left valve. L-1.5 mm. Tasch Site M-1, bed 4, Panchet Formation correlate. Mangli Village, 100 km south of Nágpur. Lower Triassic. (See Appendix for measured section and Figure 14.)

Figure 2. *Vertexia* sp. TC 40188-01, naupliid, mold of right valve. L- ~1.0 mm. Borehole No. RNM-4, depth 106 m, Panchet Formation, Andal area, Raniganj Coal Field, west Bengal. Lower Triassic. Photograph, SEM Laboratory, Geological Survey of India, Calcutta, by Shekhar Ghosh. Enlarged. (Note vertical ornamentation on posterior margin and above.)

Figure 3. *Cyzicus (Lioestheria) miculis* n. sp. TC 40079, syntype, male, internal mold of left valve. L-2.5 mm. Site RN-1-B, bed 7, downdip from RN-1-A, younger beds. Panchet Formation, Numa River, Raniganj Coal Field. Lower Triassic.

Figure 4. *Cyzicus (Euestheria)* sp. K 51/290, Geological Survey of India, Spiti District, Himachal Pradesh. Lipak River section of Kashmir (Po Series). Polygonal ornamentation on growth bands; see arrow. L-1.5 mm. Lower Carboniferous. A poor SEM photograph.

Figure 5. *Pseudoasmussiata bengaliensis* n. sp. TC 40113, syntype, male?, a dimorph, crushed valve. L-2.7 mm. Site RN-1-B, bed 2. Panchet Formation, Raniganj Coal Field. Lower Triassic.

Figure 6. *Cyzicus (Euestheria)* sp. Geological Survey of India SEM Laboratory photograph of left valve. (Note part of attached right valve.) L-5.0 mm. East Bokaro Coal Field, Mahadeva Formation, top of Luga Hill, Supra-Panchet. Upper Triassic.

Plate 31

SOUTH AMERICA

Argentina, Brazil

Figures 1, 2. *Cyzicus (Euestheria) forbesi* (Jones). (Argentina.) **1:** Syntype, British Museum (BM-50521, no. 8 on slab), external mold of left valve. L-11.6 mm. **2:** Syntype (BM 50521, no. 9 on slab), orbicular, juvenile form, internal mold of right valve. L-8.2 mm. Cacheuta, ~32 km southwest of Mendoza (lat 33°S). Upper Triassic (Rhaetic).

Figures 3, 5. *Cyzicus (Lioestheria) malacaraensis* n. sp. (Argentina.) **3:** Slab TC 70509, holotype, internal mold of left valve. L-5.8 mm. **5:** TC 70513, paratype, internal mold of left valve; detail showing coarse ornamentation. L-5.0 mm. Malacara Limestone, Malacara, Santa Cruz Province. Jurassic.

Figures 4, 6. *Leaia (Hemicycloleaia) leanzai* Leguizamón. (Argentina). Syntypes include: Cordoba PZ-1071, 1077, and 1079–1080. **4:** Lectotype, Cordoba PZ-1071, external mold of right valve; note depression of medial sector of valve creating illusion of a third rib. L-8.3 mm. **6:** Paralectotype, Cordoba PZ-1080, remnant of mold of overlying valve and an internal mold of a left value. L-10.0? mm (Figures 4 and 6 have been enlarged from Leguizamón's smaller photographs.) Tasa Cuna Formation, lat 30°46′S, long 65°18′W, northwest of Cordoba Province. Lower Permian.

Figure 7. *?Acantholeaia* cf. *regoi* Almeida. (Brazil). Departamento Geologia, Petrografia e Mineralogia (D.G.P.) VII-130, internal mold of left valve. L-4.8 mm. (Photograph from Mendes, 1954, Pl. 14, Fig. 12.) Paraná State, road from Ponta Grossa to Prudentopolis, km 78 + 600 m. Upper Permian.

Figure 8. *Cyzicus (Lioestheria)* cf. *regularis* (Reed). (Brazil). TC 90503, internal mold of right valve, probably a female valve. L-5.6 mm. Upper Rio do Rasto, along Highway BR-116, Serra do Espigão, Santa Catarina State. Upper Permian.

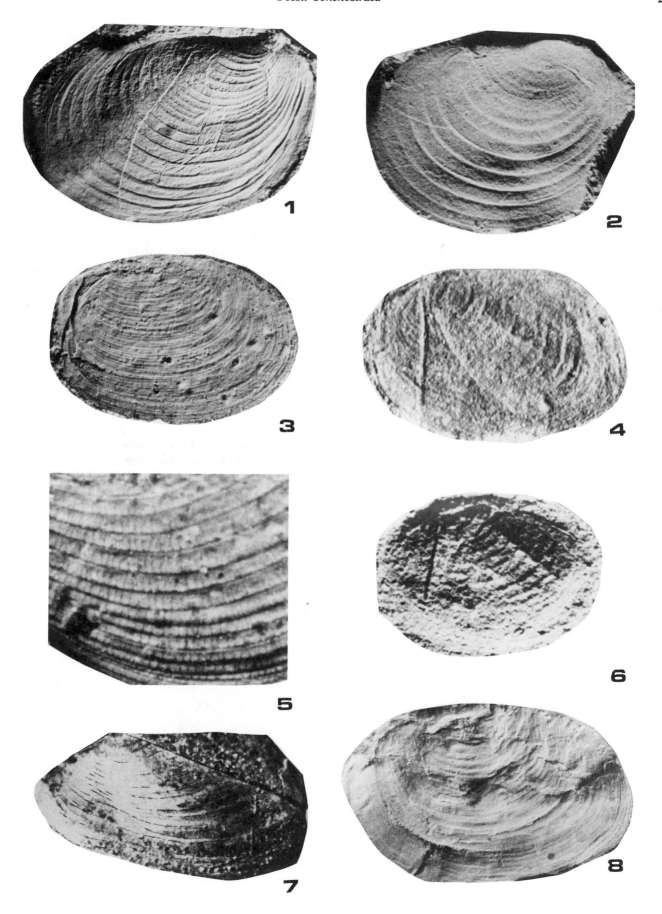

Plate 32

SOUTH AMERICA

Brazil

Figures 1, 2. *Palaeolimnadiopsis* sp. **1:** Universidad do São Paulo, USP No. 7-987 (=Tasch Reference No. 90025a). Internal mold of left valve. L-5.4 mm. **2:** Detail of recurved posterior sector of 1. Bahía State, Lower Cretaceous.

Figure 3. *Palaeolimnadiopsis* sp. 2, Universidad Federal do Minas Gerais (UFMG) No. CO-1. Internal mold of left valve, an unusually long species with recurved posterior growth bands. L-4.2 cm. Santa Maria Formation, Rio Grande do Sul State. Upper Triassic. (Photograph by Ioco Katoo.

Figure 4. *Palaeolimnadiopsis freybergi* Cardoso. UFMG No. 001. Holotype, internal mold of right valve; recurvature of all growth bands beyond the seventh band from umbo. L-~15.2 mm. Areado Formation, Paraíba State. Lower Cretaceous.

Figure 5. *Palaeolimnadiopsis* sp. UFMG No. 2522, 2523 (formerly *Pteriograpta* cf. *reali* Teixeira 1960). Internal mold of right valve. L-20.0 mm. Sousa Formation, Paraíba State. Upper Jurassic–Lower Cretaceous. (Photograph by Ioco Katoo.)

Figure 6. *Cyzicus (Lioestheria)* cf. *regularis* (Reed). TC 92774, internal mold of right valve (male?). L-5.0 mm. Tasch Station 3, bed 4B. Upper-Middle or Upper Rio do Rasto, along Highway BR-116, Serra do Espigão, Santa Catarina State. Upper Permian.

Figure 7. *Cyzicus (Euestheria)* sp. 1. TC 90520, internal mold of left valve. L-3.3 mm. Tasch Station 3, bed 3. Upper Rio do Rasto Formation, locality, and age as in 6 above.

Figure 8. *Palaeolimnadiopsis subalata* (Reed). Mold of right valve. L-10.5 mm. (Photograph from Mendes, 1954, Pl. 13, Fig. 5). Poço Preto, Upper-middle Rio do Rasto Formation, Santa Catarina State. Upper Permian.

Plate 33

SOUTH AMERICA

Brazil

Figures 1, 2, 4, 6. *Macrolimnadiopsis* pauloi Beurlen, 1954 Universidad Federal do Rio
   Grande do Sul (UFRGS), topotypes. **1:** MP-M-27, a female valve. L-17.0 mm. **2:**
   MP-P-33, a male valve. L-11.0 mm. **4:** MP-M-38, portion of valve showing sculpture,
   enlarged. **6:** MP-M-28, a female valve. L-14.0 mm. (Photographs courtesy of Dr. Pinto
   and Ms. Purper; see Pinto and Purper, 1974.) Pastos Bons Formation, Muzinho, Piauí
   State. Upper Jurassic.
Figures 3, 5. *Gabonestheria brasiliensis* n. sp. **3:** TC 60500a, holotype, internal mold of left
   valve which provides a good view of the size and shape of the umbonal projection with
   a pointed terminus. Note placement of the origination of projection and compare to
   Cornia. L-2.1 mm. Same locality and section as no. 5 below. **5:** TC 60501, paratype,
   internal mold of right valve which provides a better view of the dorsal margin. L-2.6
   mm. Lower Rio do Rasto Formation, São Pascoal, 400 km southeast of Curitiba, Santa
   Caterina State. Upper Permian.
Figure 7. *Estheriina (Estheriina)* sp. 1. TC 90067a, internal mold of left valve. L-5.8 mm.
   Note apparent superposition of umbo relative to rest of valve, an estheriinid characteris-
   tic. Lower Rio do Rasto Formation, same locality and age as in 3, above.
Figure 8. *Leaia (Hemicycloleaia) pruvosti* (Reed). External mold of two bicarinate valves.
   L-10.5 mm. (Photograph from Mendes, 1954, Pl. 13, Fig. 8.) Upper valve. 1 km from
   railroad Station at Poço Preto. Late Middle Rio do Rasto Formation. Upper Permian.

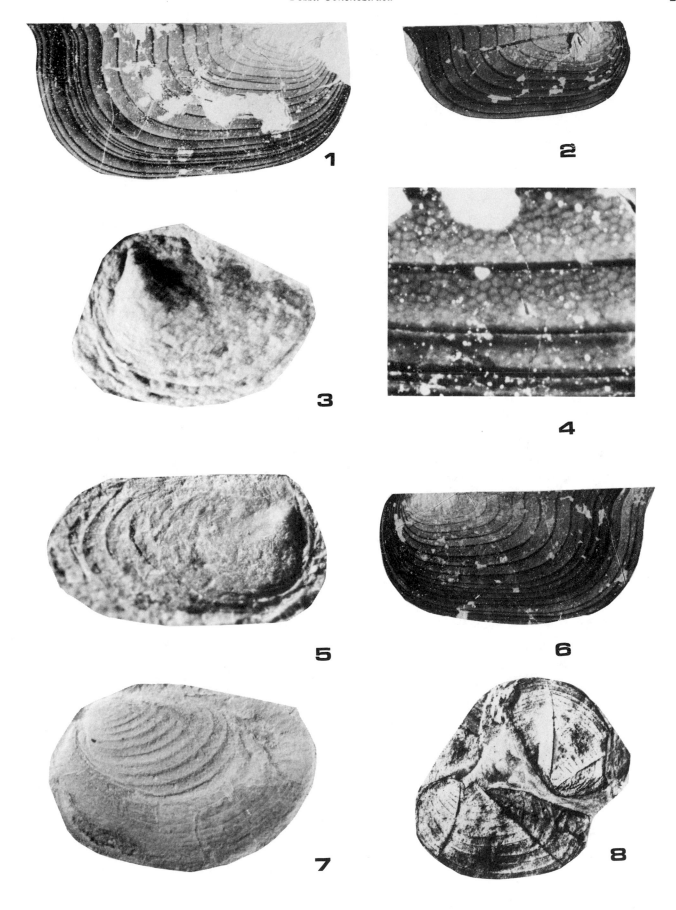

P. Tasch

Plate 34

SOUTH AMERICA

Brazil

Figure 1. *Cyzicus (Euestheria)* sp. 2. TC 60000, external mold of right valve. L-4.8 mm. Tasch Station 2, bed 4B, along highway BR-116, Serra do Espigão, Santa Caterina State, Upper Rio do Rasto Formation. Upper Permian.

Figures 2, 4. Monoleaia unicostata (Reed). **2:** TC 60550, internal mold of right valve. Beta angle 25°, L-4.7 mm. Tasch Station 3, bed 4. Same locality and age as 1 above. **4:** DGM No. 1423 (Rio de Janeiro). Internal mold of left valve. Beta angle 22°, L-2.0 mm Poço Preto, Santa Catarina. Middle Permian.

Figure 3. *Cyclestherioides (Cyclestherioides) pintoi* n. sp. TC 90333, holotype, L-7.1 mm. (=Station 17, Bigarella and others, 1967, Fig. 21.) Estrada Nova Formation, Santa Catarina. Middle Permian.

Figure 5. *Estheriina* sp. [=*Estheriina (Nudusia)* sp., *non Cyzicus (Euestheria) azambujai* Pinto.] MP-M-99 (3), UFRGS. Note elevation of smooth terminal segment above dorsal margin. Enlarged. Santa Maria Formation, Passo das Tropas, Arroio Cançela, Rio Grande do Sul State. (Photograph courtesy of Pinto.) Upper Triassic.

Figure 6. *Estheriina (Estheriina) bresiliensis* Jones. British Museum I-6827. Syntype (paralectotype), internal mold of paired left (lower) and right (upper) valves. L-7.0 mm. Along São Francisco Railroad to Salvador, near Salvador, Bahía State. Lower Cretaceous.

Figure 7. *Estheriina (Estheriina) expansa* Jones. British Museum I-3475. Holotype, internal mold of right valve. L-7.2 mm. Ihlas Formation. Railroad cutting, 89 km from Salvador, Bahía, near Pojúca, Bahía State, Brazil. Lower Cretaceous.

Figure 8. *Lioestheria* sp. Pinto and Purper [=*Cyzicus (Euestheria)* sp.] MP (UFRGS) No. MP-M-39. Topotype, internal mold of right valve. L-2.1 mm. (Photograph courtesy of Pinto.) Pastos Bons Formation, Muzinho, Piauí State. Upper Jurassic.

P. Tasch

Plate 35

SOUTH AMERICA

Brazil

Figures 1, 2. *Estheriina (Estheriina) bresiliensis* Jones. **1:** British Museum I-6828. Syntype (lectotype), mold of right valve. L-4.6 mm. Note prominent umbonal area more convex than rest of valve. *2:* BM-I-6822B, paralectotype, mold of right valve with sharp demarcation of umbo from rest of valve. L-5.7 mm. Along São Francisco Railroad to Salvador near Salvador (Bahía), Bahía State. Lower Cretaceous.

Figures 3, 4, 5. *Cyzicus (Lioestheria) mawsoni* (Jones). **3:** BM-I-6805, syntype, portion of right valve showing some of the original brownish coloration (i.e., dried haemolymph) and cancellate ornamentation in the lower sector. L-5.5 mm. **5:** Detail of cancellate ornamentation of 3, enlarged. **4:** BM-I-6804. Lectotype, portion of left valve and its internal mold. The most complete specimen of the syntype, with umbo raised above the short dorsal margin. L-5.2 mm. São Francisco Railroad, at 12–13 km and 73–74 km from Salvador (Bahía). Lower Cretaceous.

Figure 6. *Aculestheria novojilovi* Cardoso. Universidad do São Paulo, DGP-7-985, holotype. Internal mold of left valve showing a prominent spinous projection posteriorly; dorso-anterior recurvature (no projection visible on holotype as in paratype which displays both projections). L-6.7 mm. (Photograph courtesy of Roberto Cardoso.) Candeias Formation, 5 km from Santa Amaro, Bahía State. Upper Jurassic–Lower Cretaceous.

Figure 7. *Cornia* sp. TC 90089. Internal mold of left valve showing a prominent subcentral node. L-4.47 mm. Tasch bed 1A (Papp's bed 15), Lower Rio do Rasto Formation, São Pascoal, Santa Caterina State.

Figure 8. *Cyzicus (Euestheria) azambujai* (Pinto). UFRGS, MP-238. Holotype, umbonal area bears growth bands not visible in photograph. Enlarged. (Photograph courtesy of Pinto.) Santa Maria Formation, Passo das Tropas, Arroio, Cançela, Rio Grande do Sul State. Upper Triassic.

Plate 36

SOUTH AMERICA

Brazil and Peru

Figure 1. *Cyzicus (Euestheria)* sp. 2 (Brazil.) Universidad Federal do Minas Gerais (UFMG), CO-4189-2, internal mold of left valve. L-5.7 mm. (Photograph courtesy of Ioco Katoo.) "Belvedere," Santa Maria Formation, Rio Grande do Sul State. Upper Triassic.

Figure 2. *Echinestheria?* sp. (Brazil.) UFMG, CO-4192-B, internal mold of incomplete right valve and underlying mold of left valve. L-1.15 mm. Note prominent, compressed umbonal spine characteristic of *Echinestheria.* Same locality and age as 1 above. (Photograph courtesy of Ioco Katoo.)

Figure 3. *Graptoestheriella* cf. *fernandoi* Cardoso. (Brasil.) UFMG, FS-45, fragment of internal mold showing characteristic topography of an estheriellid. Enlarged. Botucatú Formation, São Paulo State. Lower Cretaceous. (Photograph courtesy of Ioco Katoo.)

Figure 4. *Cyzicus (Euestheria) ariciensis* (Jones). (Peru.) British Museum I-6815, Lectotype, internal mold of left valve. L-9.8 mm. Bed F, Moro de Arica, Arica (Department of Arequipa), southern Peru, lat 18°25′S, long 70°15′W. Carboniferous.

Figure 5. *Pseudoasmussiata katooae* n. sp. (Brazil.) UFRGS-4186, No. 2, internal mold of right valve. L-3.1 mm. (Photograph courtesy of Ioco Katoo.) "Belvedere," Station 4 on Location Map (See Appendix), uppermost Santa Maria Formation, Rio Grande do Sul State. Upper Triassic.

Plate 37

SOUTH AMERICA

Bolivia, Brazil, Uruguay

Figures 1, 4. *Cyzicus (Lioestheria)* sp. (Bolivia.) **1:** Internal mold of right valve, antero-ventral sector eroded; note inset umbo rising above dorsal margin. L-2.26 mm. **4:** Detail of 1, showing lioestheriid ornamentation. Enlarged. Ipaguazu Formation, Gulf Oil Corp., Y.P.F.B. Carandacti No. 2, depth 2592 m. Permian–Tartarian.

Figure 2. *Echinestheria semigibosa* Cardoso. (Brazil.) [=*Cornia semigibosa* (Cardoso).] DGP, Universidad do São Paulo (USP)-7-989, syntype, fragmented internal mold of left valve showing faint embryonic ribs and a small umbonal spine. Because *all* other specimens lack incipient ribs, it has been designated an apparent corniid. L-1.1 mm. (Photograph courtesy of Roberto Cardoso.) Motoco Formation. Upper Triassic.

Figure 3. *Graptoestheriella* sp. (Brazil.) Universidad Federal do Minas Gerais-unnumbered. Carbonized molds of right valve (bottom) and left valve (top), from one of two cores taken at depth 236.3-237.8 m (Photograph courtesy of Ioco Katoo.) Enlarged. Rio Passagen, Brejo dos Freires, Uiraúma, Paraíba State. Lower Cretaceous.

Figures 5, 6. *Cyzicus (Lioestheria)* sp. (Uruguay.) **5:** Fragment of internal mold of lioestheriid. Enlarged. **6:** Detail of cancellate ornamentation of 5. Enlarged. (Photograph and data courtesy of Rafael Herbst, Corrientes, Argentina.) Tacuarembó Formation, lower part; near city of Tacuarembó. Jurassic. [=*C. (Lio.) ferrando.*]

Plate 38

SOUTH AMERICA

Brazil, Colombia, Venezuela

Figure 1. *Cyzicus (Lioestheria) bigarellai* n. sp. (Brazil.) TC-90006. Holotype, external mold of left valve. L-4.7 mm. Estrada Nova Formation, Tasch Field No. 17.16, Santa Caterina State. Middle Permian.

Figure 2. *Cyzicus (Euestheria)* sp. 1 (Venezuela.) Will Maze Field No. 231D, mold of left valve; note subterminal umbo with tapered nipple-type terminal segment. Enlarged. Near La Villa del Rosario, Tincoa Formation?, Sierra de Perija, northwest Venezuela. Lower Jurassic.

Figures 3, 4. *Estheriella?* sp. (Colombia.) **3:** Will Maze Field No. C2-79-3, mold of left valve; arrow points to eroded remnants of fringe of radial costae; rest of fringe eroded. L-5.9 mm. **4:** Detail of eroded radial costae (see arrow). Enlarged. Bocas Formation near Bucaramanga, Santander massif. Triassic–Lower Jurassic.

Figures 5, 6, 7. *Estheriina (Estheriina) astartoides* Jones. (Brazil.) **5:** BM-I-6917 (on original attached label, I-6816). Syntype, mold of right valve, a new photograph of a poorly preserved specimen. L-2.5 mm. **6:** Reproduced from Jones, 1897a, Pl. 8, Fig. 8c, a drawing. Compare new photograph to Jones' drawing. **7:** BM-I-6817. Syntype, the other dimorph designated here, a probable male, internal mold of right valve. L-2.6 mm. (All other specimens of the syntype are fragmentary.) Between Pojuca and São Thiago, Bahía State. Lower Cretaceous.

Figure 8. *Cyzicus (Euestheria)* sp. 2. (Venezuela.) Will Maze Collection. Crushed internal mold of right valve in a coquina. L-4.0 mm. Site 231 D, Tincoa? Formation, Sierra de Perija, northwest Venezuela.

Plate 39

SOUTH AMERICA

Venezuela

Figures 1, 2. *Cyzicus (Lioestheria) mazei* n. sp. **1:** Will Maze Collection No. 213A, syntype, internal mold of left valve. L-3.7 mm. **2:** No. 213B, syntype, internal mold of left valve, lower left on photograph; internal mold of right valve, upper figure on photograph. Length of lower left valve, L-2.5 mm. La Villa del Rosario, Tincoa? Formation, Sierra de Perija. Lower Jurassic.

Figure 3. *Cyzicus (Lioestheria)* cf. *pricei* (Cardoso). Will Maze Collection No. 230A-D, internal mold of right valve. L-4.6 mm. La Quinta Formation?, Sierra de Perija, northwest Venezuela. Lower Cretaceous?

Plate 40

MITCHELL'S TYPES

All "F" numbers are on deposit at the Australian Museum, Sydney

Figures 1, 2. *Leaia (Hemicycloleaia) mitchelli* Etheridge, Jr. **1:** F 25426, a broadly ovate dimorph; external mold of left valve. H-5.8 mm. **2:** F 25487A. Holotype, subovate left valve; arrow points to embryonic rib that appears occasionally. H-5.7 mm.

Figure 3. *Leaia (Hemicycloleaia) collinsi* Mitchell. F 25421. Holotype, right valve with characteristic terminal umbo and small alpha angle. L-6.3 mm.

Figure 4. *Leaia (Hemicycloleaia) pincombei* Mitchell. F 25469. Holotype, right valve; note slightly sinuate dorsal margin. L-6.6 mm.

Figure 5. *Leaia (Hemicycloleaia) mitchelli?* Etheridge, Jr. [Formerly *Quadriradiata quadriradiata* (Mitchell).] F 25465A, external mold of right valve and internal mold of left valve. L-6.5 mm. Compare this new photograph of Mitchell's type with his retouched photograph (see text).

Figure 6. *Cycloleaia discoidea* (Mitchell). F 25419. Holotype, external mold of left valve; note longitudinally ovate configuration and larger anterior sector of valve than in *L. (H.) mitchelli* Etheridge, Jr. (cf. 2 above), L-5.5 mm.

Figure 7. *Leaia (Leaia) oblonga* Mitchell. F 25423. Holotype, right valve; suboblong valves are characteristic. L-7.8 mm.

Figure 8. *Leaia (Leaia) oblongoidea* n. sp. F 25420. Holotype, right valve; note characteristic elongate, rectangular shape; cf. 7 above. L-6.6 mm. (See text, Australia, for description.).

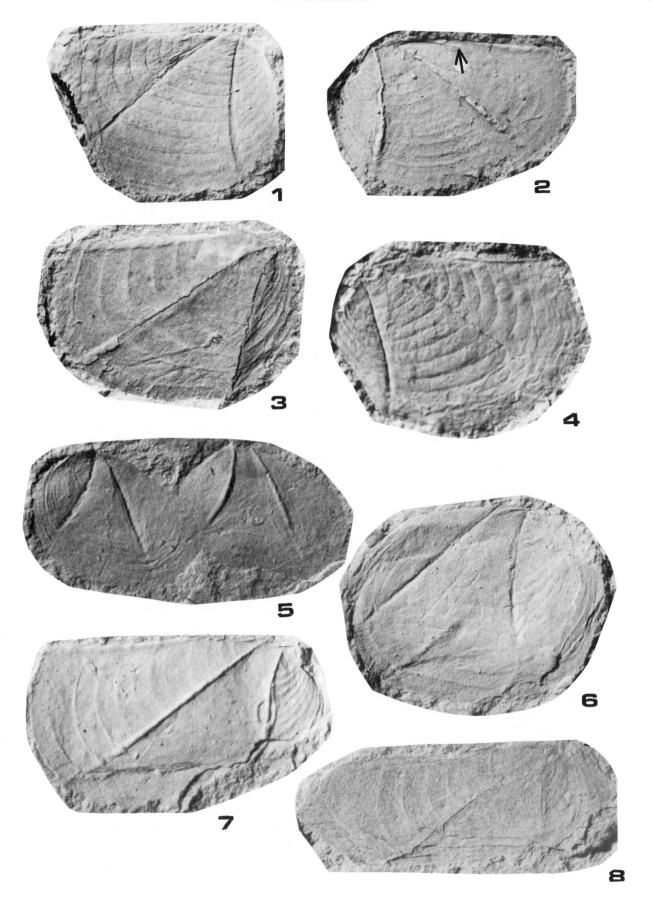

Plate 41

MITCHELL'S TYPES

Figure 1. *Leaia (Hemicycloleaia) mitchelli* Etheridge, Jr. (formerly *Leaia intermediata* Mitchell). F 25459A, Mitchell's holotype, a left valve, morphologically and metrically close to *L. (H.) mitchelli* Etheridge, Jr. (Pl. 41, Fig. 3); arrow points to the faint variable embryonic rib, L-6.0 mm. Note the valve crease which Mitchell's retouched photograph (1925, Pl. 1, Fig. 3) inadvertently added to the embryonic rib.

Figure 2. *Leaia (Hemicycloleaia) latissima* Mitchell. F 25466. Holotype, external mold of right valve. L-8.0 mm. Note ovate configuration and short, straight dorsal margin.

Figure 3. *Leaia (Hemicycloleaia) sulcata* (Kob.). F 25461. Holotype, external mold of right valve. L-7.3 mm. Note shallow valve fold near dorsal margin; i.e., the white streak in photograph was misinterpreted as a third rib by Mitchell.

Figure 4. *Leaia (Hemicycloleaia) mitchelli* Etheridge, Jr. (formerly *belmontensis*) Mitchell. F 25429. Holotype, a complete external mold of a right valve overlies a fragment of an underlying valve in such a manner that the dorsal margin of the former appears to be a third rib. L-4.9 mm. A tilt of the head 15° to the left will show the correct configuration of the overlying valve.

Figure 5. *Leaia (Hemicycloleaia) compta* Mitchell. F 25424. Holotype, internal mold of left valve; the striking feature, aside from from the prominent umbo, is curvature of the posterior rib, which is usually straight. L-7.0 mm.

Figure 6. *Leaia (Hemicycloleaia) etheridgei* (Kob.) (formerly *Leaia* sp.) F 25427. Holotype, external mold of left valve. L-4.6 mm. Note mold of a probable second antenna of a conchostracan, interpreted originally and subsequently as a third rib.

Figure 7. *Leaia (Hemicycloleaia) elliptica* Mitchell. F 25463. Holotype, external mold of left valve. L-5.3 mm.

Figure 8. *Leaia (Hemicycloleaia) ovata* Mitchell. F 25456. Holotype, external mold of left valve. L-7.0 mm.

Plate 42

MITCHELL'S TYPES

Figure 1. *Palaeolimnadia (Palaeolimnadia) wianamattensis* (Mitchell). F 25490. Lectotype, right valve with large umbo and few growth bands. L-3.3 mm.

Figure 2. *Cyzicus (Euestheria) trigonellaris?* (Mitchell). F 25484. Holotype, right valve, with eccentric-ovate configuration. L-6.2 mm.

Figure 3. *Cyzicus (Euestheria) lata* (Mitchell). F 25472. Lectotype, left valve with oblique-subovate configuration. L-6.6 mm.

Figure 4. *Cyclestherioides (C.) lenticularis* (Mitchell). F 25481. Holotype, left valve, with sub-round, prominent umbo directed anteriorly. L-1.7 mm.

Figure 5. *Cyzicus (Euestheria) novocastrensis* (Mitchell). F 25469. Holotype, right valve showing steep incline from umbo anteriorly. L-7.1 mm.

Figure 6. *Cyzicus (Euestheria) ipsviciensis* (Mitchell). F 25488. Lectotype, right valve, polygonal ornamentation visible between growth lines above antero-ventral sector. L-5.2 mm.

Figure 7. *Palaeolimnadia (Palaeolimnadia) glabra* (Mitchell). F 25475. Holotype, right valve, with large umbonal area slightly oblique anteriorly. L-3.2 mm.

Figure 8. *Leaia (Hemicycloleaia) paraleidyi* Mitchell. F 25454. Holotype, mold of left valve with a characteristically smaller alpha angle. L-6.0 mm.

Plate 43

MITCHELL'S TYPES

Figure 1. *Cornia coghlani* (Etheridge, Jr.) F 25724. Lectotype, obliquely ovate valve with prominent umbonal node, internal mold of right valve. H-1.5 mm.

Figure 2. *Cyzicus (Euestheria) obliqua* (Mitchell). F 25481. Holotype, transversely subovate valve with subterminal umbo; external mold of left valve. H-3.5 mm.

Figure 3. *Cyzicus (Euestheria) belmontensis* (Mitchell). F 25473. Holotype, mold of right valve, greatest height behind subterminal umbo; polygonal ornamentation. H-4.0 mm.

Figure 4. *Cyzicus (Lioestheria)* sp. [Formerly *"bellambiensis"* (Mitchell).] F 25485. Elongate valves with subterminal umbo; hachure-like marks between growth lines. H-~6.0 mm.

Figure 5. *Palaeolimnadia (Grandilimnadia) glenleensis* (Mitchell). F 25482. Syntype. Internal mold of left valve; note comparatively large umbo and transverse ovate configuration. H-~2.0.

Plate 44

Figures 1, 2. *Lioestheria lallyensis* Depéret and Mazeran 1912, emended. Permian. No. 1789. Numbers are reversed on Kozur and others, 1981, Plate, Figure 1; not No. 13 but No. 3. **1:** "Holotype": an incomplete valve with a so-called radial element and supposedly with a "hemispherical bulge" or node not visible, and a large umbo free of growth lines. Note that the curved structure on the umbo is segmented and, on the reader's left of the structure, there are apparently remnants of growth bands. Note the inaccurate line drawing (Kozur and others, 1981, p. 1440, Fig. 1/1) herein shown in Figure 21, 1. **2:** A paratype (not No. 3 but No. 13) described as having an umbo free of growth bands; however, note on the crushed umbo remnants of white lines similar to those that commonly demark growth bands.

Figures 3, 6. *Lioestheria lallyensis* Depéret and Mazeran, emended, Kozur and others. **3:** From Obora, Boskovicer Furche, ČSSR (Martens, 1982, Pl. XIV, Fig. 4), specimen MNG-3506-4-1, described as a left valve with "radial element" on the larval shell—an apparent second antenna. **6:** Compare *"Estheria" triangularis* Fritsch 1901 (Pl. 161, Fig. 1) showing fossilized appendages exposed by erosion on left valve, s-scape, II, biramous 2d antenna. (Cf. Figs. 6 and 3 and Fig. 21, 5 herein.)

Figures 4, 5. *Euestheria autunensis* Depéret and Mazeran, Raymond 1946. Permian. (In Depéret and Mazeran, 1912, Pl. 5, Fig. 4). **4:** Note that this is a coquina of flattened valves. **5:** Enlargement of the highest valve on the reader's left in 4 to show that growth bands extend high on the umbo and the rounded anterior sector.

Figure 7. *"Euestheria" autunensis* Depéret and Mazeran, Raymond, 1946. Permian. Number 1792. Left valve (in Kozur and others, 1981, Plate, Figure 4). Note prominent ornamentation of contiguous granules, one of the characteristics of the euestheriid valve. (This is contrary to the contention of Kozur and others [1981] that *"Euestheria"* could not be distinguished from *Lioestheria.*)

Figures 8, 9. *Lioestheria lallyensis* s.s. Depéret and Mazeran emended. **8:** The original figure for the species. (Cf. Kozur and others, 1981, p. 1440, text-Fig. 1/2, and Fig. 8 herein). The left valve on bottom of the photograph, submedial, is flattened into a turreted configuration. **9:** Enlarged view. Note the white bands (=growth bands) high up on the umbo in contradiction to one feature of the Kozur and others emended description.

Plate 45

ANTARCTICA

Photographs arranged in rows; read legend from left to right

Row 1. Left: Blizzard Heights, section traced along this slope. Right: Tasch Station 0 around bend and continued (arrow); a conchostracan-bearing interbed below basalt sill. Lower Jurassic.

Row 2. Left: Blizzard Heights, Tasch Station 1 (photographed by David Elliot), ~33 m west of Station 0; a fossiliferous interbed (see measured sections, Appendix. Right: Ohio Range (Mercer Ridge), northwest view; arrow indicates *Leaia* zone, Mt. Glossopteris Formation (Middle to Late Permian). (For details of boundary normal faults, see Long, 1965, Figure 22 and text.).

Row 3. Left: Overview of Storm Peak and environs. Right: Ohio Range (*Leaia* Zone—the section measured and traced). (See Appendix for Storm Peak section details.)

Row 4. Left: Storm Peak, Upper Flow interbed (photograph by Dietmar Schumacher). Right: Mauger Nuntak section. (See Appendix for David Elliot's section data to which Tasch added the fossil assemblage.)

Plate 46

ANTARCTICA, INDIA, SOUTH AMERICA

Figures 1, 2. Antarctica. Carapace Nunatak. **1:** Looking southeast. Several sections were measured directly below the capping basalt and the fossiliferous interbed traced several hundred metres along its length. **2:** Tasch Station 2. (Alpine pick points to conchostracan-bearing interbed.) (See Plate 49 also.)

Figure 3. India. Panchet Formation, Raniganj Basin, Lower Triassic, along the Numa River, Tasch Station RN-1A (see Appendix for biota, measured sections, and location of this and other sites).

Figures 4, 5. South America. Brazil, Rio do Rasto Formation, Upper Permian, along Highway 116, Serra Espigão, Santa Catarina, Tasch Station 3, diastem section. **4:** Collecting bags indicate conchostracan-bearing beds. **5:** Continuation of the same section as in 4. (See Appendix for measured sections.)

Figure 6. South America. Brazil, São Pascoal. Rio do Rasto Formation, Upper Permian. A rich conchostracan-bearing site with both *Cornia* and *Gabonestheria* genera. Collecting bags indicate successive fossiliferous units. (See Appendix for measured section.)

Plate 47

AUSTRALIA

New South Wales (Permian)

Figure 1. Conchostracan-bearing bed at entrance to John Darling Colliery, Womara, New South Wales. Bank 2.60 m above road level. Station 1, bed 4; *Glossopteris* flora; bed contains three species of leaiid conchostracans and a pelecypod.

Figures 2, 3. 2: Station 3, beds 3, 4; *Glossopteris* flora and a single *Leaia* species. Outcrop on top and behind a large pebble conglomerate quarry, Floraville Road (see 3 legend that follows). **3:** Belmont Quarry, north of Maxwell Street on West Quarry Road, a thick pebble conglomerate.

Figures 4, 5, 6. 4: Merewether Beach, Station 14, near sewer outlet, Merewether. Section with Dudley seam (not seen in photograph). Shell band shows height of high tide level which covers the fossiliferous outcrop seen in 5 and 6. **5:** Sampling the two conchostracan-bearing ironstone beds; bed 8 (upper ironstone). **6:** Station 14, bed 3, lower ironstone-bearing fossils.

Figure 7. Station 6a–c (composite section), bank along north shore of Lake Macquarie. Tracing the conchostracan-bearing bed along the double pipeline at three substations, in fossiliferous cherts.

Plate 48

AFRICA

Lesotho, Republic of South Africa, Angola

Figures 1, 2. 1: Lesotho. Kop, Mofoka's Store, Masero District, southwest side of ridge capped by Cave Sandstone. Arrow points to site of conchostracan-bearing section (Tasch Station 4). **2:** Massive, capping Cave Sandstone in contact with a bed underlain by Cave Sandstone. Conchostracans were found in lower part that permit correlation of this section with that in 3 below.

Figure 3. Republic of South Africa. Barkly Pass. (Tasch Station 2A.) Arrow points to Johann Loock (sedimentologist, University of Orange Free State, Bloemfontein), who is standing on the conchostracan-bearing bed which correlates with that shown in 2 above.

Figure 4. Republic of South Africa. Siberia. (Tasch Station 1A.) A composite section. Outcrop along Greyling's Pass, west of Siberia. The section is capped by the massive Cave Sandstone and, farther down the pass, the next lower of three Cave Sandstone units is exposed.

Figure 5. Lesotho. Thabaning, Mafateng District. (Tasch Station 3.) Numerous successive conchostracan-bearing beds which contain, among others, *Cornia* and *Palaeolimnadia.*

Figures 6, 7. Northern Angola, view of escarpment. **6:** Upper Stage of the Cassanje Series, so-called Phyllopod Beds divided into three zones by Oesterlen; the lowest zone seen in the photograph is conchostracan bearing, and this zone is spread over 2000 km^2. (Photograph and field data by Dr. Phillip M. Oesterlen.) **7:** Detail of lower Phyllopod (conchostracan) Bed in the lowest zone in 6.

Plate 49

ANTARCTICA

Figures A. B. Antarctica. Carapace Nunatak (lat 76°43′S, long 159°27′E), southern Victoria Land. **A:** Aerial map (courtesy of the Division of Polar Programs, National Science Foundation). Taken looking north, reversed here to show the southern view of the Kirkpatrick basalt sill. **B:** Detail of basalt sill and contact with the conchostracan-bearing shale. Carapace Sandstone; less than 1 m thick and traceable discontinuously over 307 m of the southeast-southwest–trending exposure. (White blotches are due to paint used to delete ink notations made on the aerial map.) (Cf. Pl. 46, 1 herein.)

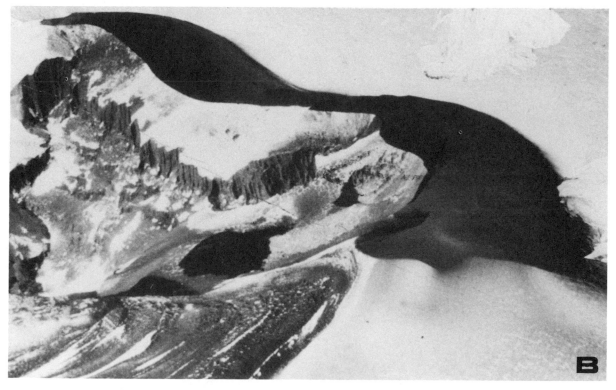

APPENDIX

Antarctica
  Blizzard Heights
  Storm Peak
  Carapace Nunatak

Australia
  Newcastle Coal Measures

India
  Mangli
  Raniganj Basin
  Sironcha Reserve Forest

Evolutionary Trends (Newcastle Coal Measures)

Table 1. Australia

Table 2. Australia

Measured Sections

ANTARCTICA

Blizzard Heights

Station 0. *Measured at termination of outcrop at east end around bend (southeast); and because interbed's basal contact was covered, section was offset at east end. (See Fig. 3B and 4.)*

Bed No.		Thickness (cm)
7-9.	Basalt, vein quartz below, and underlain by basalt.	--
6.	Laminated, baked, and carbonaceous argillite, conchostracans, *Cyclestherioides (Cyclestherioides) alexandriae* n. sp.	0.75
5.	Baked argillite with pseudo-banding; lower 1.27 cm with fish scales and poorly preserved conchostracans.	3.17
4.	Massive, blocky, weathered argillite; conchostracans at 1.27 cm below top of bed: *Glyptoasmussia meridionalis* n. sp; and fish scales.	6.34
3.	Blocky argillite, weathered dark brown, with fish scales and conchostracans *Estheriina (Nudusia) stormpeakensis* n. sp., Cyzicus *Lioestheria) antarctis* n. sp., *Cyzicus (Lioestheria) longulus* n. sp., *Cyzicus (Lioestheria)* cf. *brevis* Shen, *Estheriina (Nudusia)* sp. 2, *Cornia* sp. 1, *Cornia* sp. 2	5.08
2.	Fossiliferous argillite with ostracods, fish scales, and conchostracans; *Cyzicus (Euestheria) ellioti* n. sp., *Estheriina (Estheriina)* sp. 2, *Cyzicus (Euestheria) castaneus* n. sp., *Cyzicus (Euestheria)* sp., *Cyzicus (Euestheria) beardmorensis* n. sp.	5.08–5.71
1.	Dark, baked, blocky argillite, weathered dark brown; upper 0.86 cm, a laminated layer with conchostracans *Cyzicus (Euestheria)* sp., *Glyptoasmussia meridionalis* n. sp. (Sample 1A taken at same position as bed 1 but offset to east end and then some 2.44 m westward, where gradational contact with underlying basalt was exposed	5.08

Note: *Cyzicus (Euestheria) ellioti* n. sp. and *Cyzicus (Euestheria) beardmorensis* n. sp. from D.E. Site 23.59AA is equivalent to Tasch Station 0, bed 2. *Cornia* sp. 2 occurs in D.E. Site 23.58C, equivalent to Tasch Station 0, bed 3. *Estheriina (Nudusia)* sp. 2 occurs in D.E. Site 23.59B, equivalent to Tasch Station 0, bed 3.

Total Measured Thickness _____

25.50–26.13 cm
0.25–0.26 m

Station 1. *33 or more m west of Station 0.*

Bed No.		Thickness (cm)
--	Tuff, light buff to brown, massive, jointed, upper part agglomerate.	--
9.	Baked argillite, overlain by buff clay and thin tuff; contact appears conformable.	3.80
8.	White-weathering argillite, easily crumbled; conchostracans: *Estheriina (Nudusia)* sp. 1, *Cyzicus (Euestheria) minuta* (von Zieten).	3.15
7, 6.	(7) Blocky, white-weathering argillite, trace fossil; coprolite with fish scale (Tasch, 1976, Pl. 1, Figs. 3, 4, 8). Conchostracans: *Cyzicus (Lioestheria) malangensis* (Marlière), *Cyzicus (Lioestheria) antarctis* n. sp., *Cyzicus (Euestheria) minuta* (von Zieten). Bed 7 grades down to bed 6. (6) A yellow-brown, thin-bedded argillite with conchostracans: *Glyptoasmussia prominarma, Cyzicus (Euestheria) beardmorensis, Cyzicus (Euestheria) juracircularis, Cyzicus (Euestheria) castaneus, Cyzicus (Euestheria) minuta* (von Zieten).	3.80
5.	Conchostracan-bearing argillite with *Cyclestherioides (Sphaerestheria)* cf. *aldensis* Novojilov; *Cyzicus (Euestheria) minuta* von Zieten), *Cyzicus (Euestheria) castaneus* n. sp., and fish scales	1.30
4. 1.	Light-gray, medium-bedded argillite. Conchostracans: *Cyzicus (Lioestheria) antarctis* n. sp., *Cyzicus (Euestheria) minuta* (von Zieten), *Cyzicus (Euestheria) beardmorensis* n. sp., *Cyzicus (Euestheria formavariabalis* n. sp., *Glyptoasmussia meridionalis* n. sp., *Estheriina (Estheriina)* sp.	9.00

251

3. Blocky, light-gray argillite, trace-fossil braided structure (Tasch, 1976, Pl. 1, Fig. 1) and conchostracans: *Cyzicus (Lioestheria) malangensis* (Marlière). — 9.00

2. Conchostracan-bearing argillite (sampled as upper 2B and lower 2A units: 2B. *Cyzicus (Euestheria) formavariabalis* n. sp. 2A. *Cyzicus (Lioestheria) malangensis* (Marlière) *Cyclestheriodes (Cyclestherioides) alexandriae* n. sp. *Cyzicus (Euestheria) ellioti* n. sp. — 7.60

1. Tan to khaki, baked clay with conchostracan, *Cyzicus (Euestheria) castaneus* n. sp.; sharp contact with bed 2, and conformable contact with underlying amygdoidal basalt; contact surface of very low relief. — 28.00

Total Measured Thickness ———————
65.65 cm
0.65 m

Storm Peak

Lower Flow

Station 0 *(Tasch Number) Reconstructed section based on field data communicated by David Elliot, supplemented by Tasch's field work, collections, and study of the biota at Stations 1 and 2, as well as analysis of Elliot's fish slabs and conchostracan-bearing slabs.*

Bed No.		Thickness (cm)
4.	Baked, laminated shale.	3.81
3.	Fish bed, baked, laminated shale, with the fossil fish *Oreochima ellioti* Shaeffer; slabs with fish fossils, 11–12 mm thick; conchostracans on same bedding as fish fossils; *Cyzicus (Euestheria) ichthystromatos* n. sp. (cf. Station 1, bed 4).	11.73
2.	Baked, laminated shale with conchostracans: *Estheriina (Estheriina)* sp. 2, *Cyzicus (Euestheria) crustapatulus* n. sp.	6.75
1.	Contact with tuff, basal bed. Below this bed and within tuff, wood fragments and palynomorphs; *Classopollis anasillos* Filatoff, *Classopollis classoides* Pflug, and *Callialasporites trilobatus* Balme (see Tasch and Lammons, 1978, Pl. 1, Figs. 12–14). (Cf. Sta. 1, bed 1.)	5.71

Total Measured Thickness ———————
27.60 cm
0.28 m

Lower Flow

Station 1. *Approximately 50 m east of Station 0.*

Bed No.		Thickness (cm)
7.	Contact with green tuff that grades upward into a veined amygdoidal basalt.	--
6.	Hard, thin-bedded argillite with conchostracans on upper and lower surfaces, in direct contact with tuff; conchostracans: *Cyzicus (Euestheria) castaneus* n. sp.	5.08
5.	Hard, thin-bedded argillite with conchostracans on several levels: *Cyzicus (Euestheria) castaneus* n. sp.	3.81
4.	Fish bed. Thin, undulating, crustlike argillite with *Oreochima ellioti* Shaeffer; unidentified insect and conchostracan: *Cyzicus (Euestheria) rhadinis* n. sp. (upper surface).	5.71
3.	Blocky argillite with conchostracans poorly preserved.	8.52
2.	Hard argillite with fossil wood.	5.08
1.	Fossil wood horizon—whole logs; palynomorphs: *Classopollis anasillos* Filatoff, *Classopollis classoides* Pflug, and *Callialasportites trilobatus* Balme. This horizon is underlain by basalt.	30.5–61.0

Total Measured Thickness ———————
58.7–89.2 cm
0.59–0.89 m

Lower Flow

Station 2. *Approximately 11.1 m West of Station 0.*

Bed No.		Thickness (cm)
--	Basalt cliff; at base, debris underlain by tuff. Bed 4 in contact with base of tuff	--
4*.	Hard argillite, both surfaces with palynomorphs: *Patinosporites* sp. and conchostracans: *Cyzicus (Euestheria) castaneus* n. sp., *Cyzicus (Euestheria) ichthystromatos* n. sp.	3.17
3.	Hard argillite with conchostracans on both surfaces: *Cyzicus (Euestheria) castaneus* n. sp.	7.62
2.	Blocky, hard argillite with conchostracans: *Cyzicus (Euestheria) castaneus* n. sp.	6.37
1.	Thin-bedded argillite with ostracodes and conchostracans in upper contact with bed 2. Same conchostracan species as bed 2. Palynomorphs: *Ischyosporites marburgensis* De Jersey, *Chasmatosporites* cf. *C. rimatus* Nilsson. Underlain by red-weathering basalt.	1.89

Total Measured Thickness ———————
~19.0 cm
0.19 m

*Bed numbers reversed here; in the field, top bed was Number 1. See Tasch and Lammons, 1978, legend for Pl. 1, Station 0, bed 0 (= bed 1, a typesetting error).

Upper Flow, Slump Block

Bed No.   Thickness (cm)

11. (Sampled as upper, middle, and lower):
Upper: soft, tan, massive tuffaceous clay mudstone; slope covered with tuffaceous debris.
Middle: massive, buff-clay mudstone, weathers blocky; extensively jointed; beetle elytra and partial insect wing. Conchostracan: *Glyptoasmussia australospecialis* n. sp.; trace fossils: conchostracan valve galleries (Tasch, 1976, Pl. 1, Figs. 5 and 6.)
Lower: blocky, buff-weathering, clay mudstone; probable insect; conchostracans: *Cyzicus (Lioestheria) disgregaris* n. sp., *Cyzicus (Euestheria) crustapatulus* n. sp., *Palaeolimnadia (Grandilimnadia)* cf. *minutula* n. sp.   30.48

10. Hard, laminated shale (undulate lamination); cryptalgal laminate; ostracodes and conchostracans on top and bottom surface: *Cyclestherioides (Cyclestherioides) alexandriae* n. sp., *Estheriina (Nudusia) stormpeakensis* n. sp. Palynomorphs: *Chasmatosporites* cf. *C. rimatus* Nilsson.   4.44

9. Blocky, light-gray, hard mudstone, laminated near base; conchostracans on bottom layer on several levels: *Asmussia* sp. 1.   8.25

8. Thin-bedded, laminated (undulant lamination) to hard, platy shale, weathered buff; conchostracan in lower one half.   5.08

7. Light gray, platy and laminated, hard argillite that grades upward; conchostracan: *Palaeolimnadia (Grandilimnadia)* cf. *glenleensis* (Mitchell).   5.08

6. White-weathered, laminated mudstone; charophyte; trace fossil: braided structure Type C (Tasch, 1976, Pl. 1, text-fig. 1a–c); conchostracans (lower one-half): *Estheriina (Nudusia) stormpeakensis* n. sp., *Cyzicus (Lioestheria) disgregaris* n. sp., *Palaeolimnadia (Grandilimnadia) minutula* n. sp., *Cyzicus (Euestheria) crustapatulus* n. sp., *Palaeolimnadia (Grandilimnadia)* cf. *glenleensis* (Mitchell).   24.76

5. Very thin-bedded, laminated to platy shale; weathers in thin brown sheets; conchostracans (lower): *Cyzicus (Euestheria) crustapatulus* n. sp., *Estheriina (Estheriina)* sp. 3.   17.14

4. Light-gray claystone, laminated mudstone; conchostracans: *Pseudoasmussiata defretini* n. sp., *Palaeolimnadia (Grandilimnadia) minutula* n. sp.   3.81

3. Lithology same as bed 4 above.   5.08

2. Gray, laminated mudstone with several very thin shale partings between beds 2 and 3; conchostracans: *Cyzicus (Euestheria) transantarctensis* n. sp., *Palaeolimnadia (Grandilimnadia)* cf. *glenleensis* (Mitchell).   8.89

1. Massive claystone/mudstone; upper third, buff, grading down to white (lower one-third). Base not exposed; conchostracans: *Palaeolimnadia (Grandilimnadia)* sp. 2, *Estheriina (Nudusia) stormpeakensis* n. sp., *Cyzicus (Euestheria)* sp. 1; trace fossil: braided structure, type C (see bed 6 above for reference). (Other trace fossil in Upper Flow debris; borings in conchostracan valve ([Tasch, 1976, Pl. 1, Fig. 7].) Conchostracans: *Glyptoasmussia* cf. *luekiensis* Defretin-Le Franc and *Estheriina (Nudusia)* sp.   5.08

Total Measured Thickness ——————
         118.09 cm
         1.18 m

Carapace Nunatak

Station 2. *Approximately 300 m east of Station 1. The fossiliferous beds were traceable for 12 m. (See Fig. 6.)*

Bed No.   Thickness (cm)

-- Capping basalt sill   --

2. Laminated to massive black shale with plant debris and abundant palynomorphs. (Sampled in 3 units.)
Unit C (upper): Massive, black shale in direct contact with base of the flow. Palynomorphs: *Araucaracites australis, Undulatosporites* sp., *Cibotiumspora juriensis, Classopollis simplex, Laricondites triquetrus;* carbonized wood; notostracans (caudal filaments).

2. *Unit B (middle):* Laminated shale; *Araucaracites australis;* carbonized needles; brown and carbonized insect fragments; syncarids, mayfly nymph, ostracodes, notostracans, isopods.
Unit A (lower): Laminated black shale with *Araucaracites australis* and carbonized notostracans.   53.6–60.9

1. Massive black shale with plant debris; fish scales and teeth; abundant palynomorphs: *Classopollis chateaunovi* Reyre, *Exesipollenites* cf. *E. tumulus* Balme; carbonized needles and carbonized, paired beetle elytra; conchostracans: *Cyzicus (Lioestheria) disgregaris* n. sp.   7.6–15.2

-- Basalt underlain by Carapace Sandstone   --

Total Measured Thickness ——————
         61.2–76.1 cm
         .61–.76 m

Station 1' (Central). *Approximately 150 m West of Station 2.*

Bed No.		Thickness (cm)
--	Capping basalt sill.	--
1.	Laminated, hard shale that grades down to a massive shale. Biota include syncarid, carbonized and buff wood fragments; carbonized needles; mayfly nymph; notostracan; fish scale; ostracodes (in packets and as strew) and conchostracans: *Cyzicus (Lioestheria) disgregaris* n. sp.	15.2–60.9
--	Lower basalt sill.	--
	Total Measured Thickness	
		15.2–60.9 cm
		0.15–0.61 m

It is obvious that the ~15.2-cm figure above corresponds to the larger thickness of Station 2, bed 1, and that the ~60.9-cm figure above corresponds to the larger thickness of Station 2, bed 2.

Station 1 (South). *Approximately 150 m southwest of Station 1'.*

Bed No.		Thickness (cm)
--	Capping basalt sill	--
3.	Fissile shale in sharp contact with capping sill. Biota include scalariform tracheids, carbonized wood fragments, carbonized needles, ostracodes, mayfly nymph, and conchostracans: *Cyzicus (Lioestheria) disgregaris* n. sp.	15.2–20.3
2.	Massive black shale with carbonized beetle elytra (Tasch, 1973), carbonized wood fragments, brown plant fragments, scalariform tracheids, notostracans, syncarids, ostracodes, mayfly nymph, and conchostracans: *Cyzicus (Lioestheria) disgregaris* n. sp.	
1.	Laminated shale in sharp contact with underlying sill. Palynomorphs: abundant triletes; carbonized leaf fragments, needles, and wood; brown plant fragments; syncarids; mayfly nymph; beetle elytra and other insects; notostracans and conchostracans: *Cyzicus (Lioestheria) disgregaris* n. sp. (beds 1 and 2)	10.2–15.2
--	Lower basalt sill* underlain by Carapace Sandstone**	--
	Total Measured Thickness	
		25.4–35.5 cm
		0.25–0.35 m

*8.35 m thick; **30 to 36 cm thick.

Station 0. *Approximately 7.0 m south of Station 1.*

Bed No.		Thickness (cm)
--	Capping basalt sill	--
4.	Fissile shale: scalariform tracheids and conchostracans: *Cyzicus (Lioestheria) disgregaris* n. sp.	
3.	Same as bed 4. *Araucaracites australis* Cookson, *Classopollis* sp., and *Inaperturopollenites* Balme.	
2.	Same as bed 4. Same suite of palynomorphs as bed 3 plus scalariform tracheids; carbonized plants, notostracans, ostracodes, syncarids, and conchostracans: *Cyzicus (Lioestheria) disgregaris* n. sp.	
1.	Same as bed 4. *Classopollis* sp., *Araucaracites australis* Cookson; carbonized needles and conchostracans: *Cyzicus (Lioestheria) disgregaris* n. sp. (beds 1–4)	15.2–20.3
--	Lower basalt sill underlain by Carapace Sandstone.	--
	Total Measured Thickness	
		15.2–20.3 cm
		0.15–0.20 m

(Palynomorphs: *Klukisporites neovariegatus* Filatoff and *Contignisporites cooksoni* (Balme) Dettmann, also occur in Station 0 slabs but bed is uncertain).

AUSTRALIA

Newcastle Coal Measures

Station 1. *Entrance to John Darling Colliery, Wommara, New South Wales, Australia. Bank to road level.*

Bed No.		Thickness (cm)
6.	Hard, tripolitized white chert (possibly tuffaceous), overlain by soil cover. Sharp contact with bed 5.	0.48
5.	Light-gray to blue-gray, hard, massive chert with *Glossopteris* flora.	0.60
4.	Thin, platy, argillaceous limestone bearing brown carbonaceous plants, *Glossopteris* flora, and conchostracans (leaiids) in black, shaly, minifacies; conchostracans 6.35 mm below top of bed: *Leaia (Hemicycloleaia) magnumelliptica* n. sp., *Leaia (Hemicycloleaia) mitchelli, Leaia (Hemicycloleaia) immitchelli* n. sp., and a pelecypod.	0.02
3.	Same as bed 4 but relatively blocky and soft; no fossils.	0.03
2.	Same as bed 3 and covered; no fossils.	?
1.	Covered. (Debris of chert and argillaceous limestone, the latter with *Glossopteris* flora; also leaiid conchostracans that eroded from bed 4.)	?
	Leveled Thickness	
	(From uppermost bedrock to road level.)	~2.60 m

Station 2. *Warner's Bay Road and the Esplanade, 0.2 km north of Warner's Bay Road New South Wales, Australia. Slump dump bank, hence only hard, lower, in situ beds were measured. Old road on top of bank and Esplanade at base. Basal section covered and slumped.*

Bed No.		Thickness (m)
--	Covered upper section.	
2.	White silty argillite at base; barren; argillaceous slab 80 mm thick	0.08
1.	Carbonaceous slabs with conchostracans: *Leaia (Hemicycloleaia) mitchelli,* Etheridge, Jr.; *Leaia (Hemicycloleaia) immitchelli* n. sp.; *Leaia (Leaia) oblongoidea* n. sp.; *Leaia (Hemicycloleaia) ovata* Mitchell; *Leaia (Hemicycloleaia) compta* Mitchell; *Leaia (Hemicycloleaia) elliptica* Mitchell; *Leaia (Hemicycloleaia) collinsi,* Mitchell; *Palaeolimnadia (Palaeolimnadia) wianamattensis* (Mitchell).	0.14–0.19
--	Covered basal section. A fragment of bed 2 found here.	--
	Levelled Thickness	
	(From Old Road to Esplanade)	4.03 m

Station 3. *Belmont quarry (tuff/gravel), New South Wales, Australia. Floraville Road, behind and on top of large pebble conglomerate quarry; north of Maxwell Street on West Quarry Road. Accurate measurement of this section was difficult.*

Bed No.		Thickness (m)
6.	Pebble conglomerate.	2.7–3.3
5.	White weathering, tuffaceous shale.	1.0–1.3
4, 3.	Upper unit: brown weathering, thin, carbonaceous argillite, bearing *Glossopteris* flora and *Leaia (Hemicycloleaia)* cf. *collinsi* Mitchell, 2.5 cm Lower unit: plant fossils in blocky clay. Between upper and lower units a thin (0.1–0.3 cm) arenaceous minifacies.	0.66
2, 1.	Khaki-green, plastic and blocky clay; lower portion, coally bed, brown-black weathering, with plant fossils.	0.66
	Estimated Total Thickness	
		5.92–8.30 m

Station 4. *Kimble Hill. Parish Kahibah, Portion 311, New South Wales, Australia. Mt. Hutton area, at bend in Violettown Road, 0.48 km southwest of Wilson Road South. (Due to extensive cover between outcropping rocks, it was not possible to measure this section. The cherts contain several conchostracan species. The composite section below, in view of the covered sectors, can only be regarded as a possible interpretation.)*

Bed No.		Thickness
7.	Blocky, gray-black chert, fossiliferous: *Palaeolimnadia (Palaeolimnadia) wianamattensis* (Mitchell); *Leaia (Hemicycloleaia) ovata* Mitchell; *Leaia (Hemicycloleaia) mitchelli* Etheridge, Jr.; *Leaia (Hemicycloleaia) immitchelli* n. sp.; *Leaia (Hemicycloleaia) compta* Mitchell.	--
6.	Covered	--
5.	Arenite underlain by argillite; nonfossiliferous.	--

4.	Covered.	
3.	Brown-black, conchostracan-bearing chert: *Palaeolimnadia (Palaeolimnadia) glabra* (Mitchell); *Palaeolimnadia (Palaeolimnadia) wianamattensis* (Mitchell); *Leaia (Hemicycloleaia) mitchelli* Etheridge, Jr.; *Leaia (Leaia) oblongoidea* n. sp., and an insect wing.	--
2.	Covered.	
1.	Brown-black, conchostracan-bearing shale and nodular chert: *Palaeolimnadia (Palaeolimnadia) wianamattensis* (Mitchell); *Leaia (Hemicycloleaia) ovata Mitchell.*	--
	Thickness Not Determined	--

Station 5. *James Street Quarry. 1.13 km southwest of Station 4 (west of Floraville Road); east of slope off James Street, Mt. Hutton Area, near Tingara Heights, New South Wales, Australia.*

Bed No.		Thickness (m)
22.	*Thin, platy chert, weathers brown (limonitic).*	0.40
21.	*White weathering, tripolitized, black chert, thin to medium bedded; basal portion arenaceous*	0.64
20.	*Yellow-white, silty argillite; conchoidal and spherical weathering.*	0.20
19.	*Blocky, hard, brown-weathering chert.*	0.50
18.	*B. Blocky, massive chert, white-brown weathering with abundant plant fossils on upper surface; conchostracan fossils about 7.8 cm below surface; plants. Palaeolimnadia (Palaeolimnadia) glabra (Mitchell); Leaia (Leaia) oblonga Mitchell; Leaia (Leaia) oblongoidea n. sp.; Leaia (Hemicycloleaia) collinsi Mitchell; Cycloleaia discoidea (Mitchell); Leaia (Hemicycloleaia) etheridgei (Kobayashi); Leaia (Hemicycloleaia) paraleidyi Mitchell.*	0.15
	A. Blocky and platy; lower portion, thin chert with insect fossil (beetle elytra) and conchostracan, Leaia (Hemicycloleaia) sp.	0.11 0.26
17.	Thin bedded, platy to laminated, brown-weathering shale with abundant limonitic conchostracan fossils: *Leaia (Hemicycloleaia) mitchelli* Etheridge, Jr.; *Leaia (Hemicycloleaia) immitchelli* n. sp.; *Leaia (Hemicycloleaia) collinsi* Mitchell; *Leaia (Hemicycloleaia) paraleidyi* Mitchell; *Leaia (Leaia) oblongoidea* n. sp.; *Leaia (Hemicycloleaia)* cf. *etheridgei* (Kobayashi). (Lateral sampling was undertaken over a distance of 10 m with collections made every 1.6 m. This sampling yielded a few additional species [see Fig. 11].)	0.09
16.	Thin, platy chert and shale.	0.36

15. Thin-bedded, laminated, hard argillite, brown weathering, with plant fossils. 0.18
1. Plastic, green-white weathering clay, weathers in small, blocky, pieces. 0.50
13. Blocky, white-weathering siltstone. 0.80
12. Hard, blocky, green-gray clay. 2.08
11. (Levelled down to top of coal bed 10.) Plant-bearing, brown-red weathering argillite. 2.02
10. Carbonaceous, plastic clay; coally, interfingers with bed 11. 0.13
9. White-brown stained, massive siltstone with plant fossils. 0.56
8. Platy to plastic, brown-red, plant-bearing argillite with plant fossils; basal few millimetres coally. 1.38
7. Coally beds, about 7.0 cm each, separated by 0.5 cm of tan, plastic clay. 0.10–0.12
6. White, limonite-stained, silty, plastic clay. 0.62
5. Black to brown, coally, plastic clay; weathers black and brown 0.15–0.18
4. Plant-bearing, hard, argillite; weathers brown; limonitic stain and limonitic replacement of plant fossils; sharp contact with underlying hard chert. 0.47
3. Massive, white-weathering, blue-gray chert, base covered. --
2. Covered. --
1. Red-brown–weathering siltstone at base of slope; covered in part. --
Thickness of Measured Section (Beds 22 to 4) 11.24 m
Total Thickness Undetermined

Station 6. *Warner's Bay, New South Wales, Australia. Bank on north shore. 6.6 to 13.7 m from the intersection of Berkeley Street and Moulden Street and the Esplanade, along a double pipe line. (Composite section of three substations, 6a, b, and c.)*

Substation 6c.

Bed No.		Thickness (m)
8.	Covered. Hand level from top of bed 7 to Esplanade Road.	9.60
7.	Hard, blue chert, weathers brown.	0.25
6.	White-weathering tuffaceous clay.	0.26
5.	B. Hard, blue to buff chert with basal portion brown-weathering, bearing plant fossils; *Leaia (Hemicycloleaia) paraleidyi* Mitchell; ~33.0 m west of offset.	0.33

Substation 6b.

A. (Offset to northeast of section 6a.) Band of blue chert, bearing conchostracans; about 0.27–0.30 m above bed 2 (coally bed). *Leaia (Hemicycloleaia) mitchelli* Etheridge, Jr.; *Leaia (Hemicycloleaia) paraleidyi* Mitchell. 0.01 0.34

Substation 6a.

4. Platy, white-weathering argillite, plant bearing. 0.15
3. Buff weathering, brown, plant-bearing, silty argillite. 0.11
2. Coally, carbonaceous brown-black siltstone. 0.22
1. Tuffaceous, white-buff, brown-weathering mudstone. (Basal siltstone, not seen at this station but farther east along the bank, separated by an undetermined covered interval.)
Total Measured Thickness _____ ~11.30 m

Station 7. *Mt. Hutton Area, New South Wales, Australia. 2.41 km north-northwest of Station 5. James Street Creek, in creek bed, downslope to creek, but above creek floor. North on Floraville Road; northwest on Warmouth Road coming from Station 5; northeast on Warner-Charlestown Road along which, starting from Esplanade Road, the distance is 0.6 km to the power substation. Three substation outcrops were studied: 7A (youngest), 7B, and 7C (creek bed, oldest).*

Substation 7A. *Quarry 0.16 km on Warner-Charlestown Road to Burton Road, and 0.32 km to substation 7A, wooded area.*

Bed No.		Thickness (m)
--	Rock debris in old quarry, an undetermined interval above the Upper Pilot coal seam; beyond slabs of hard chert with blue-black interiors embedded in soil, a deep soil cover; slabs, white-brown weathering, with conchostracan fossils: *Leaia (Hemicycloleaia) immitchelli* n. sp.; *Leaia (Hemicycloleaia) ovata* Mitchell; *Leaia (Leaia) oblonga* Mitchell; *Leaia (Hemicycloleaia) belmontensis* Mitchell; *Palaeolimnadia (Palaeolimnadia) glabra* (Mitchell).	0.02–0.05

Substation 7B. *Northeast of power substation. Upper Pilot coal seam underlain by a lower, nonfossiliferous chert at road level, and an overlying siltstone also overlain by a chert capped by a conglomerate which continues to the southwest of the outcrop in successive layers.*

Levelled from Warner-Charlestown Road to creek floor, 13.1 m (subtract 1.05 m thickness of creek bed below). Beds along slope from base to Warner-Charlestown Road; nonfossiliferous siltstone, tuffaceous shale, siltstone, and covered interval. 12.05

Substation 7C. *Creek Bed.*

4. Hard, blocky, massive chert, dense blue, gray-buff weathering; plants in carbonaceous debris. 0.38

3. Thin-bedded, carbonaceous black chert with plants and ribbed conchostracans: *Leaia (Hemicycloleaia)* cf. *mitchelli* Etheridge, Jr.; *Leaia (Hemicycloleaia) belmontensis* Mitchell; *Leaia (Hemicycloleaia) elliptica* Mitchell; *Leaia (Hemicycloleaia) immithcelli* n. sp.; *Leaia (Leaia) oblonga* Mitchell.   0.18

2. Thin, platy, black-blue chert; basal 0.01 as in bed 3; black chert with insect and plant fossils.   0.30

1. Brown, blue-gray, platy argillite exposed in quarry floor, the rest of bed covered.   0.19

Total Measured Thickness _____
(Substations 7A, B, C)   15.82 m

Station 8. *Pacific Highway, northeast turn off to Violettown Road (formerly Main Road on Map 18 referred to below) coming from Belmont, New South Wales, Australia. Outcrop floors in road ditch on right side of Violettown Road near where the latter is intersected by the Pacific Highway. Distance from Belmont Post Office to Station 8: 3.54 km. (Map 18, Coordinates 3D, Robinson's Newcastle Street Directory, 4th Edition.)*

Bed No.		Thickness (m)
15.	White, tuffaceous sandy shale.	1.56
14.	Siltstone, weathers white.	1.44
13.	White tuffaceous shale.	1.60
12.	Covered; possibly same as bed 13 but covered by gravels and soil.	2.92
11.	Hard, massive black chert with plant fossils.	0.06
10.	Platy to thin-bedded brown, fossiliferous shale with plant fossils and *Leaia (Hemicycloleaia) mitchelli* Etheridge, Jr.	0.20
9.	Black, light gray, nonfossiliferous argillite.	0.28
8.	Insect/Conchostracan fossil bed; thin-bedded to platy, white-weathering argillite with plant fossils. Bed 8 subdivided into four fossiliferous subunits, a (youngest)-d (oldest): a-upper 2.0 cm: *Leaia (Hemicycloleaia) etheridgei* (Kobayashi); b-.29 cm below "a": *Leaia (Hemicycloleaia) belmontensis* Mitchell, *Palaeolimnadia (Palaeolimnadia) wianamattensis* (Mitchell), *Palaeolimnadia (Palaeolimnadia) glabra* (Mitchell); c-5-6 cm thick: *Palaeolimnadia (P.) glabra* (Mitchell), *Leaia (Hemicycloleaia) etheridgei* (Kobayashi); d-basal, 1.2 cm thick: *Palaeolimnadia (P.) wianamattensis* (Mitchell), *Palaeolimnadia (P.) glabra* (Mitchell), *Leaia (Hemicycloleaia) compta* Mitchell, *Leaia (Hemicycloleaia) mitchelli* Etheridge, Jr., *Leaia (Hemicycloleaia) paraleidyi* Mitchell. (Insect wing in debris.	0.46
7.	Brown-weathering, thin-bedded shale with carbonaceous fragments of plants.	2.15

(From base of insect/conchostracan bed to top of siliceous log bed [bed 6 below], covered, except for upper part of bed 7 and log bed.)

6. Siliceous log bed (upper part); white weathering, massive argillite with plant material; blue-black to brown interior.   0.42

5. Partly covered from base of bed 6 to top of bed 4; hard, platy argillite, interior black.   0.58

4. Bed 5 covered to top of bed 3.   0.20

3. Hard, blocky argillite.   0.68

2. Hard, tan argillite. (This bed goes under cover on Violettown Road and is found again on Pacific Highway in contact with underlying conglomerate.)   0.63

1. Conglomerate, underlies bed 2; larger pebbles at base, smaller pebbles in contact with bed 2.   --

Total Measured Thickness _____
   13.18 m

Station 9B. *Between Saros Street, Master Street, and Wommara Avenue, Wommara, New South Wales, Australia. Upper slope to Wommara Avenue. Slump bank opposite two houses; near section shatter and fill. Only lithologies could be determined. A composite section. (See Map 19, Robinson's Newcastle Street Directory, 4th Edition.)*

Bed No.		Thickness (m)
6.	Siltstone.	--
5.	Tuffaceous shale.	--
4.	Chert.	--
	(Offset)	
3.	Conchostracan-bearing chert (debris in house roadway): *Leaia (Hemicycloleaia)* sp. undet. (but not *mitchelli*); chert underlies the fossiliferous layer.	--
2.	Nonfossiliferous chert.	--
1.	Plant-bearing argillite.	--

Thickness Undetermined _____

Station 9A. *Darling Street, Wommara, New South Wales, Australia. School fence bank, covered. (Map 10, Robinson's Newcastle Street Directory, 4th Edition).*

Bed No.		Thickness (m)
4.	Covered.	--
3.	Conchostracan-bearing shale: *Leaia (Hemicycloleaia) ovata* Mitchell; *Leaia (Hemicycloleaia) magnumelliptica* n. sp.; *Leaia (Hemicycloleaia) belmontensis* Mitchell; *Leaia (Hemicycloleaia) mitchelli* Etheridge, Jr.	0.25–0.51
2.	White tuffaceous shale.	--
1.	Limonitic siltstone; underlying bed covered.	2.03

Thickness Measured _____
   2.28–2.54 m

(Total Thickness Undetermined)

Station 10. *Elizabeth Street, south to top of Victoria Street (off Floraville Road), Floraville, New South Wales, Australia. (Map 18, Robinson's Newcastle Street Directory, 4th Edition.) (Above the ditch section of Station 10, there is an estimated 61 m of conglomerate and siltstone.*

Bed No.		Thickness (m)
2.	Slabs of brown-weathering chert with plant and conchostracan fossils. Conchostracan-bearing horizon, 76–114 mm thick and 128 mm below top of bed: *Leaia (Hemicycloleaia) mitchelli* Etheridge, Jr.; *Leaia (Hemicycloleaia) immitchelli* n. sp.	0.33
1.	Blue chert, nonfossiliferous.	0.05
	Measured Thickness	
		0.38 m
	Total Thickness of Ditch Section (Levelled from top of Bed 2 to Floraville Road.)	1.50 m

Station 11. *Southwest of Windale, New South Wales, Australia. Windale Road going northeast 0.40 km to Floraville Road from the outcrop; past intersection of Violettown Road and Windale Road. (Map 18, Robinson's Newcastle Street Directory, 4th Edition.*

Bed No.		Thickness (m)
8.	Soil with blocky chert fragment, nonfossiliferous.	0.86
7.	Conchostracan-bearing chert: *Leaia (Hemicycloleaia) mitchelli* Etheridge, Jr.	0.04
6.	Conchostracan-bearing chert, blocky, upper 25 mm with fossils: *Palaeolimnadia (Palaeolimnadia) wianamattensis* (Mitchell); *Palaeolimnadia (Palaeolimnadia) glabra* (Mitchell); *Leaia (Hemicycloleaia) ovata* Mitchell.	0.06
5.	Light-gray, hard chert, nonfossiliferous	0.13
4.	Hard, light-gray chert, nonfossiliferous	0.02
3.	White weathering, thin, platy shale.	0.14
2.	Blocky, plant-bearing argillite with brown interior and carbonaceous fragments	0.04–0.10
1.	Gray-tan weathering, thin, platy argillite, covered to top of basal conglomerate; 50–75 mm above conglomerate, siltstone	3.81
	Total Measured Thickness	
		5.10–5.16 m
	(Levelled from top of soil to loose pebble conglomerate at base.)	5.12 m

Station 12. *Tingira Heights, New South Wales, Australia. Near transmission line; continuation of Burton Road; straight up the hill in creek bed bank (South Creek). (Map 18, Robinson's Newcastle Street Directory, 4th Edition.) (Composite Section.)*

Substation 12a.

Bed No.		Thickness (m)
6.	Conglomerate.	--
5.	Siltstone. Chert fragments in roadbed, nonfossiliferous. (Below bed 5.)	--

Substation 12b.

Bed No.		Thickness (m)
4.	Hard, blue-black chert, weathers brown, with carbonaceous fragments and plants; conchostracan-bearing: *Leaia (Hemicycloleaia) pincombei* Mitchell; *Leaia (Hemicycloleaia) mitchelli* Etheridge, Jr.; *Palaeolimnadia (Palaeolimnadia) glabra* (Mitchell); and insect wing.	0.13
3.	Siltstone, covered.	2.66
2.	Log bed, steeply dipped on road and under trees.	--
1.	Siltstone and underlying conglomerate	--
	Measured Thickness	
		2.79 m
	Total Thickness Undetermined	

Station 13. *Kahibah, New South Wales. Australia. (See Newcastle-Lake Macquarie Shire Map.) Chert (tuff) quarry behind street outcrop. From entrance to the quarry on Tingira Drive, to Croudace Bay Road, 0.32 km and 1.93 km to Violettown Road. (Map 31, Robinson's Newcastle Street Directory, 4th Edition.)*

Bed No.		Thickness (m)
4.	Overlying conglomerate. Thickness undeterminable.	--
3.	Tuff (chert) interval, barren.	2.78
2.	Chiefly banded tuffaceous argillite, fossiliferous, subdivided into sampling units: 2a (oldest) through 2h (youngest). h-hard, plant-bearing argillite, brown-white weathering, with a few very thin coally streaks. g-*Leaia (Hemicycloleaia)* cf. *compta* Mitchell. f-Barren. e-*Palaeolimnadia (Palaeolimnadia) wianamattensis* (Mitchell). d-Barren. c-*Leaia (Hemicycloleaia)* cf. *sulcata* (Kobayashi); *Palaeolimnadia (Palaeolimnadia) wianamattensis* (Mitchell). b-*Leaia (Hemicycloleaia) mitchelli* Etheridge, Jr.; *Leaia (Hemicycloleaia) immitchelli* n. sp.	

a-*Leaia (Hemicycloleaia) kahi-bahcnsis* n. sp; *Cycloleaia discoidea* (Mitchell); *Leaia (Hemicycloleaia)* cf. *etheridgei* (Kobayashi); *Leaia (Hemicycloleaia) mitchelli* Etheridge, Jr.; *Leaia (Hemicycloleaia) immitchelli* n. sp., *Leaia (Leaia) oblonga* Mitchell. (Levelled from base of bed 4 to base of 2a, thickness ~4.0 m. Units 2a–2h measured individually.)   1.22

1. From base of 2a to quarry floor, unfossiliferous   0.03
0. Plant-bearing tuffaceous shale. (Levelled from quarry floor to top of conglomerate band in road approach to quarry.)   ~10.30

Total Measured Thickness —————
~14.33 m

Total Levelled Thickness   14.60 m

Station 14. *Merewether Beach, near sewage outlet, Merewether, New South Wales, Australia. (Section measured below Dudley seam.)*

Bed No.		Thickness (m)
16.	Dudley coal seam	--
15.	Black shale, white weathering.	0.66
14.	Gray siltstone.	0.14
13.	Black shale and siltstone.	0.27
12.	Hard, cherty ironstone.	0.03
11.	Black silty shale with ironstone lenses in the middle.	0.07
10.	Black shale with ironstone lenses in middle. *Glossopteris* flora at base of shale and a leaiid conchostracan (subsequently lost), Tasch Collection. *Leaia (Hemicycloleaia) compta* Mitchell (probably occurred here), Mitchell Collection	0.44
9.	Gray siltstone. (Sea level at high tide.)	1.06
8.	Ironstone. *Cyzicus (Euestheria) lata* (Mitchell), *Cyzicus (Euestheria) novocastrensis* (Mitchell) (Tasch Collection) Mitchell's types of these and the following species also occurred in this bed: *Cyclestherioides (Cyclestheriodes) lenticularis* (Mitchell), *Cyzicus (Euestheria) obliqua* (Mitchell), *Cyzicus (Euestheria) trigonellaris* (Mitchell).	0.09
7.	Siltstone platform. Total thickness of platform not determined; covered below shale. Levelled thickness from lower 0.66 m of platform to sea level.	2.00
6.	Black shale, argillites, and siltstone.	0.12
5.	Ironstone.	0.02
4.	Black shale and siltstone.	0.26
3.	Ironstone with *Cyzicus (Euestheria) lata* (Mitchell).	0.02–0.06
2.	Black siltstone and argillite.	0.33–0.50

1. Green-stained, gray siltstone, at sea level at low tide.   0.20

Total Measured Thickness —————
5.71–5.94 m

Levelled Thickness   6.25 m*

*The difference here can account for the unmeasured covered portion of the siltstone platform.)

INDIA (See Fig. 12)

Mangli

3. Station M-1*. *Red shale pit, Mangli Village, ~91 m above the* Glossopteris *Bed, Nerduda River. (Red-streaked gray Kampti sandstone surrounds the area but no bedrock exposed near the pit.)*

Bed No.		Thickness (cm)
9.	Yellow shale, some red banded, few *Cyzicus (Euestheria) mangaliensis* (Jones). Overlain by soil.	5.80
8.	Red-banded hard shale with yellow (limonitic) streaks. Upper slab: red banded, limonitic streaks, no fossils, ~3.0 cm. Lower slabs: seasonally banded, silty shale, slightly micaceous, no fossils, ~7.0 cm.	10.0
7.	Hard, red-yellow argillite. Three units: Upper: 2.8 cm, *Cornia panchetella* n. sp., *Pseudoasmussiata indicyclestheria* n. sp. Middle: hard red argillite, 2.8 cm. Lower: same as middle, 2.6 cm. (*Note: plant fossils in float from beds 7–9.) (Offset 2.9 m southeast to shale pit.)	11.2
6.	Massive, hard red-yellow–streaked shale; Upper, 3.2 cm, barren. Conchostracans at base of this unit in a band 4.0 cm thick, *Cyzicus (Euestheria) mangaliensis* (Jones), *Cornia panchetella* n. sp.	7.20
5.	Hard red shale with yellow streaks, banded in part. Upper units A and B. *Cyzicus (Euestheria) mangaliensis* (Jones) in unit 6.0 cm thick. Lower unit banded red shale with limonitic streaks. *Cyzicus (Euestheria) mangaliensis* (Jones) in unit 8.4 cm thick.	14.4
4.	Massive-bedded red shale with yellow streaks, conchostracans throughout. *Cyzicus (Euestheria) mangaliensis* (Jones) abundant in basal 0.2 m.	13.6
3.	Medium-bedded red-yellow argillite with conchostracans: *Cyzicus (Euestheria) mangaliensis* (Jones), *Cornia panchetella* n. sp.	11.6
2.	Massive-bedded, hard, red argillite; limonitic bands. Upper units A and B. *Cyzicus (Euestheria) mangaliensis* (Jones), *Cornia panchetella* n. sp. Lower unit, barren (float from upper unit also collected).	18.0

1. Partly covered red argillite; yellow bands; conchostracans in limonitic band, the upper 10 cm:

    1.0 cm below top of bed, abundant and well-preserved conchostracans: *Cyzicus (Euestheria) mangaliensis* (Jones).

    2.3 cm below top: *Cyzicus (Euestheria) mangaliensis* (Jones).

    3.0 cm below top: barren.

    3.8 cm below top: *Cyzicus (Euestheria) mangaliensis* (Jones).

    6.1 cm below to as far down as possible to reach, actual bottom of bed 1 undetermined. 6.1

    Total Measured Thickness —————————

    97.90 cm
    ~0.98 m

(Kampti Sandstone overlies top of bed 9 and a nearby well, 2.4 m below bed 1; exposes red and yellow shale to depth of water surface, 22.7 m below soil level. Thus, the same type of lithology extends downward.)

Raniganj Basin

D[1]-Station RM-1. *Gorgangi Hill on Machkanda nala; lat 23°35'20"N, long 86°50'40"E.*

Bed No.		Thickness (m)
7.	Capping buff sandstone.	4.81
6.	Red shale bands intercalated in white weathering shales; some pelletiferous, massive to blocky, and some small, hard calcareous nodules. Poorly preserved conchostracan fossils in khaki siltstone (Field Unit A), 0.7 m thick; carbonized plants and worm burrows.	3.30
5.	Same lithology as fossiliferous unit in bed 6; plant fossils and poorly preserved conchostracans (Field Unit B).	1.16
4.	Khaki siltstone as in fossiliferous beds above (Field Unit C): *Cyclestherioides (Cyclestherioides) machkandaensis* n. sp.	1.00
3.	Shale to basal sandstone. In a coarse micaceous siltstone layer, *Cyzicus (Euestheria) basbatiliensis* n. sp.	5.20
2.	Plastic, smooth, red shale with very poorly preserved conchostracans.	0.03
1.	Basal sandstone, some 0.41 m exposed, total thickness not determined.	--
	Total Measured Thickness —————————	
		15.50 m

D[1] - Station RN-2*. *Numa River, tributary of Dawudi River; lat. 23°41'50"N, long 80°59'40"E. (Measured in southwest direction; gentle dip to southeast. Thick soil cover. 18 m to southeast, worm burrow.)*
*RN-2 is ~79.3 m above the Permian–Triassic boundary

Bed No.		Thickness (cm)
7–9	Bed 9, partly covered; 7–9 light khaki and tan, micaceous, all beds barren.	109.0
6.	Khaki siltstone, conchostracan-bearing: *Cyzicus (Euestheria) raniganjis* n. sp., *Cyclestherioides (Sphaerestheria)* sp. This bed extends about 1.61 m.	0.9
5.	Dark, micaceous siltstone, carbonized plant fossil.	28.3
4.	Top and bottom of same slab of khaki siltstone bears *Cyzicus (Euestheria) raniganjis* n. sp. and some other poorly preserved conchostracans.	2.6
3.	Gray siltstone, poorly preserved conchostracans.	7.7
2.	Brown siltstone with worm burrow.	14.6
1.	Brown siltstone to river bed. Barren.	162.5
	Total Measured Thickness —————————	
		325.6 cm
		3.26 m

Note: Distance between RN-1 and RN-2 is 365.8 m upstream. **Cyzicus (Euestheria) dualis* n. sp., apparently belongs to one of the fossiliferous beds of RN-2.

D[1] - Station RB-2. *Along Basbatili nala to creek bed; lat 23°17'10"N, long 86°53'55"E.*

Bed No.		Thickness (cm)
7.	Top conchostracan bed with poorly preserved conchostracans, overlain by coarse, brown sandstone; fossils in khaki micaceous siltstone. (This bed is traceable downdip to a soil bank that overlies the successive clam-shrimp-bearing beds.) Carbonized plant debris and, downdip, worm burrow in khaki siltstone that is separated from bed 1 by a 50 cm gap.	276.0
6.	Khaki clay and silt (bank of creek). Barren.	93.1
5.	Khaki shale with hematitic conchostracan: *Cyzicus (Euestheria) basbatiliensis* n. sp.	5.0
4.	Khaki-brown micaceous siltstone with sparse carbonized plant debris.	10.0
3.	Khaki-green plastic shale with poorly preserved conchostracans.	1.0
2.	Khaki-to-green siltstone and shale.	22.0
1.	Khaki-green, plastic shale with poorly preserved conchostracans, a few centimetres above creek level.	3.5
	Total Measured Thickness —————————	
		410.6 cm
		4.11 m

D^1 - Station RN-1-A. *Composite Section. Numa River; lat. 23°41'55"N, long 86°57'30"E, dip to southeast. (RN-1 is about 46.3 m above Permian-Triassic boundary.)*

Bed No.		Thickness (cm)
	Soil cover and slump.	
4.	a. Conchostracan-bearing siltstone, fossils poorly preserved.	3.6
	b. Nonfossiliferous siltstone.	6.4
	c. Conchostracan-bearing khaki siltstone; fossils poorly preserved.	2.0
	d. Same as "c" with *Cyclestherioides (Sphaerestheria)* sp.	8.0
		20.0
3.	Conchostracan-bearing bed in siltstone; fossils poorly preserved, 15 cm below top of bed.	21.1
2.	Barren siltstone weathering white to top of mudcrack at river level.	29.4
1.	Mudcrack bed at river level, barren.	5.0
	Total Measured Thickness	
		75.5 cm
		0.75 m

(Note: Distance between RN-1-A and RN-1-B [downdip] is 7.2 m.)

D^1 - Station RN-1-B.

Bed No.		Thickness (cm)
	Soil cover and slump.	
10.	Gray, brown to khaki siltstone with *Cyclestherioides (Sphaerestheria)* sp. (Downdip, 6.8 cm from the top, a thin conglomerate to top of underlying conchostracan-bearing bed (i.e., pebble[s] in laminated siltstone).	28.6
9.	Conchostracan-bearing siltstone; fossils poorly preserved.	1.7
8.	Barren siltstone, except for poorly preserved conchostracans at the base.	8.8
7.	Siltstone, barren.	7.5
6.	Siltstone with carbonized plant fossils.	1.5
5.	Siltstone, barren.	4.5
4.	Laminated siltstone with *Cyzicus (Lioestheria) miculis* n. sp.	2.0
3.	Siltstone, barren.	10.3
2.	Conchostracan-bearing siltstone, with *Cyzicus (Euestheria) basbatiliensis* n. sp. and *Pseudoasmussiata bengaliensis* n. sp.	2.0
1.	Siltstone, to river level, barren.	28.0
	Total Measured Thickness	
		95.0 cm
		0.95 m

Sironcha Reserve Forest and Environs

D^2-Station K-1. *2.5 km east-southeast of Kadamba Village (Sirpur Taluka, Adilabad District, Andhra Pradesh).*

Bed No.		Thickness (cm)
10.	Top bed overlain by soil; red, thin, hard argillaceous limestone; bearing fossil insects: Heteroptera, Diptera, Coleoptera, Blattaria, and plant fossils.	9.0

9.	Hard, gray argillaceous limestone, fossiliferous; insects: Blattaria, beetle elytra, insect wings, carbonized plants; basal, hard red shale parting, and hard, ferruginous/silicified slab with conchostracans poorly preserved (cf. same lithology in K-7 section).	10.0
8.	White to gray calcareous argillite with fossil wood, seeds, insect wings, fish, and conchostracans: *Estheriina (Nudusia) adilabadensis* n. sp.	5.6
7.	Hard argillaceous limestone with insect wings (Blattaria?).	6.0
6.	Calcareous marl with *Pseudoasmussiata andhrapradeshia* n. sp., and *Cyzicus (Euestheria) crustabundis* n. sp. in upper part.	9.0
5.	Hard argillaceous limestone, insect wings.	12.5
4.	Hard, gray argillaceous limestone, upper part, 5.0 cm; lower part, calcareous argillite and marl, 1.0 cm	6.0
3.	Upper unit: light-gray, banded argillaceous limestone with insects: Coleoptera, Homoptera, Blattaria, Diptera, Neuroptera, also suborder Hemiptera and some unknown forms; carbonized plants and conchostracans: *Estheriina (Nudusia) adilabadensis* n. sp.	11.0
2.	Thin-bedded, silicified fossiliferous limestone with insects. Insect wings: Diptera and Coleoptera; upper unit, 20.5 cm. Lower unit, hard calcareous argillite or argillaceous limestone, with insects: Coleoptera and Blattaria, 3.5 cm.	24.0
1.	Basal bed at creek level (Thoalla Madugu Vorre), siltstone with crystalline calcite. (Bed is thicker but concealed below creek floor.)	~33.0
	Total Measured Thickness	
		126.1 cm
		1.26 m

D^2-Station K-2. *Approximately 500 m upstream from K-1 and base of section, an estimated 4.57 m above K-1.*

Bed No.		Thickness (cm)
9.	Top bed in soil; hard, thin laminated calcareous argillite; light-white to gray buff; undulating laminae; somewhat crystalline, barren.	26.5
8.	Thin, gray-brown weathering, hard argillaceous limestone; upper unit with ostracodes and insect wings: Coleoptera, Blattaria, Neuroptera, Ephemerida, and unknowns; 7.0 cm. Lower unit: plant fossils, fish scales, and insects: pronota, elytra; 5.6 cm.	12.6

7. White to buff argillaceous limestone. Upper unit: white buff; thin, platy argillite with insects: paired wings, Blattaria, and fish scales; 2.0 cm. Lower unit: white buff, thin-bedded argillaceous limestone with plants and fish scales; 10.6 cm. — 12.6
6. Thin, platy, buff-white argillite, barren. — 6.4
5. Internally pitted, gray, thin-bedded limestone, recrystallized in zones; carbonized plant fragments. — 11.7
4. Thin, light-gray limestone with crystalline partings; barren. — 9.7
3. Gray geodic limestone, fossiliferous. Sampled in three units: Upper unit with *Estheriina (Nudusia) bullata* n. sp.; 6.4 cm. Middle unit, thin-bedded limestone, barren; 3.5 cm. Lower unit, thin-bedded, with abundant insects: Coleoptera, Neuroptera, Diptera, Hemiptera, Mecoptera, and unknowns; 7.7 cm. — 17.6
2. Thin-bedded limestone. Upper unit with *Paleolimnadia (Grandilimnadia)* sp. and insects: Blattaria, Homoptera, Ephemerida, Orthoptera, and lower unit with *Estheriina (Nudusia) indijurassica* n. sp. and insects: Homoptera, Blattaria, Coleoptera, and unknowns. — 9.1
1. Massive limestone in creek bed, base not exposed; upper portion, insects: Hemiptera, Coleoptera. — 20.0

Total Measured Thickness —— 126.6 cm
1.27 m

10. Argillaceous limestone with coconut calcite partings. — 35.0
9. Banded, gray-tan, argillaceous limestone with coconut calcite partings and several layers of conchostracans: *Cyzicus (Euestheria) crustabundis* n. sp. — 9.6
8. White buff, marly limestone with thin, hard calcite bands; barren. (*Offset II, 15.1 m upstream.) — 24.2
7. Medium to massive-bedded, white buff, slightly crystalline limestone. Upper portion with chert pebbles at base of unit and probable fish scale, 8.3 cm. Lower portion, chert fragments, 24.7 cm. — 33.0
6. Buff calcareous argillite; barren. — 22.0
5. Massive, hard limestone, buff-gray; bedrock of lower creek unit; dips below bed 6; barren. Fish scales, upper portion. (*Offset III, 18.3 m upstream.) — 50.0
4. Upper Creek. Massive limestone, reticulate, buff; barren. — 44.5
3. Thin-bedded limestone, white buff, portion reticulate; with honeycomb weathering; barren. — 95.4
2. Massive white buff limestone, reticulate; barren. — 60.5
1. Thin to medium-bedded white argillite with *Cyzicus (Euestheria) crustabundis* n. sp. Measured from base of bed 2 to creek bed (base not determined). — 87.4

Total Measured Thickness —— 671.5 cm
6.71 m

D²-Station K-4. *From the stream section on eastern side of tank, 1.2 km southeast of Metapalle. Composite Section.*

Bed No.   Thickness (cm)
16. Along bank, massive limestone, reticulate, thin calcareous partings, parallel and vertical; honeycomb weathering. Upper unit: buff-white, interior slightly recrystallized with *Cyzicus (Euestheria) crustabundis* n. sp. and possible fish scale; middle and lower units, same as upper, but barren. — 60.2
15. Medium-bedded limestone, white buff, reticulate, recrystallized interior, barren. — 22.8
14. Medium-bedded white buff limestone, hard and argillaceous, barren. — 70.5
13. Thin-bedded white-gray limestone, barren. — 18.0
12. Buff to tan limestone, banded, platy, with mudcracks. (*Offset I, 14.2 m upstream, barren). — 20.0
11. Buff to gray limestone, with several generations of conchostracans: *Cyzicus (Euestheria) crustabundis* n. sp. and insects: Blattaria. — 18.4

D²-Station K-5. *Chalma Wagu, southeast 9, Chitur; Forest Map No. 9/1, Conpt 245, Sironcha Reserve Forest; about 1.2 km southeast of Chitur Village; lat 18°46′50″N, long 80°06′35″E.*

Bed No.   Thickness (cm)
Soil cover with polished pebbles.
12. Buff, massive limestone, deeply pitted, with fish scales in upper unit only; middle and lower units, barren. — 77.7
11. Combined ripplemark-mudcrack bed; thin-bedded, gray and buff limestone bearing the indicated structures. Amplitude of ripples (cm): 4, 10, 25.5, 33.5; direction of flow, S 10°E. — 9.0
10. Buff-gray limestone, thin bedded, with insects: Blattaria, Mecoptera(?), and conchostracans; *Cyzicus (Euestheria) crustabundis* n. sp.; very thin chert bands. Upper unit, barren; middle, fish scale and the above indicated fauna; lower, barren, small mudcracks, 14.1 cm thick. — 33.3

9. Upper surface, with large mudcracks. Buff-gray, medium-bedded limestone with *Cyzicus (Euestheria) crustabundis* n. sp. Twenty-two layers traceable along the weathered face of the bed. Two large units comprise the Upper and Lower Ripple layers. Upper Ripple unit, 6.0 cm to crest from top of bed (mudcrack surface), 7.0 cm to trough; amplitude 1.0 cm. Lower Ripple unit with fish scale on ripple surface; to crest, 10.7 cm from top of bed to trough, 12.5 cm; amplitude, 1.8 cm. (Lowest trough reversed to become the upper trough due to change in current direction in 10 yr [based on 10 laminae between].) Rate of sedimentation, 1.1 cm/yr.   24.5

8. Thin layer of argillaceous limestone, small fragment of probable plant fossil.   10.0

7. Mudcrack surface; buff-gray, medium bedded limestone, with crystalline calcite crust, 12.0 cm thick; fish fragments in one layer, 4.6 cm below top; mudcrack bed 6.5 cm below top (small mudcracks).   13.3

6. Carbonized debris in marly limestone (white gray), debris sparse. Top unit, carbonized plant; middle, fish scale; bottom unit, barren.   27.8

5. Conchostracan-bearing bed; medium-bedded limestone, seasonally banded. 2.5 cm above base, *Cyzicus (Euestheria) crustabundis* n. sp.; 3.0 cm above base, carbonized wood; 4.1 cm above base, *Cyzicus (Euestheria) crustabundis* n. sp. 5.0 cm above base, unidentified conchostracans, poorly preserved, and carbonized wood; beetle elytra and Blattaria. Uppermost level of bed has a plant fragment.   11.5

4. Thin, calcareous, argillaceous partings at base of which are estheriids; below are coconut calcite partings. Fossil conchostracans, *Cyzicus (Euestheria) crustabundis* n. sp. from top of bed occurs at 2.0 cm, 2.6 cm, and so on, below. Base of bed with fossil wood impressions.   7.4

3. Coconut calcite partings with calcareous marly surfaces bear faint estheriid impressions that are unidentifiable.   31.7

2. Seasonally banded, hard argillitic limestone with conchostracans. From top of bed to 14.0 cm below top, conchostracans and banding: 1.1 cm below top, undetermined estheriids; 5.0 cm below top, thin argillaceous layers with fish fossils and conchostracans, *Cyzicus (Euestheria) crustabundis* n. sp. and fossil wood; base of bed, barren   34.6

1. Black, platy carbonaceous shale with ostracodes; true thickness undetermined; at base, upper portion of interbedded calcareous shale and thin, hard, calcareous limestone. Small vertebrate bone embedded in carbonaceous shale and calcareous limestone 12.1 cm below base of bed 2.   ~37.6

Total Measured Thickness _____

~346.2 cm
~3.46 m

D²-Station K-5-A. *Offset 110.7 m upstream from Station K-5. Bank section overlies directly K-5, bed 1, found under river sand.*

Bed No.   Thickness (cm)

2. Below less than 1 m of soil, hard, thin, platy argillaceous, calcareous limestone. Also hard, red ferruginous siltstone and sand (cf. K-1, bed 9, that also has a slab with estheriids. Poorly preserved conchostracans in both instances).   60.6

1. Hard, gray, red-banded limestone; mudcracks in contact with the limestone; barren.   33.8

Total Measured Thickness _____

94.4 cm
0.94 m

D²-Stations K-6 and K-6-A. *Composite section. Uske Wagu, East Branch, about 0.8 km northwest of Variam Village. Sironcha Reserve Forest, lat 18°44'00"N, long 80°04'30"E.*

Bed No.   Thickness (cm)

12. Hard, buff limestone, medium-bedded, slightly recrystallized; base slightly silty. Barren   18.7

11. Covered interval. Thickness by levelling.   120.0

10. Covered interval. Thickness by levelling.   16.3

9. Hard, buff limestone, silty, banded, barren.   18.5

8. Hard, buff limestone, slightly recrystallized, silty, banded, barren.   14.9

7. Upper portion, hard, buff-gray limestone with chert fragments, barren. Lower portion, same, but slightly recrystallized, barren.   8.2

6. Large mudcracks on upper surface; samples in three units: Upper: buff-brown, cherty to hard limestone, fish scale. Middle: hard buff limestone, carbonized plant and wood, fish scales, etc. Lower: gray-buff, hard limestone, carbonized fragments, chert fragments. Thirty-nine laminae counted (all units). Several mudcrack and one ripple-mark bed. Upper to middle unit, 12 cm below top. Middle mudcrack bed to bottom, 15.2 cm below top; ripple marks, 6.0 cm below top.   30.2

5. Thin-bedded, platy limestone. Three units: Upper, 2.5 cm below top, carbonized wood, fish scales, fins, etc. Middle, clay on siltstone bank, barren. Lower, cherty, platy, brown limestone, barren.   38.4

4. Thin, platy, banded limestone with estheriids, *Cyzicus (Euestheria) crustabundis* n. sp. to basal hard limestone bed* (below river sand). (*Offset to east bank where better preserved and exposed.)   48.2–56.2
*Offset. Station K-6-A (i.e., K-6 continued downward on east bank).

3. Mudcrack surface, thin, hard limestone and buff-gray, silty.   8.0

2. Medium-bedded, hard limestone, buff, silty.   27.0

1. Basal bed in creek (bottom exposed by digging); hard buff argillite and argillaceous limestone, massive. Three units. Upper; brown, silty, barren. Middle; chert fragments, fish fragments. Lower (base); fish scale, carbonized.   62.2

Total Measured Thickness _____
410.6–418.6 cm
4.11–4.19 m

D²-Station K-7. *Kotá Formation, southwest of Kotá Village, near Pranhita River. Because of slump and soil creep, the true thickness could not be adequately determined. Thickness measured (within the stated limitations) by levelling from soil to top of bed 14, 7.13 m and from that bed top to river level, 2.92 m, or a total thickness of 10.05 m.*

Bed No.   Thickness (cm)
Soil, uppermost portion with abundant fossil wood, overlies bed 14.

14. Thin-bedded to massive, and banded, brown calcareous limestone overlies a ferruginous band and coconut calcite partings. Poorly preserved conchostracans in banded ferruginous layer. (Cf. Station K-1, bed 9, K-5-A, bed 2.)   ~82.4

13. Buff-gray, medium-bedded, silty, marly limestone; calcite veining. Barren.   22.6

12. Surface mudcracked (medium-size cracks); hard, thin-bedded, buff limestone with coconut calcite partings forming crust over mudcracks. Barren.   5.7

11. Sequence of thin-layered successive calcite partings.   54.5

10. Mudcrack surface, side bears cherty partings.   6.0

9. Hard cherty limestone with conchostracans, poorly preserved.   16.6

8. Mudcrack bed, hard, banded, buff limestone; surface with conchostracans: *Cyzicus (Euestheria) crustabundis* n. sp., 3.0 cm below top of bed.   11.4

7. Covered unit. Marl-coated calcite with thin shale below.   38.1

6. Upper portion, marl, 11.5 cm thick, grading down to banded, hard limestone with *Estheriina (Estheriina) pranhitaensis* n. sp.; banded limestone directly underlies a nodular bed with conchostracans, which also occur at base of bed 6. These conchostracans are poorly preserved.   ~71.4

5. Massive, buff-gray limestone down to mudcrack surface.   56.2

4. Mudcrack surface underlain by banded limestone with poorly preserved conchostracans; bed carbonized from top to bottom. Upper portion, nodular limestone and coconut calcite partings.   42.0

3. Hard, thin-bedded limestone, buff; pinches out southwest in a distance of less than 1 m. Barren.   5.5

2. Mudcrack surface; massive, buff-gray limestone with fish scales and fish skeletons; some scales carbonized.   30.0

1. Marly bed, mostly covered. Complete thickness not determined due to soil creep downslope from top of bed 14.   ~292.0

Total Measured Section _____
734.4 cm
7.34 m
Based on levelling, corrected to   10.05 m

Note: This Kotá Formation type section, with a top ferruginous banded layer bearing poorly preserved conchostracans, permits correlation between K-7 and K-5-A. The new conchostracan species, *Cyzicus (Euestheria) crustabundis* n. sp., occurs in K-7, K-5, and K-4, as well as K-1, and indicates a correlation between north and south outcrops of the Kotá Formation (See Fig. 12, D²-Kotá.)

Evolutionary Trends and Morphologic Aspects

NEWCASTLE COAL MEASURES, AUSTRALIA

INTRODUCTION

Available data on the fossil conchostracans from the New South Wales (Australia) Newcastle Coal Measures did not previously allow for analysis of evolutionary trends and morphologic aspects, as well as related themes: recurring species, time-restricted occurrences, and systemic boundaries. However, the stratigraphic placement of Mitchell's conchostracan types, as well as other fossil data (Fig. 26), and the establishment of a traceable conchostracan horizon (Fig. 9) has provided the necessary context for consideration of such relationships.

Evolutionary Trends. (Palaeolimnadia). Two palaeolimnadiids are known from the Newcastle Coal Measures: *Palaeolimnadia (Palaeolimnadia) glabra* (Mitchell) and *Palaeolimnadia (Palaeolimnadia) wianamattensis* (Mitchell). They occur together at Tasch Station 11, bed 2. *Palaeolimnadia (Grandilimnadia) glenleensis* (M.) made its first appearance in the Blackwater Group of the Bowen Basin—a correlate of the Newcastle Coal Measures.

Before exploring the relationship of these three species, two types of data need to be reviewed: (1) morphological distinctiveness and (2) distribution.

Morphological Distinctiveness. The umbonal area is the most distinguishing and variable characteristic of the three palaeolimnadiids noted above. In *P. (P.) glabra* (M.), the umbonal area is large, eccentrically placed on the valve, and larger than its equivalent in *P. (P.) wianamattensis* (M.). In *wianamattensis (M.),* the umbonal area is not eccentrically disposed. Both *glabra* and *wianamattensis* bear few growth bands. By contrast, *Palaeolimnadia (G.) glenleensis* M. has a considerably reduced and subterminal umbo and also bears more growth bands than the other two species.

It is apparent that *P. (G.) glenleensis* (M.) is either an offshoot of a common ancestor of the three species or evolved from one of the other two. The basis for favoring the second alternative is discussed below.

Distribution. Viewed stratigraphically, *P. (P.) glabra* (M.) and *P. (P.) wianamattensis* (M.) occur in beds underlain by the Upper Pilot seam, whereas *P. (G.) glenleensis* (M.) occurs in beds equivalent to those in the Newcastle Coal Measures that are underlain by the Fassifern, the next higher coal seam above the Upper Pilot (See Fig. 26). Thus, *P. (G.) glenleensis* (M.) is geo-

logically later in its first appearance than the other two species. (It was originally described from the Triassic.) These factors, together with the umbonal change in *glenleensis* (M.), may be taken to indicate derivation of *glenleensis* (M.) from one of the earlier appearing two species.

Considered from the geographic distribution of *glabra* (M.) and *wianamattensis* (M.), both species occur at three Tasch stations, 4, 8, and 11, whereas *"wianamattensis"* appears at Station 2 without *"glabra"* (M.). *P. (P.) glabra* (M.) was found without *wianamattensis* M. at three stations, 5, 7, and 12.

If the reduced umbonal area in *P. (G.) glenleensis* (M.) and the comparatively smaller umbonal area in *wianamattensis* (M.) relative to *P. (P.) glabra* (M.) is taken into account, then one can postulate a trend. Incidentally, that trend did not die out in late Paleozoic time, when *glabra* did, because *P. (P.) wianamattensis* (M.) persisted throughout the Australian Triassic, whereas *P. (G.) glenleensis* (M.) has a less complete Mesozoic record; it is known only from the Middle Triassic.

The evidence allows the conjecture that the ancestral stock of *"wianamattensis"* (M.) could have been *"glabra."* *P. (G.) glenleensis* (M.), having further reduction of the umbonal area, possibly evolved from *"wianamattensis"* (M.), whose first appearance preceded its own.

Leaia. *Leaia (Hemicycloleaia) mitchelli* (Etheridge, Jr.) is the most uniquitous and numerically abundant species in the beds above the Upper Pilot seam and below the Fassifern seam. It occurs at 10 of 14 stations.

Foreshortening of the ribs of *Leaia (H.) mitchelli* (E.) gave rise to the new species *Leaia (H.) immitchelli* n. sp. (Pl. 23, Figs. 2, 10). The new species overlaps *mitchelli* in occurrences, appearing together with it at six Tasch stations and alone only at Station 1.

To determine the possible origin of *L. (H.) mitchelli* E., an inquiry is needed into its first appearance and association with other ribbed conchostracans. Its first appearance in the Coal Measures was at Tasch Stations 1, 4, 7, 9, 12, and 13, in association with *Leaia (Hemicycloleaia) compta* M., among other leaiids (See Fig. 9). Mitchell (1925) recorded this last-named species from beds below the Dudley Seam (Tasch Station 14; see Appendix) and at Croudace Hill. Novojilov (1956a, p. 32), on the basis of Mitchell's paper, thought *L. (H.) compta* M. and *mitchelli* E. were one species. Mitchell also originally thought (1925, p. 445) that *compta* M. might be a variety of *mitchelli,* but later changed his mind and treated *compta* as a distinct species. This writer's study of the holotypes indicated that the most outstanding difference separating *compta* M. from *mitchelli* E. was the *eta* angle; i.e., the angle formed by its curved posterior rib and a tangent to it drawn to the dorsal margin. That angle and also other features noted (see emended taxonomy of *L. (H.) mitchelli* E. and *L. (H.) compta* M.) indicate that *compta* M. is a distinct species, as Mitchell thought.

The concurrence of Mitchell, Novojilov, and Tasch on the similarities shared by *L. (H.) mitchelli* E. and *compta* M., appears to be a clue to the origin of *mitchelli* E., especially in light of the vertical distribution. It seems reasonable to infer that *compta* M., the geologically older species, gave rise to *mitchelli* E. by modification of its originally curved posterior rib into the straight one of *mitchelli* E., and by alteration of that rib's position, which gives *mitchelli* E. smaller alpha and beta angles. Otherwise, several ratios of the two species are close, for example, H/L, a/H, and c/L, as well as d_1 and d_2.

Leaia (Leaia) oblonga (Mitchell) and *L. (L.) oblongoidea* n. sp. are closely related species found in beds above the Upper Pilot seam. The former is oblong in configuration; the latter, elongate-rectangular, with a longer posterior sector and a slightly curved dorsal margin. Because of their nonoverlapping distribution, it is difficult to determine whether these species descended from a common unknown progenitor or whether one gave rise to the other. Their distribution is as follows: *L. (L.) oblonga,* Belmont quarry only (=Tasch Station 3); *L. (L.) oblongoidea* n. sp. occurred at Tasch Stations 2, 4, and 8. These two species are one example of the evolution of ribbed valves in the Coal Measures.

As noted earlier, *L. (H.) compta* M. appears to have given rise to *L. (H.) mitchelli* E. In turn, the closeness of almost all parameters, as well as the angles *beta, delta,* and *gamma* between *mitchelli* E. and *L. (H.) paraleidyi* M. indicates the likely derivation of *paraleidyi* M. from *mitchelli* E. by a decrease in the *alpha* angle (81.0° ±2.0° in *mitchelli* E. to 73°–74° ±2.0° in *paraleidyi* M. All three of these species show close resemblances. The loci or sequence in the genetic code governing ribs, their position on the valve, and their curvature, apparently were the active evolutionary site(s) through the time represented (from the geologically oldest Bar Beach Formation, through to Eleebana and the Catherine Hill Bay Formations, the geologically youngest). The so-called "incipient ribs," though occasional in occurrence, also denote flexibility in the code for ribbedness. Failure of such ribs to spread throughout the *mitchelli* (E.) population can be attributed to natural selection.

Recurring Species. For recurring species in the Coal Measures, minor variations occur in several angles and parameters. *L. (H.) paraleidyi* M., discussed above in a different context, and *L. (H.) elliptica* M. are both illustrative of this observation. For *paraleidyi* M., from older to younger beds, the *alpha* angle and the form ratio varied slightly: 74°, 74° to 73°; H/L - .630 to .683.

The same holds for *elliptica* M., *alpha* angle 80°–83°; H/L - .561 to .566.

Time Restricted Occurrences. Among the ribbed types in the Coal Measures are numerous species that appear in a single time frame and then do not recur; i.e., above a given coal seam. Species in this category are *Cycloleaia discoidea* (M.), *Leaia (H.) etheridgei* Kobayashi, *Leaia (L.) oblonga* M., *Leaia (L.) oblongoidea* n. sp., *Leaia (H.) kahibahensis* n. sp., *Leaia (L.) quadrata* M., *Leaia (H.) collinsi* M., and *Leaia (H.) magnumelliptica* n. sp. All of these species are restricted to the Croudace Bay Formation (see measured sections, Newcastle Coal Measures, Fig. 9). Two ribbed species were found to be confined to the Eleebana Formation, *Leaia (H.) latissma* M. and *Leaia (H.) pincombei* M.

Systematic Boundaries. (Carboniferous–Permian). As can be observed in Fig. 26, there is a considerable time gap in Australian conchostracan occurrences from the Carboniferous Lower Visean–Upper Namurian of the Canning Basin to the Permian (Tartarian) of the Sydney Basin. Very distinct changes did occur in the conchostracan fauna during this apparently barren interlude. A few Carboniferous genera, but no species, persisted into the Permian. Three Carboniferous genera, *Rostroleaia, Monoleaia,* and *Limnadiopsileaia* did not recur in the Australian Permian. Two genera did: *Leaia* and *Cyzicus.* Yet, the first group of genera persisted elsewhere. *Rostroleaia* was originally described from the Upper Permian (Kazanian) of the Urals (USSR) (Novojilov, 1952), *Monoleaia* from the Rio do Rasto Formation (Upper Permian) of Brazil (Reed, 1929), and *Limnadiopsileaia* from the Wellington Formation (Leonardian) of Kansas (Tasch, 1962). (See also "South America," this monograph, for a Brazilian *Monoleaia* species in the recent Tasch collection.) Only species of *Leaia* are known from locality M15 in the Newcastle Coal Measures of the Sydney Basin. Age-equivalent deposits in the Bowen Basin have *Palaeolimnadia* as well as *Leaia* (Fig. 27). No leaiids are known to have survived into the Lower Triassic anywhere in the world.

Species of *Cyzicus* appear in the Lower Triassic of the Canning Basin and in equivalent horizons in the Bonaparte and Carnarvon basins, as well as in the Knocklofty Formation of Tasmania. In the last, there are species of *Estheriina* and *Palaeolimnadiopsis* which are known from the Blina Shale of the Canning Basin and in the Bonaparte Basin, but are unknown in the Sydney Basin. Triassic conchostracans are unknown so far from Antarctica, despite their plentiful occurrence in the Australian Triassic.

During our field work in the Antarctic Triassic, deliberate search was made for conchostracan fossils with only negative results from the limited number of sites studied. No conchostracans are known from the Jurassic of Australia but they are found on other Gondwana continents: India (Tasch and others, 1975), Antarctica (Tasch, 1979), South America (Cardoso, 1962), central and west Africa (Haughton, 1963) and southern Africa (Lesotho; Ellenberger, 1970; Tasch, 1980) (Fig. 24A). In Australia, by Late Triassic time (Wianamatta Group), only palaeolimnadiids, corniids, and cyziciids are known. There followed an apparent Jurassic hiatus of conchostracans, and by Lower Cretaceous time, a single species is known, *Cyzicus (Lioestheria) branchocarus* Talent. The absence of Jurassic forms in Australia points up the question of the origin of the Cretaceous cyziciid species.

Cyziciids occur in the Jurassic beds of the Transantarctic Mountains (Queen Alexandra Range) at several localities (Storm Peak, Blizzard Heights, Manger Nunatak), in the Indian Jurassic Kotá Formation (Tasch and others, 1975), the interbeds of the Jurassic Drakensberg lavas, and in the Cave Sandstone sequence of Lesotho in southern Africa (Ellenberger, 1970; Stockley, 1947; Tasch, 1980). Because India was in Jurassic proximity with Antarctica and Australia, according to Curray and Moore (1974) (Frontispiece), that may have provided a prior dispersal track for Australian Cretaceous cyziciids. There was more possibility for dispersal because the bight of Australia was proximate to Antarctica in Jurassic time, according to Sproll and Dietz (1969), if, in fact, there were no indigenous conchostracan inhabitants of Jurassic freshwater bodies in Australia.

The problem of Mesozoic conchostracans in Antarctica, India, and Australia has been considered on several occasions (Tasch and others, 1975; Tasch and Jones, 1979b, among others). The thrust of this interpretation established a tie between palaeolimnadiid and estheriniid conchostracan bioprograms in the Lower Triassic of western Australia (subsurface deposits, Media Well No. 1) of the Canning Basin, the Blina Shale, and the Bowen Basin of eastern Australia. The same relationship also exists in Jurassic equivalents of Antarctica (Transantarctic Mountains) and India (Kotá Formation). If the continents of India–Antarctica–Australia were all in Mesozoic contact or relatively close proximity (Tasch, 1980), that would have allowed for conchostracan egg dispersal among them.

The absence of Jurassic conchostracans in Australia, but not in the proximate Gondwana continents, poses a problem, because Jurassic continental beds are present in Australia. What conditions existed in Australia during Jurassic time to inhibit or thwart hatching of conchostracans (assuming their presence) in contrast to the recurrent appearance of conchostracan populations in Australia during Triassic time, as well as their appearance in the Cretaceous?

There were numerous Jurassic lakes and swamps in eastern Australia (for reference, see below), while western Australia was chiefly marine. Erosion of the Paleozoic mountain ranges supplied enormous quantities of detritals (quartz-rich sands) which filled many eastern and southern basins, and at other times covered the Great Artesian Basin (Hind and Helby, 1969). Extensive fluvatile drainage existed. Toward the end of the Paleozoic there were marine incursions and volcanism restricted to some eastern basins. Jurassic tholeiitic dolerite intrusions were extensive in Permian-Triassic basins of Tasmania and other Australian mainland basins (Clarence-Moreton Basin) which had widespread basalt flows (McElroy, 1969). There were also coal deposits in the Great Artesian Basin and elsewhere (Laseron and Brunnschweiler, 1969; Brown and others, 1968).

It now seems quite probable that the absence of Jurassic conchostracans in Australia is real and not merely an artifact of inadequate collections. There is now a fair body of surface and subsurface information on both the marine and nonmarine Jurassic of the continent, and the latter has never yielded conchostracans.

In addition to the possibly extensive shallow marine barrier in Western Australia between eastern Australia, and Antarctica where Jurassic conchostracans existed, the substrates of the Jurassic continental pools of Australia may have had unsuitable coarse detrital substrates or perhaps a humic acid barrier in the peaty swamps. Certainly, there was an immense amount of decaying vegetation, as indicated by the Walloon Coal Measures (McElroy, *in* Packham, 1969). Thus far, the Australian Jurassic lacustrine biota consists of plants (White, 1981), fish (Lake Talbragar) (Woodward, 1895), and freshwater mussels (Victoria lakes) (Laseron and Brunnschweiler, 1969).

TABLE 1. AUSTRALIAN MESOZOIC SPECIES BY BASIN AND AREA.

Included are Species described by Tasch, *in* Tasch and Jones, 1979a, 1979b, 1979c; Tasch, 1975; and literature citations. These are in addition to localities given under description of Mitchell's types below and Mesozoic species described in this memoir.

===

TRIASSIC
Numbers correspond to those on Locality Map, Fig. 7.

Canning Basin
26. *Cyzicus (Lioestheria) australensis* (Novojilov). Blina Shale, Bell's Ridge near Lennard River, 100 km southeast of Derby; lat 18°S, long 124.5°E, near Joseph Bonaparte Gulf, Western Australia. Lower Triassic.
27. *Estheriina (Nudusia) blina* Tasch, Blina Shale (see No. 26). Lower Triassic.
29. *Cyzicus (Lioestheria) erskinehillensis* Tasch. Blina Shale, Erskine Hill, 100 km southeast of Derby. Lower Triassic.
30. *Palaeolimnadia (Palaeolimnadia) wianamattensis* (Mitchell). Blina Shale, Wongil Ridge, 12 km east-southeast of Yeeda Station near Joseph Bonaparte Gulf, Western Australia. Lower Triassic.
32. *Palaeolimnadia (Grandilimnadia)* sp. undet. 1. Blina Shale, WAPET Meda No. 1 well (for details, see No. 37 below, cuttings from 192–195 m.
33. *Palaeolimnadia (Grandilimnadia) medaensis* Tasch. Meda No. 1 well, cuttings at 179–182 m (for locality, see No. 37 below).
37. *Cyzicus (Lioestheria)* sp. undet. 5. Blina Shale, M52, northeast bank of Lake Jones (lat 19°27′17″S, long 126°06′22″E) and WAPET Meda No. 1 well (lat 17°24′00″S, long 124°11′30″E), 56 km east of Derby on the Lennard shelf. Lower Triassic.
38. *Cyzicus (Lioestheria)* sp. undet. 3. Blina Shale, BMR Locality M29, Chilpada Chara, Minnie Range, Cornish 1:250,000 Sheet. Lower Triassic.
39. *Cyzicus (Lioestheria)* sp. undet. 4. Blina Shale (for locality, see No. 38 above).

Sydney Basin
60, 61. *Cornia coghlani* (Etheridge, Jr.). Dents Bore, Holt Sutherland Estate, George's River. Lower Triassic.
71. *Palaeolimnadia (Palaeolimnadia) wianamattensis* (Mitchell). Brookvale, Beacon Hill near Manly, New South Wales (Australian Museum F55873).
(Additional Sydney Basin not on Index Map. Geological Survey of New South Wales, 1980 Bulletin, v. 26 p 412):
   Cornia coglani (?) Bulgo Sandstone. Lower Triassic.
   Cornia (?) *coghlani* (?) Stanwell Park Claystone. Lower Triassic.
   Cornia (?) *coghlani* (Etheridge, Jr.). Dooralong Shale, Narrabeen Group. Lower Triassic.

Carnarvon Basin
68. *Cyzicus (Euestheria) minuta* (?) (von Zieten). Kockatea Shale, 7.2 km south of Redbluff, near Kalbari, Western Australia. Lower Triassic. (Cockbain, 1974).

Joseph Bonaparte Gulf
40. *Palaeolimnadia (Grandilimnadia) arcoensis* Tasch. ARCO-AAP Petrol No. 1 well, lat 12°48′30″S, long 128°26′50″E, offshore, 265 km, west-southwest of Darwin, depth 3204.8 m. Lower Triassic.
41. *Cyzicus (Euestheria) bonapartensis* Tasch, (same as 40 above), well core 8, depth 3204.8 m.
42. *Cycizus (Euestheria)* sp. undet. 1 (same as 40 above), well cuttings, 3352.6 m.
43 *Cyzicus (Euestheria)* sp. undet. 3. (Same as 40 above.) 3217–3230.7 m.
44. *Cyzicus (Euestheria)* sp. undet. 2 (same as 40 above), well cuttings, 3352.6 m.
45. *Palaeolimnadia (Grandilimnadia) profunda* Tasch. (Same as 40 above.) depth 3422.7 m junk basket core.
46. *Cyzicus (Euestheria) dickensi* Tasch. NMR locality 619, 6 km southeast of Port Keats Mission (lat 14°15′53″S, long 129°33′37″E). Lower Triassic shales, 6 km south of Mount Goodwin.

Bowen Basin, Queensland
51. *Cyzicus (Lioestheria)* sp. undet. 1. BMR locality A27, 2.4 km south-southwest of Taurus Homestead, Duaringa. Base of Rewan Group. Lower Triassic.

52. *Cyzicus (Lioestheria)* sp. undet. 2. BHP Blackwater No. 2 bore. (Same locality as 55 below.) Core slice at 197.5 m, ? base of Rewan Group. Lower Triassic.
53. *Cyzicus (Lioestheria)* sp. undet. 3. BHP Blackwater No. 2 bore (same locality as 55 below), core slice at 101.4 m, Rewan Group. Lower Triassic.
54. *Palaeolimnadia (Palaeolimnadia) wianamattensis* (Mitchell). (Same locality as 55 below.) Split core at 102.9 m.
55. *Palaeolimnadia (Grandilimnadia)* sp. undet. 1. BHP Blackwater No. 2 bore, 2 km southeast of Blackwater, Queensland, Rewan Group, split core at 102.9 m. Lower Triassic.
56. *Estheriina (Nudusia) circula* Tasch. (Same locality as 57 below.) Base of Rewan Group. Lower Triassic.
57. *Estheriina (Nudusia) rewanensis* Tasch. BMR locality B 63, 4.8 km west-northwest of Taurus Homestead, Duaringa (No. 57); also Canning Basin, Meda No. 1 well, Blina Shale, depth 216.4 m (see No. 37).

Port Hacking, New South Wales

62, 63, 64, 65. *Cornia coghlani* (Etheridge, Jr.). Port Hacking, Heathcote, Narrabeen and Liverpool; all near Sydney (cited in Etheridge, Jr., 1888, p. 4). [Cf.*Cyzicus (Lioestheria) fictacoghlani* n. sp.] Cremorne Bore No. 2, Sydney, depth 643.7 m, Narrabeen Group. Lower Triassic. (Described herein.)

Ipsvich, Queensland

66. *Palaeolimnadia (Grandilimnadia)* sp. Denmark Hill. Upper Triassic. (Australian Museum 39291.)
78. *Cyzicus* cf. *mangliensis* (?) Incomplete specimen impossible to identify. Same locality as 66 above. (Australian Museum 28833.)

Tuggerah Lake, N.S.W.

79. *Cornia coghlani* (Etheridge, Jr.). Tuggerah Formation, Narrabeen Group near Newcastle. (H. G. Raggatt, 1969, p. 405–406.) West margin of Tuggerah Lake, depth 191.1 m (Australian Museum F 51455.)

Tasmania

(See Tasch, 1975, vol. 109, for species descriptions.)
80. *Cyzicus (Lioestheria)* sp. 1. Tinderbox, about 20 km south of Hobart.
81. *Palaeolimnadiopsis tasmani* Tasch. Tinderbox. Same locality as 80.
82. *Cyzicus (Lioestheria)* sp. 2. Knocklofty, about 3 km northwest of Hobart.
83. *Cyzicus (Lioestheria)* sp. 3. Knocklofty. Same locality as 82 above.
85. *Palaeolimnadia (Palaeolimnadia)* cf. *wianamattensis* (Mitchell). Cascades, about 4 km southwest of Hobart.
86. *Palaeolimnadia (Palaeolimnadia)* sp. 1. Cascades. Same locality as 85 above.
87. *Palaeolimnadia (Grandilimnadia)* sp. 1. Cascades and Old Breach, about 12 km north of Cascades. Same locality as 85 above.
88. *Palaeolimnadia (Palaeolimnadia) banksi* Tasch. *(nom. corr.)* Poatina, Ross Sandstone, southwest of Launceston.
89. *Palaeolimnadia (Palaeolimnadia) poatinis* Tasch. Same locality as 88 above.

TABLE 2. SCHEDULE OF AUSTRALIAN FOSSIL CONCHOSTRACAN LOCALITIES.

Paleozoic-Mesozoic. To be read in conjunction with Locality Map, Fig. 7. (Numbers 15–25 and other grouped numbers refer to several species from the same locality.)

Map Numbers

15–25. Grant Range No. 1 well, Anderson Formation, 90 km south-southeast of Derby, near J. Bonaparte Gulf, western Australia (=WA). Carboniferous (Late Visean to possibly early Namurian).
26–27. Bell's Ridge, near Lennard River, 100 km southeast of Derby, 18°124.5°E, near J. Bonaparte Gulf, Northern Territory (NT), Blina Shale, Canning Basin, Lower Triassic.
28–29. Erskine Hill, about 30 km northwest of Lennard River, 100 km southeast of Derby, near J. Bonaparte Gulf, WA, Blina Shale, Canning Basin. Lower Triassic.

Bonaparte Gulf, WA, Blina Shale, Canning Basin. Lower Triassic.

30. Wongil Ridge, 12 km east-southeast of Yeeda Station, lat 17°38′30″S, 123°45′00″E, near J. Bonaparte Gulf, WA, Blina Shale, Canning Basin. Lower Triassic.

31–36. Meda No. 1 well, 56 km east of Derby on Lennard Shelf at lat 17°24′00″S, long 124°11′30″E, Western Australia (=WA), Blina Shale, Canning Basin. Lower Triassic.

37. Lake Jones. Mount Bannerman 1:250,000 Sheet. Lat 19°27′17″S, long 126°06′22″E, WA, Blina Shale, Canning Basin. Lower Triassic.

38–39. Chilpada Chara, Minnie Range. Cornish 1:250,000 Sheet. 23.3 km north-northwest of Culvida Soak. Lat 20°01′35″S, long 126°50′49″E, WA, Blina Shale, Canning Basin. Lower Triassic.

40–45. ARCO-AAP Petrol No. 1 well. Lat 12°47′30″S, long 128°26′50″E. 265 km west-southwest of Darwin, WA. Lower Triassic, nonmarine unit C.

46. South of Mount Goodwin, 6 km southeast of Port Keats Mission. Lat 14°15′53″S, long 129°33′37″E. NT. Lower Triassic.

47. Narrien Range, northwest of Emerald, Drummond Basin, Raymond Formation, Queensland. Lower Carboniferous.

48–50. Clermont 1:250,000 Sheet. Bowen Basin, Queensland, 4.2 km south-southwest of Winchester Homestead, Blackwater Group. Upper Permian.

51. Duaringa 1:250,000 Sheet. Bowen Basin, Queensland, 2.4 km south-southwest of Taurus Homestead about 18 km south-southwest of Blackwater, Rewan Group (Sagittarius Sandstone). Base of Lower Triassic.

52–55. Blackwater, Duaringa, Bowen Basin, Queensland. 2 km southeast of BHP Blackwater No. 2 Bore, Rewan Group. Lower Triassic.

56–57. Duaringa, Bowen Basin, Queensland, 4.8 km west-northwest of Taurus Homestead, Rewan Group. Lower Triassic.

58. Koonwarra, 4.2 km east of, and 155 km northeast by road, of Melbourne, Victoria. Lower Cretaceous (Valanginian–Aptian).

59, 71. Brookvale, Beacon Hill, Sydney Basin, northeast of Sydney, New South Wales (=NSW). Hawkesbury Sandstone. Middle Triassic.

60. Moore Park, Surry Hills, Sydney, Sydney Basin, NSW. Lower Triassic.

61. Dents Creek, Holt Sutherland Estate, near Georges River and Sydney. Narrabeen Group, Sydney Basin. Lower Triassic.

62. Port Hacking, near Sydney, Sydney Basin, NSW. Lower Triassic.

63. Heathcote, near Sydney, Sydney Basin, NSW. Middle-Triassic?

64. Narrabeen, near Sydney, Sydney Basin, NSW. Narrabeen Group. Lower Triassic.

65. Liverpool, near Sydney, Sydney Basin, NSW. Hawkesbury Sandstone. Middle-Triassic.

66. Denmark Hill, Ipsvich Coal Measures, Ipsvich, near Brisbane, Queensland. Upper Triassic.

67. Charleston, between Newcastle and Lake Macquarie, Newcastle Coal Measures. Permian (Tartarian).

68. Redbluff, 12 km south of Kalbari, southern Carnarvon Basin, WA. Lower Triassic.

69. Bowning near Yass, NSW. Permian.

70–70A. Newcastle, NSW, Newcastle Coal Measures. Upper Permian.

72. Tuggerah Lake, near Newcastle, NSW. Narrabeen Group. Lower Triassic.

73. Weoroona, south of Blackwater, headwaters of Sirius Creek, Queensland. Upper Permian.

74. Glenlee, near Hawkesbury River and Sydney, NSW. Lower Triassic.

75. Hornsby, near Sydney, NSW. Middle Triassic.

76. South of Bulli, near Sydney, NSW. Upper Permian.

77. Garie Beach, 10 km south-southwest of Port Hacking, near Sydney, NSW. Lower Triassic.

78. Benelong, near Dubbo, NSW. Upper Permian.

79. Buchan Caves, Snowy Mountains, Victoria. Lower Middle Devonian.

80–81. Tinderbox, south of Hobart, Tasmania. Lower Triassic.

82–83. Knocklofty, northwest of Hobart, Tasmania. Lower Triassic.

84. No entry.

85–87. Cascades, southwest of Hobart, Tasmania. Lower Triassic.

88–89. Poatina, southwest of Launceston, Tasmania. Lower Triassic.

90. Old Beach, northwest of Hobart, Tasmania. Lower Triassic.

91–92. Deleted.

REFERENCES CITED

Almeida, F.F.M. de, 1950, Acantholeaia un novo Génera de Leaiadea: Rio de Janeiro, Brazil, Divisão de Geologia e Mineralogia Notas Preliminares e Estudos, no. 51, p. 3–6, pl. 1, fig. 1–4, text fig. 2-4.

Audouin, V., 1837, Communications, *in* Société Entomologique de France, Annales, v. 6, p. 5–516: Entomologie Bulletin, 1er Trimestre, Session 1, p. ix–xi.

Baird, W., 1849, Monograph of the family Limnadiidae, a family of entomostracous Crustacea: Zoological Society Proceedings, v. 17, p. 84–90, pl. 11.

——, 1859, Description of some recent Entomostraca from Nágpur, collected by the Reverend S. Hislop: Zoological Society Proceedings, v. 27, p. 232, pl. 63, fig. 1.

Ball, H. W., undated, Report on Conchostraca from the Lethem Shales, British Guiana, *in* White, E. I., Report to the Geological Survey of British Guiana: British Museum (Natural History).

Ball, H. W., Borns, H. W., Jr., Hall, B. A., Brooks, H. K., Carpenter, F. M., and Delevoryas, T., 1979, Biota, age, and significance of lake deposits, Carapace Nunatak, Victoria Land, Antarctica: Fourth International Gondwana Symposium (India), Papers, I, p. 166–175.

Ballance, P. F., and Watters, W. A., 1971, The Mawson Diamicite and the Carapace Sandstone Formations of the Ferrar Group at Allen Hills and Carapace Nunatak, Victoria Land, Antarctica: New Zealand Journal of Geology and Geophysics, v. 14, no. 3, p. 512–527.

Barnard, K. H., 1931, A revision of the South African Branchiopoda (Phyllopoda): South African Museum Annual, v. 29, p. 255–256.

Besairie, H., 1952, Les Formations du Karroo á Madagascar: Alger, 19th Congrés Géologique International, Symposium sur Les Séries de Gondwana, p. 181–186.

Beurlen, K., 1954, Um novo género de Conchostráceo de familia Limnadiidae: Rio de Janciro, Brazil, Divisão de Geologia e Mineralogia Notas Preliminares e Estudos, no. 83, p. 1–7, text figs. 1–3.

Beyrich, W. H., 1857, Protokoll der Juni–Sitzung: Berlin, Zeitschrift der Deutschen Geologischen Gesellschaft, v. 9, p. 374–377.

Bigarella, J. J., Pinto, I. D., and Salamuni, R., eds., 1967, Brazilian Gondwana Geology: Curitiba, International Symposium on Gondwana Stratigraphy and Paleontology, Excursion No. 3, Guidebook, 122 p.

Bishop, J. A., 1968, Aspects of post-larval life history of *Limnadia Stanleyana* King (Crustacea: Conchostraca): Australian Journal of Zoology, v. 6, p. 885–895.

Blanford, W. T., 1860, On the rocks of the Damuda Group and their associates in Eastern and Central India, as illustrated by the re-examination of the Rániganj Field: Bengal, Journal of Asiatic Society, v. 29, p. 352–358.

Bock, W., 1953, American Triassic Estheriids: Journal of Paleontology, v. 27, no. 1, p. 62–76, pl. 11-13.

Bond, G., 1952, The Karroo System in Southern Rhodesia: Alger, 19th Congrés Géologique International, Symposium sur Les Séries de Gondwana, p. 209–223.

——, 1955, The Madumabisa (Karroo) Shales in the Middle Zambesi Region: Transactions and Proceedings of the Geological Society of South Africa, v. 58, p. 71–99, pl. IX-XIV.

——, 1964, A new phyllopod from the Karroo of Northern Rhodesia: Arnoldia Miscellaneous Publication, v. 1, no. 15, 4 p., 1 text fig.

Borges, A., 1952, Le Systéme du Karroo au Mozambique (Afrique Orientale Portugaise): Alger, 19th Congrés Géologique International, Symposium sur Les Séries de Gondwana, p. 232–250.

Borns, H. W., Hall, B. A., Ball, H. W., and Brooks, H. K., 1972, Mawson Tillite, Victoria Land, east Antarctica: Reinvestigation continued: Antarctic Journal of the United States, v. 7, p. 106–107.

Brongniart, A., 1820, Mémoire sur le Limnadia, nouveaux genre des Crustacés: Mémoire Musée Histoire Naturelle, v. 6, p. 83–93.

Bronn, H. G., 1850, Uber *Gampsonyx fimbriatus* Jordan aus der Steinkohlen Formation von Saarbrücken und vom Murg-Thal: Stuttgart, Neues Jahrbuch für Mineralogie, Geognosie, Geologie, und Petrefakten-Kunde, p. 575–583.

Brown, D. A., Campbell, K.S.W., and Crook, K.A.W., 1968, The Geological Evolution of Australia and New Zealand: Pergamon Press, 409 p.

Burmeister, K.H.K., 1861, Reise durch die la Plata-staaten mit besonderer culturzustand der Argeninischen republik: Halle, H. W. Schimidt, v. 1, p. 77.

Cardoso, R. N., 1962, Alguns Conchostráceos Mesozoicos do Brasil: Boletim da Sociedade Brasileira dê Geologia, v. 11, no. 2, p. 21–32, 1 pl. 6 text figs.

——, 1965, Sôbre a Ocorrencia no Brasil de Afrograptidae e Monoleiolophinae, Conchostráceos Carenados: Rio de Janeiro, Brasil, Divisão de Geologia e Mineralogia Boletim, no. 221, 35 p., Est. I–III.

——, 1966, Cochostráceos do Grupo Bahía: Brasil, Boletim do Instituto de Geologia, Escola Federal de Minas de Ouro Preto, v. 1, no. 2, 76 p., Est. I–III.

——, 1971, Contribicão ao estudo da Formacão Areado: Estratigrafia e descricão dos Filopodos fosseis: Brasil, Arquivos do Museu de História Natural, Universidade Federal do Minas Gerais, v. 1, p. 8–43, Est. I–II, text figs. 1–11.

Carpenter, F. M., 1969, Fossil insects from Antarctica: Psyche, v. 76, no. 4, p. 418–425, fig. 1.

Chakravorty, S. K., and Ghosh, S. C., 1973, Cited in, New vertebrate fossil horizon in North Karanpura Coalfield: Geological Survey of India News, v. 4, no. 4, p. 4–5.

Chang Wen-tang, Chen Pei-chi, Shen Yan-bin, 1976, Fossil Conchostraca of China: Peking, Science Press, 325 p., 138 pl.

Chen Pei-ji, 1982, Jurassic conchostracans from Mengyin District, Shandong: Acta Palaeontologica Sinica, v. 21, no. 1, p. 138–139, 2 pl.

Cockbain, A. E., 1974, Triassic conchostracans from the Kockatea Shale: Geological Survey of West Australia, Annual Report 1973, p. 104–107.

Cohee, G. V., Glaessner, M. P., and Hedberg, H. D., eds., 1978, Contributions to the geologic time scale: Tulsa, American Association of Petroleum Geologists, p. 271, fig. 2.

Colbert, E. H., 1973, Continental drift and the distribution of fossil reptiles, *in* Tarling, D. H., and Runcorn, S. K., eds., Implication of Continental Drift to the Earth Sciences: New York, Academic Press, p. 395–412.

——, 1981, The distribution of tetrapods and break-up of Gondwana, *in* Cresswell, M. M., and Vella, P., eds., Gondwana five: Selected papers and abstracts of papers presented at the 5th International Gondwana Symposium: Rotterdam, A. A. Balkema, p. 277–282.

Cox, J. C., 1881, Notes on the Moore Park borings: Australia, Proceedings Linnean Society of New South Wales, v. 5, p. 273–281, 18 pl.

Curray, J. R., and Moore, D. G., 1974, Sedimentary and tectonic processes in the Bengal deep-sea fan and geosynclines, *in* Burke, C. A., and Drake, C. L., eds., The Geology of Continental Margins: New York, Springer-Verlag, p. 617–627, 18 text-figs.

Daday de Deés, E. (Jenō), 1910, Monographie systematique des phyllopodes anostracés: Paris, Annales des Sciences Naturelles Zoologie, v. 9, no. 11, p. 91–492, 84 text figs., 5 figs. in appendix.

——, 1915, Conchostracés: Paris, Annales des Sciences Naturelles Zoologie, v. 20, p. 39–330.

——, 1923, Conchostracés, pt. 2: Paris, Annales des Sciences Naturelles Zoologie, v. 10, no. 6, p. 255–390.

——, 1925, Conchostracés, pt. 3: Paris, Annales des Sciences Naturelles Zoologie, v. 8, p. 143–184.

——, 1926, Conchostracés, pt. 3 continued: Paris, Annales des Sciences Naturelles Zoologie, v. 9, p. 1–81.

—— 1927, Conchostracés, pt. 3 finish: Paris, Annales des Sciences Naturelles Zoologie, v. 10, p. 1–112.

Daguin, F., 1929, Étude stratigraphique et paléontologique du Carbonifère de la rive droite de l'oued Guir: Notes et Mémoires Service Mines et Carte Géologique Maroc.

David, T.W.E., 1907, The geology of the Hunter River Coal Measures, New South Wales, Pt. 1, General geology and the development of the Greta Coal Measures: Geological Survey of New South Wales Memoir, p. 332.

Defretin, S., 1950, Sur quelques Estheria du Trias Francais á facies Germanique et de l'Hettangien: Société Géologique du Nord Annales, v. 70, p. 214–227, pl. 8 and 9.

——, 1953, Quelques conchostracés du Nord-Cameroun: Bulletin de la Société Géologique de France, v. 6, no. 6, p. 679–690.

——, 1958, Remarques à propos de la Note de N.I. Novojilov sur quelques conchostracés Chinois et Africains: Société Géologique du Nord Annales, v. 67, p. 244–260, pl. 18.

Defretin-Le Franc, S., 1965, Étude et révision de Phyllopoda Conchostracés en provenance U.R.S.S.: Annales de la Société Géologique du Nord, p. 33.

——, 1967, Étude sur les Phyllopodes de Bassin du Congo: Musée Royal de l'Afrique Central, Tervuren, Belgique, Annales, series 8th, Sciences Géologiques, no. 56, 119 p., 14 pl., 22 text figs.

——, 1969, Les Conchostracés Triasiques du Groenland Oriental. Notes on Triassic stratigraphy and paleontology of north-eastern Jameson Land (east Greenland): Meddelelser om Grønland, v. 168, no. 2, pt. 3, p. 124–136, 2 pl.

Defretin, S., and Fauvelet, E., 1951, Présence de Phyllopodes Triasiques dans la région d'Argana-Bigoudine (Haut-Atlas Occidental): Division Mines et de la Géologie, Service Géologique. Notes et Mémoires No. 85, Rabat, Morocco, Fortin-Moullot, p. 129–135, 1 pl.

Defretin, S., Durand Delga, M., and Lambert, A., 1953, Faunules du Trias Superieur dans le Nord-Constantinois (Algérie): Société d'Histoire Naturelle. l'Africa du Nord, v. 44, p. 185–195.

Defretin, S., Joulia, F., and Lapparent, A. F., 1956, Les Estheria de la region D'Agadès (Niger): Bulletin de la Société Géologique de France, v. 6, no. 6, p. 679–690.

de la Beche, H. T., 1832, Handbuch der Geognose bearbeitet von H. von Decken: Berlin, Duncher und Humblot, p. 453.

Deleau, P., 1945, Les bassins carbonifères du Sur Oranais (Algérie): Les bassins de Columb Béchar et le bassin du Guir: Bulletin de la Société Géologique de France, v. 15, p. 625–632.

Depéret, C., and Mazeran, P., 1912, Les Estheria du Permien d'Autun: Société d'Histoire Naturalle d'Autun Bulletin, v. 25, p. 165–173, pl. 5, text figs., 1–4.

de Souza, A., Sinelli, O., and Gonçalves, N.M.M., 1971, Nova ocorrência fossilifera na Formacão Botucatú, in Anais, Congresso Brasil de Geologia, 25th, p. 281–295.

Diener, K., 1915, The Anthracolithic faunae of Kashmir, Kanaur, and Spiti: Palaeontologia Indica, n.s., Memoir 5, part 2, p. 113, 115–116.

Dietrich, W. O., 1939, Trias in Nord Adamawa: Zentralblatt für Mineralogie, Abteilung B, no. 2, p. 60–62.

Doumani, G. A., and Tasch, P., 1965, A Leaiid Conchostracan Zone (Permian) in the Ohio Range, Horlick Mountains, Antarctica, in Hadley, J. B., ed., Geology and Paleontology of the Antarctic: American Geophysical Union, Antarctic Research Series, v. 6, no. 1299, p. 229–239.

Duan Wei–wu, 1982, Middle Jurassic–Early Cretaceous conchostracans from Northeastern Lianoning: Acta Palaeontologica Sinica, v. 21, no. 4, p. 497–504, pl. 2, fig. 1.

Dunker, W. (B.R.H.), 1846, Monographie der norddeutschen wealdenbildung. Ein beitrag zur geognosie und naturgeschichte der vorwelt: Braunschweig, Oehme und Muller, pt. 1, 83 p.

Du Toit, A. L., 1937, Our Wandering Continents: Edinburgh, Oliver and Boyd, 332 p.

Ellenberger, P., 1970, Les niveaux paléontologiques de première apparition des mammifères primordeaux en Afrique du Sud et leur ichnologie. Establissement de Zones stratigraphiques détaillées dans le Stormberg du Lesotho (Afrique du Sud) (Trias Supérieur a Jurassique) in 2nd Gondwana Symposium, South Africa, July to August, 1970; Proceedings and Papers, International Union of Geological Sciences; Council for Scientific and Industrial Research, Scientia, Pretoria: Pietermaritzburg, The Natal Witness (Party) Limited, p. 343–457.

Elliot, D. H., 1970, Jurassic tholeiites of the central Transantarctic Mountains, in Gilmour, E. H., and Stradling, D., eds., Proceedings Second Columbia River Basalt Symposium: Eastern Washington State College, p. 301–325.

Emery, K. O., Uchupi, E., Phillips, J., Bowin, C., and Mascle, J., 1975, Continental margin off western Africa: Angola to Sierra Leone: American Association of Petroleum Geologists Bulletin, v. 59, no. 12, p. 2209–2265. (p. 2257).

Etheridge, R., Jr., 1888, The invertebrate fauna of the Hawkesbury–Wianamatta Series (beds above the Productive Coal Measures) of New South Wales: Memoirs of the Geological Survey of New South Wales, Paleontology: Sidney, Australia, Charles Potter, Government Printer, v. 1, p. 4–8, pl. 1, figs., 1-10.

——, 1893, On Leaia mitchelli, Etheridge, fil., from the Upper Coal Measures of the Newcastle District: Australia, Linnean Society of New South Wales Proceedings, 2nd series, v. 7, p. 307–310.

Fedorov, L. V., Ravich, M. G., and Hofmann, J., 1982, Geologic comparison of southeastern peninsular India and Sri Lanka with a part of east Antarctica (Enderby Land, MacRobertson Land, and Princess Elizabeth Land), in Craddock, C., ed., Antarctic Geoscience, Symposium on Antarctic geology and geophysics, August 22–27, 1977, International Union of Geological Sciences, Madison, University of Wisconsin Press, p. 73–78.

Feistmantel, O., 1890, The geological and paleontological relations of the coal and plant bearing beds of the Paleozoic and Mesozoic Age in Australia and Tasmania: Memoirs of the Geological Survey of New South Wales, Paleontology, Sidney, Australia, Charles Potter, Government Printer, v. 3, p. 73–75, pl. 29.

Feys, R., 1960, Sur Estheria tenella et les Estheria du Permian Inférieur: Bulletin de la Société Géologique de France, 7th series, v. 2, p. 610–620.

Frakes, L. A., and Crowell, J. C., 1971, The positon of Antarctica in Gondwanaland, in Quam, L. O., ed., Research in th Antarctic: Washington, D.C., American Association for the Advancement of Science, Publication No. 93, p. 731–745.

Fritsch, A., 1901, Fauna der Gaskohle und der Kalksteine der Permformation Bohmens: Prag, Selbstverlag, v. 4, no. 3, p. 65–101, 25 abteilung, 11 tafel.

Furon, R., 1950, Géologique de l'Afrique: Paris, Payot, 350 p.

Gansser, A., 1964, Geology of the Himalayas: New York, Interscience Publishers (John Wiley and Sons), 289 p., Fig. 28.

Geinitz, H. B., 1876, Beitrage zur Geologie und Palaeontologie der Argentinischen Republik, II Palaeontologisher Theil, II Abtheilung: Cassel, Verlag von Theodor Fischer, p. 3, tafel 1, fig. 1-6.

Ghosh, S. C., 1973, New find of conchostracan horizons in East Bokaro Coalfield: Geological Survey of India News, v. 4, no. 7, p. 4–5, cover photographs 1–4, text fig. p. 4.

Ghosh, S. C., and Shah, S. C., 1976, Estheriella taschii sp. nov., a new Triassic conchostracan from the Panchet Formation of East Bokaro Coalfield, Bihar: Calcutta, Asiatic Society Monthly Bulletin, v. 5, no. 3, p. 7–8.

——, 1977, Estheriella taschi sp. nov., a new Triassic conchostracan from the Panchet Formation of East Bokaro Coalfield, Bihar: Journal of Asiatic Society, v. 19, no. 1–2, p. 14–18.

Giebel, C. G., 1857, Palaeontologische Untersuchungen: Zeitschrift für Naturwissenschaften, Halle, v. 10, p. 301–317, pl. 2, text fig. 6, 7.

Gislén, T., 1936, Contributions to the ecology of Limnadia: Lund, Sweden, Lund Universitet, C.W.K. Cleerup, Arrsskrift, Avd. 2, v. 32, no. 9, p. 3–19.

Goldfuss, A., 1840, Petrefacta Germaniae: Leipsig, Zwieter Thiel, p. 118, pl. 113, fig. 5.

Greigert, J., and Pougnet, R., 1967, Essai de description des formations géologiques de la Republique du Niger: Mémoirs Bureau Recherches Géologique Mineralogie, v. 48, p. 112.

Gressitt, J. L., 1961, Problems in the zoogeography of the Pacific and antarctic insects: Pacific Insects Monograph, v. 2, p. 1–94.

Grey, G., 1871, Remarks on some specimens from South Africa (Communicated with notes by Professor T. Rupert Jones, F.G.S.) [abs.]: The Geological Society of India, Proceedings, v. 27, pt. 1, p. 49–52.

Gunn, B. M., and Warren, G., 1962, Geology of Victoria Land between the Mawson and Mulock Glaciers, Antarctica: New Zealand Geological Survey Bulletin, no. 71, 157 p.

Gurney, R., 1931, Branchiopoda, in Reports of an expedition to Brazil and Paraguay 1926–27: London, Zoological Journal of the Linnean Society, v. 37, no. 252, p. 263–275.

Guthörl, P., 1934, Die Arthropoden aus dem Karbon und Permian des Saar-Nahe-Pfalz Gebietes: Berlin, Preussiche Geologische Landesanstalt, Abhandlungen, n.s., no. 164, p. 3–219, pl. 1-30.

Halle, T. G., 1913, The Mesozoic flora of Graham Island, *in* Der Schwedischen Sudpolar Expedition 1901–1903, Band 3, Lieferung 14, 123 p., 9 pl.

Harrington, H. J., 1961, Geology of parts of Antofagasta and Atacama provinces, northern Chile: American Association of Petroleum Geologists Bulletin, v. 45, no. 2, p. 169–197.

Haughton, S. H., 1924, The fauna and stratigraphy of the Stormberg Series: South African Museum Annals, v. 12, p. 326–330.

—— , 1963, The stratigraphic history of Africa south of the Sahara: New York, Hafner Publishing Co., 365 p.

—— , 1969, Geological history of southern Africa: Capetown, Geological Society of South Africa, 535 p.

Hedberg, H. D., 1964, Geologic aspects of origin of petroleum: American Association of Petroleum Geologists Bulletin, v. 48, p. 1796.

Herbst, R., 1965, La Flora fosil de la formacion Rosa Blanca, Provincia Santa Cruz, Patagonia, Con consideraciones geológicas y estratigráficas: Universidad Nacional de Tucamen, Opera Lilloana, v. 12, 107 p.

Herbst, R., and Ferrando, L. A., 1985, *Cyzicus (Lioestheria) ferrandoi* n. sp. (Conchostraca, Cyzicidae) de la Formacion Tacuarembo (Triasico Superior) de Uruguay: Argentina, Parana, Revista de la Associacion de Ciencias Naturales del Litoral, v. 16, no. 1, p. 29–47.

Hind, M. C., and Helby, R. J., 1969, The Great Artesian Basin in New South Wales, *in* Packham, G. H., ed., The geology of New South Wales: Sydney, Geological Society of Australia, p. 481–497.

Hislop, S., 1857, Geology of Nágpur State: Bombay, Asiatic Society Journal, v. 5, p. 58–76; 148–150.

Hislop, S., 1861, On the age of the fossiliferous thin-bedded sandstone and coal of the province of Nágpur, India: The Quarterly Journal of the Geological Society of London, v. 17, p. 346–354.

Hislop, S., 1862, Remarks on the geology of Nágpur: Bombay, Asiatic Society Journal, v. 6, p. 194–204; 207.

Hislop, S., and Hunter, R., 1854, On the geology of the neighborhood of Nágpur, Central India: The Quarterly Journal of the Geological Society of London, v. 10, p. 470–473.

—— , 1855, On the geology of the neighborhood of Nágpur, central India: The Quarterly Journal of the Geological Society of London, v. 11, p. 345–383.

Holub, V., and Kozur, H., 1981, Revision einiger Conchostracen-Faunen des Rotliegenden und biostratigraphische Auswertung der Conchostracen des Rotliegenden: Innsbruck, Austria, Geologie und Paläontologie Mitteilungen, Innsbruck, v. 11, no. 2, p. 39–94.

Hughes, T.W.H., 1877, The Wardha Valley coal-field: Delhi, India Geological Survey Memoir 13, p. 1–154.

Jack, R. L., and Etheridge, R., Jr., 1892, The geology and palaeontology of Queensland and New Guinea: Brisbane, Queensland, Australia, James Charles Beal, p. 1–768, pl. 1–68.

Jain, S. L., 1980, The continental Lower Jurassic fauna from the Kota Formation, *in* Jacobs, L. L., ed., Aspects of vertebrate history: Flagstaff, Museum of Northern Arizona Press, v. 20, 408 p.

Janensch, W., 1925, Wissenschaftliche ergebnisse der Tendaguru Expedition 1909–1912: Palaeontographica, supplement VII, v. 2, part 1, p. 125–138.

—— , 1933, Eine Estheria aus den Tendaguruschichten: Palaeontographica, supplement VII, v. 2, part 2, lieferung 2, seite 97–98.

Jones, T. R., 1862, A monograph of the fossil Estheriae: London, Palaeontographical Society, p. 1–134, pl. 1–5.

—— , 1878, Notes on some bivalved Entomostraca: Geological Magazine, n.s., Decade 3, v. 5, p. 100–102, pl. 3.

—— , 1897a, On some fossil Entomostraca from Brazil: Geological Magazine, n.s., v. 4, p. 195–202, pl. 8, fig. 1–5.

—— , 1897b, On fossil Entomostraca from South America: Geological Magazine, n.s., Decade 4, v. 3–4, p. 259–265.

—— , 1901, On the Enon Conglomerate of the Cape of Good Hope and its fossil Estheriae: Geological Magazine, n.s., Decade 4, v. 8, no. 8, p. 350–354.

Jones, T. R., and Woodward, H., 1894, On some fossil Phyllopoda: Geological Magazine, n.s., Decade 4, v. 1, no. 7, p. 289–293, pl. IX.

Kapel'ka, V., 1968, Les Conchostracés du Mésozoique Inférieur de l'Iénisséi (Sibérie Orientale), *in* Novojilov, N. I., and Kapel'ka, V., Nouveaux Conchostracés de Sibérie: Paris, Annales de Paléontologie (Invertébrés), v. 54, p. 120, pl. A, fig. 3.

Katoo, I., 1971, Conchostráceos Mesozóicos do Sul do Brasil: Contribuição à Estratigrafia das Formações Santa Maria e Botucatú [Ph.D. thesis]: Porto Alegre, Universidade Federal do Rio Grande do Sul, 87 p., 13 pl.

Knight, O. L., 1950, Fossil insect beds of Belmont, N.S.W.: Records of the Australian Museum, v. 23, no. 3, p. 251–253 and map.

Kobayashi, T., 1954, Fossil estherians and allied fossils: Journal of the Faculty of Science, Tokyo University, v. 9, section 2, part 1, p. 1–192, 30 text figs.

—— , 1973, Classification of the Conchostraca and the discovery of estheriids in the Cretaceous of Borneo: Geology and Paleontology of Southeast Asia, University of Tokyo Press, v. 13, p. 47–72, pl. 6.

—— , 1975, Upper Triassic estheriids in Thailand and the conchostracan development in Asia in the Mesozoic Era, *in* Kobayashi, T., and Toriyama, R., eds., Geology and Paleontology of Southeast Asia: University of Tokyo Press, v. 16, p. 57–90.

Kobayashi, T., and Kido, Y., 1943, Climatic effects on the distribution of living estheriids and its relation to the morphic character of their carapaces: Geological Society of Japan Journal, v. 50, p. 37.

Kozur, H. 1979, Erster Nachweis von *Vertexia tauricornis* Ljutkevĭc 1941 (Conchostraca) im Bundsandstein des Thuringer Beckens: Berlin, Zeitschrift für Geologische Wissenschaften, v. 7, p. 817–820.

—— , 1980, Die Conchostraken-Fauna der Mittleren Bernburg-Formation (Buntsandstein) und ihre stratigraphische bedeutung: Berlin, Zeitschrift für Geologische Wissenschaften, v. 8, no. 7, p. 885–903.

Kozur, H., and Sittig, E., 1981, Das *"Estheria" tenella*—Problem und zwei neue Conchostracen-Arten aus dem Rotliegenden von Sulzbach (Senke von Baden-Baden, Nordschwarzwald): Innsbruck, Austria, Geologie und Paläontologie Mitteilungen, v. 11, no. 1, p. 1–38.

Kozur, H., Martens, T., and Pacaud, G., 1981, Revision von *"Estheria" (Lioestheria) lallyensis* Depéret and Mazeran, 1912 und *"Euestheria" autunensis* Raymond, 1946: Berlin, Zeitschrift für Geologische Wissenschaften, v. 9, no. 12, p. 1437–1445.

Krommelbein, K., 1966, On "Gondwana Wealden" Ostracoda from N.E. Brazil and West Africa: Proceedings 2nd West African Micropaleontological Colloquium: Ibadan, p. 113–119.

Kusumi, H., 1961, Studies on the fossil estherids: Hiroshima, Japan, Hiroshima University, Geological Report No. 7, 88 p., 12 tables, 61 text figs., 9 pl.

Kyle, R. A., and Fasola, A., 1978, Triassic palynology of the Beardmore Glacier Area of Antarctica: Leon, Spain, Palinologia, v. 1, p. 313–318, pl. 1.

Langenheim, J. H., 1961, Late Paleozoic and early Mesozoic plant fossils from the Cordillera Oriental of Colombia and correlation of the Giron Formation: Servico Geológico Nacional, Boletim Geológico, v. 8, no. 1–3, p. 95–132, fig. 36.

Laseron, C., and Brunnschweiler, R. O., 1969, Ancient Australia: Sydney, Angus and Robertson, Ltd., p. 1–234.

Lea, I., 1856, Description of a new mollusk from the Red Sandstone near Pottsville, Pennsylvania: Academy of Natural Sciences of Philadelphia, Proceedings, v. 7, p. 340–341, pl. 4.

Leguizamón, R. R., 1975, Hallazgo del Generao *Leaia* (Conchostraco) en El Permico Argentino: Actas, Congreso Argentino de Paleontologia y Bioestratigrafia, v. 1, p. 357–369.

Leriche, M., 1913, Entomostraces des couches du Lualaba (Congo–Belge): Revue Zoologique Africaine, v. 3, p. 3–6, pl. I–II.

—— , 1920, Notes sur la Paléontologie du Congo: Revue Zoologique Africaine, v. 8, p. 67–86.

—— , 1932, Sur les premiers fossiles découverts au Nord de l'Angola dans le prolongement des couches du Lubilash et de couches du Lualaba: Académie des Sciences de Paris, Comptes Rendus, v. 195, p. 395–400.

Lightfoot, B., 1914, The geology of the western part of the Wankie Coalfield, South Rhodesia: Geological Survey Bulletin, v. 4, p. 3–49, pl. 1–41.

—— , 1929, The geology of the central part of the Wankie Coalfield, South Rhodesia: Geological Survey Bulletin, no. 15, p. 38.

Livret–Guide Excursion Géologique, 1979, Le Sud et le Sud Ouest de Madagascar, *in* Colloque Scientifique International, "L'Histoire du Gondwana vue de Madagascar": Tananarive, Democratic Republic of Madagascar, Academie Malagache, 35 p. (plus additional insert pages irregularly numbered).

Long, W. E., 1965, Stratigraphy of the Ohio Range, Antarctica, *in* Hadley, J. B., ed., Geology and Paleontology of the Antarctic: American Geophysical Union, Antarctic Research Series, v. 6, no. 1299, p. 71–116.

Lutkevich, E. M., 1929, Phyllopoda srenego Devona Severo–Zapadnay oblasti: Geologich Komitet, Izvestia, v. 48, no. 5 (English summary, p. 137–143, pl. 36, text fig. 1–21b). (Phyllopoda from the middle Devonian of the North-west Province.)

——, 1937, On some Phyllopoda from the USSR, Summary, The Triassic Estheriae of the Permian Platform: Annuaire de la Société Paléontologique de Russie, v. 11, p. 59–70, pl. 9, fig. 11.

——, 1938, Triassic Estherinae from the upper strata of the Tungussk Series: Institut Arctique, Transactions, t. 10, no. 3, p. 155–164, pl. 1, fig. 12–14.

——, 1939, Phyllopoda, in Atlas of the leading forms of fossil faunas of the USSR, Permian: v. 6, p. 190–193; 264–265, pl. 46, text fig. 1–5.

——, 1941, Phyllopods permskikh ottozheniy Europeyskoy Kargaly i ikh otno-sheniya, *in* Paleontology of the USSR, v. 5, pt. 10, no. 1 (English summary, p. 33–44, pl. 1–3). (The Phyllopods of the Permian deposits of the European part of USSR.)

Lyell, C., Elements of Geology, p. 452, fig. 490, cited in Cox, 1881.

Maillieux, E., 1933, Un conchostracé nouveau de l'Assise des grés et schistes de Wépion, (Ensien Inferieur): Musée Royal Histoire Naturelle de Belgique Bulletin, v. 15, no. 10, p. 4–5.

Marlière, R., 1950a, Ostracodes et Phyllopodes du Systéme du Karroo au Congo Belge et les régions avoisinantes: Tervuren, Belgique, Musée du Congo Belge Annales Sciences Géologiques, v. 6, ser. 8, 41 p., 2 text fig., 3 pl.

——, 1950b, Estudos de Geologia Paleontologia: Lisboa, Portugal, Ministerio das Colonias, Anais, v. 5, tome 4, part 1, p. 52–76, pl. I and II.

Martens, T., 1982, Zur taxonomic und biostratigraphie neuer Conchostraken-Funde (Phyllopoda) aus dem Permokarbon und der Trias von Mitteleuropa: Leipzig, Freiberger Forschungshefte, v. 375, p. 49–82, 16 figs., 15 pl.

McElroy, C. T., 1969, Walloon Coal Measures, *in* Packham, G. H., ed., The geology of New South Wales: Sydney, Australia, Geological Society of Australia, Inc., p. 469–471.

McGregor, A. M., 1927, A geological reconnaissance in the Mafungabusi gold belt, with appendices on the fossils by Dr. S. H. Haughton and Dr. A. C. Du Toit: Bulawayo, South Rhodesia Geological Survey Short Report No. 20, 13 p., map.

Mendes, J. C., 1954, Conchostrácos Permianos do Sul do Brasil, *in* Lange, F. W., Paleontologia do Paraná, Volume Comemorativo: Curitiba, Museu Para-naense, p. 153–164, pl. 12–14, 3 text figs.

——, 1960, Nota sôbre Conchostráceos Brasileiros da Familia Limnadiidae: Anais da Academia Brasileira de Ciencias, v. 32, no. 1, p. 75–78.

Mezzalira, S., 1974, Contribucão ao conhecimento da estratigrafia e paleontologia de Arenito Baurú: São Paulo, Brasil, Instituto Geografico e Geológico Boletim, no. 51, p. 7–9, 20–21, 120–129, 2 pl.

Mitchell, J., 1925, Description of new species of *Leaia:* Australia, Linnean Society of New South Wales Proceedings, v. 50, no. 4, p. 438–447, pl. 41–43.

——, 1927, The fossil Estheriae of Australia, Part 1: Australia, Linnean Society of New South Wales Proceedings, v. 52, no. 2, p. 105–112, pl. 2–4.

Molin, V. A., 1968, New species of Conchostraca from the Upper Permian and Lower Triassic of the European U.S.S.R.: Paleontological Journal, v. 2, no. 3, p. 368–373 (p. 369).

Molin, V. A., and Novojilov, N. I., 1965, Dvustvorchatye listonogiye permi i triasa Severa SSSR: Moskva, Nauka Press, 166 p., 12 pl.

Molnar, P., and Tapponnier, P., 1977, The collision between India and Eurasia: Scientific American, v. 236, no. 4, p. 30–41.

Morris, S. F., 1980, Catalogue of the type and figured specimens of fossil Crustacea (excl. Ostracoda), Chelicerata, Myriapoda, and Pycnogonida in the British Museum (Natural History): British Museum (Natural History) Publication No. 828, 53 p., 3 pl.

Netshajev (Netchaev), A. V., 1894, Faune du Permien supérieur de la Russie en Europe: Société Naturalistes, Université–Kazan, Transactions, t. 27, no. 4,

p. 1–503.

Newton, R. B., 1910, Note on some fossil nonmarine Mollusca and a bivalved crustacean *Estheriella* from Nyasaland: Quarterly Journal of the Geological Society of London, v. 66, App. 2, p. 229–248, 19 pl., fig. 15–18.

Novojilov, N. I., 1946, New Phyllopoda from the Permian and Triassic deposits of the Nordwick–Khatanga region: Nedra Artiki, no. 1, p. 172–202, pl. 1–3 (English summary, p. 194–202).

——, 1952, Nouveaux groupes generiques dans les Crustacés Phyllopodes de la Famille des Leaiidae: Moscow, Akademiya Nauk SSSR Doklady, v. 85, no. 6, p. 1369–1372, 3 figs.

——, 1953, Guide pour la recherche et la collection des Crustacés Phyllopodes fossiles: Academie Sciences de l'URSS., Institut Paléontologique Recherches et Études de Restes Organiques Fossiles, v. IV, 15 p., 17 figs., 3 pl.

——, 1954a, Novye bidy dvustvorchatykh listonogikh rakoobraznykh iz devona iuzhoñi sibiri: Moscow, Akademiya Nauk SSSR Doklady, v. 95, no. 1, p. 159–162.

——, 1954b, Phyllopodes bivalves—Leaiidae provenant des depôts carboniféres du Kazakhstan: Moscow, Akademiya Nauk SSSR Doklady, v. 96, no. 6, p. 1241–1244, 1 fig., 1 table.

——, 1954c, Crustacés Phyllopodes du Jurassique supérieur et du Crétacé de Mongolie: Travaux Institut Paléontologique Academie des Sciences de l'URSS, v. 48, p. 7–124, 75 figs., 17 pl. (Translated Bureau de Recherches Géologiques, Géophysiques et Minières.)

——, 1955, Phyllopodes Polygraptinae du Dévonien du cours Inférieur de la Torgalyx du Sud: Moscow: Akademiya Nauk SSSR Doklady, v. 102, no. 1, p. 153–155, 1 pl., 1 table.

——, 1956a, Dvustovorchatye listonogiye rakoobraznye, 1. Leaiidy: Moskva, Akademiya Nauk. SSSR. Trudy Paleontologicheskii Institut, v. 61, 129 p., 87 figs., 10 tables, 14 pl.

——, 1956b, Nouveau genre de Leaiides, Crustacés, Phyllopodes, Igorvarent-sovia, du Carbonifere: Akademiya Nauk. Doklady. SSSR. Paléontologique, v. 106, no. 6, p. 1087–1090, 1 fig.

——, 1958a, Crustacés bivalves de l'ordre des Conchostracés du Cretace inférieur Chinois et Africain: Lille, France, Société Géologique du Nord Extrait des Annales, 1957, v. 67, p. 235–243, pl. XVII.

——, 1958b, Conchostraca du Permian et du Trias du littoral de la Mer des Laptev et de la Toungouska Inférieure: Paris, Service de Information Geo-logique, Annales, Bureau de Recherches Géologiques, Géophysiques et Minières, v. 26, p. 15–63, pl. I–IV.

——, 1958c, Recueil d'articles sur les phyllopodes Conchostracés: Paris, Service de Information Géologique, Annales, Bureau de Recherches Géologiques, Géophysiques et Minières, v. 26, p. 1–135 (dated Moscow, 1950, published 1958).

——, 1958d, Conchostraca de la super-famille des Limnadiopseoidea superfam. nov: Paris, Service de Information Géologique, Annales, Bureau de Re-cherches Géologiques, Géophysiques et Minières, v. 26, p. 95–123, pl. I, II.

——, 1958e, Conchostraca de la famille nouvelle des Kontikiidae du Meso-zoique de Chine, d'Australie et des region polaires et du Paléozoique de la région de la Volga: Paris, Service de Information Géologique, Annales, Bureau de Recherches Géologiques, Géophysiques et Minières, v. 26, p. 85–92, pl. 1.

——, 1960, Sous-ordre Conchostraca Sars, 1846 (sic.) phyllopodes bivalves, *in* Chernysheva, N. E., ed., Osnovy Paleontologii, Chlenistonogie trilobituo-braznye i rakoobraznye: Moscou, Gosgeoltekhizdat, v. 8, p. 220–252, fig. 455a, 463, 586, pl. 13–15. (Translated Bureau Recherches Géologiques, Géophysiques et Minières.)

——, 1961, Dvustvorchatye listonogie devona: Akademiya Nauk SSSR, Trudy Paleontologicheskii Institut, v. 81, 132 p.

——, 1970, Vymershie Limnadioidei (Conchostraca-Limnadioidea): Moskva, Izdat. 'Nauka", p. 1–237, pl. 1–10, 214 text figs.

Novojilov, N. E., and Kapel'ka, V., 1960, Crustacés bivalves (Conchostraca) de la série Daido de l'Asie Orientale dans le Trias supérieur de Madygen (Kirghizie Occidentale): Lille, France, Société Géologique du Nord Annales, v. 80, no. 3, p. 177–189, pl. XI.

——, 1968, Nouveaux Conchostracés de Sibérie: Annales de Paris, Paléontologie

(Invertébrés), Masson et al., v. 54, no. 1, p. 126, pl. B, fig. 3.

Novojilov, N. I., and Varentsov, I. M., 1956, Nouveaux Conchostracés du Givétien de la Touva: Academie des Sciences de l'URSS, Comptes Rendus, v. 110, no. 4, p. 670–673.

——, 1958, Nouveaux Conchostraca du Dévonien de la Touva: Materiaux pour les Principes de Paléontologie, pt. 2, p. 41–45.

Oliveira, P. E., 1953, Sôbre um novo conchostráceo fóssil do estado da Bahía: Rio de Janeiro, Brazil, Divisão de Geologia e Mineralogia, v. 63, p. 1–13, 2 pl.

Osaki, K., 1970, On some Permian conchostracans from the Jangseong-ri, Samchuk-gun, Kangwon-do, Republic of Korea: Science Report Yokohama National University, Section II, no. 16, p. 73–80, pl. 12, fig. 1, text fig. 3.

Ozawa, Y., and Watanabe, T., 1923, On two species of Estheriae from the Mesozoic shale of Korea: Tokyo, Japanese Journal of Geology and Geography, v. 2, no. 2, p. 40–42.

Packard, A. S., 1874, Synopsis of the fresh-water phyllopod Crustacea of North America, *in* Hayden, F. V., U.S. Geological and Geographical Survey of the Territories: Annual Report, 1873, p. 613–622, 4 pl.

——, 1883, A monograph of the phyllopod Crustacea of North America with remarks in the order Phyllocarida: Hayden, F. V., Geologist, U.S. Geological Survey of the Territories, 12th Report, p. 295–590, 39 pl.

Packham, G. H., ed., 1969, The geology of New South Wales: Sydney, Australia, Journal of the Geological Society of Australia, v. 16, pt. 1, 654 p. (Republished as a book; Canberra, Australian Government Publishing Service.)

Passau, G., 1919, Découverte d'un gîte fossilifere au Kwango (Congo Belge): Société Géologique de Belge Congo Annales, v. 6, no. 42, 33 p.

Pennak, R. W., 1953, Fresh water invertebrates of the United States: New York, Ronald Press, 740 p.

Petters, S., 1981, Stratigraphy of Chad and Iullemmeden Basins (West Africa): Basel, Eclogae Geologicae Helvetiae, v. 74, no. 1, p. 139–159.

Philippi, R. A., 1887, Die Tertiaren und Quartaren Versteinerungen Chiles: Leipzig, F. A. Brockhaus, 266 p., 58 pl.

Picard, H. E., 1911, Uber den Unteren Buntsandstein der Mansfolder Mulde und seine Fossilien: Berlin, Jahrbuch Koniglich Preussichen Geologischen Landesanstalt, 1909, v. 30, no. 1, p. 576–622.

Pinto, I. D., 1956, Arthropódos da Formacão Santa Maria (Triassico Superior) do Rio Grande do Sul, com notícias sôbre alguns restos vegetais: Sociedad Brasileiro de Geologia Boletim, v. 5, no. 1, p. 75–94.

Pinto, I. D., and Purper, I., 1974, Observations on Mesozoic Conchostraca from the North of Brasil, *in* Anais do Congresso Brasileiro de Geologia, 28th: p. 305–316.

Plumstead, E. P., 1962, Fossil flora of Antarctica: Geology 2, Transantarctic Expedition, 1955–1958, Scientific Report no. 9, p. 1–140.

——, 1964, Paleobotany of Antarctica, *in* Adie, R. J., ed., Antarctic Geology: Amsterdam, New Holland Publishing Co., p. 637–654.

Pruvost, P., 1911, Note sur les Entomostracés Bivalves: Lille, France, Société Géologique du Nord Annales, v. 40, p. 60–80, 3 pl.

——, 1919, Introduction a l'étude du Terrain Houiller du Nord et du Pas-de-Calais: La Faune continentale du Terrain Houiller du Nord de la France: Paris, Ministere des Travaux Publics, Mémoire, 574 p., 51 text figs., 29 pl.

——, 1920, Découverte de *Leaia* dans le Terrain Houiller du Nord et du Pas-de-Calis. Observation sur le genre *Leaia* et des differentes espèces: Société Géologique du Nord Annales, v. 43, p. 254–280, pl. 2.

Pryor, E. A., and Mursa, V., 1968, Surface Geology, Newcastle Coalfield: Belmont, The Broken Hill Proprietary Company, Ltd. (map).

Raggart, H. G., 1969, Macroflora and Fauna, *in* Packham, G. H., ed., The Geology of New South Wales: Sydney, Geological Society of Australia, p. 405–406.

Raymond, P. E., 1946, The genera of fossil Conchostraca: An order of bivalved Crustacea: Cambridge, Harvard University, Museum of Comparative Zoology Bulletin, v. 96, no. 3, p. 218–307, pl. 1–6.

Reed, F.R.C., 1929, Novos phyllopodos fosseis do Brasil: Rio de Janeiro, Ministerio da Agricultura, Industria e Commercio, Servico Geologico Mineralogico do Brasil Boletin 24, no. 34, 17 p., 1 pl.

Reible, P., 1962, Die conchostraken (Branchiopoda, Crustacea) der Germanischen Trias: Stuttgart, Neus Jahrbuch für Geologie und Paläontologie Abhand-

lungen, v. 114, no. 2, p. 169–244, pl. 6–10.

Rennie, J.V.L., 1934, Note on an *Estheria* from the Witteberg Series: South African Journal of Science, v. 31, p. 233–235.

——, 1937, Fossils from the Lebombo Volcanic Formation: Servicos de Industria Minas e Geologia de Colonia de Moçambique, v. 1, Lourenço Marques (Moçambique), p. 14–24, pl. I, II.

Riek, E. P., 1970, Origin of the Australian insect fauna, *in* Second Gondwana Symposium (South Africa), Proceedings and Papers, International Union of Geological Sciences: Pretoria, South Africa, Council for Scientific and Industrial Research, Scientia, p. 593–598.

Rigby, J. F., and Shah, S. C., 1980, The flora from the Permian nonmarine sequences of India and Australia: A comparison, *in* Cresswell, M. M., and Vella, P., eds., Gondwana Five: Selected papers and abstracts of papers presented at the 5th International Gondwana Symposium: Rotterdam, A. A. Balkema, p. 39–42.

Rivas, O. O., and Benedetto, G., 1977, Paleontologia y edad de la Formación Tinacoa, Sierra de Perija, Edo, Zulia, Venezuela: Caracas, 5th Congreso Geologico Venezolano, Ministerio de Energía y Minas, División de Exploraciones: Memoria, t. 1, p. 15–28, 3 pl.

Robinson's Newcastle Street Directory, 4th Edition, undated: Sydney, Australia, Robinson, H.E.C., Party, Ltd., 128 p., 49 maps.

Roch, E., ed., 1953, Itineraires Géologiques dans le Nord du Cameroun et le Sud–Ouest du Territoire du Tchad, avec la collaboration por le parte pétrographique de Elizabeth et Anne Faure–Maret: Paris, Cameroun Mines Service Bulletin, no. 1, 131 p.

Rocha-Campos, A. C., and Farjallat, J.E.S., 1966, Sôbre a extensão da Formacão Botucatú na Região Meridional de Mato Grosso: Sociedad Brasileria Geologia Boletim, v. 15, no. 4, p. 93–105, figs. 6, 9.

Roux, S. F., 1960, A fossil conchostracan from the Middle Ecca (Lower Permian), Transvaal: South African Journal of Science, v. 56, no. 11, p. 282–284.

Rusconi, C., 1946a, Varias especies de Trilobites y Estherias del Cambrico de Mendoza: La Sociedade de Historia y Geographica de Cuyo Revista, v. 1, p. 239–246, pl. 1, text fig. 1–6.

——, 1946b, Acerca de la *Estheria minorpriori* (Ostracoda-*sic.*) de Mendoza: Argentine, Facultad Ciencias Fisico y National Boletin, Anais, v. 9, no. 1, p. 753–758.

——, 1948a, Algunas especies de Esterias del Triasico en Mendoza: Museo Historie Natural Mendoza (Argentine), Revista, v. 2, no. 3, p. 199–202.

——, 1948b, Apuntes sôbre el Triasico y el Ordovicio de El Challao, Mendoza: Museo Historie Natural Mendoza (Argentine), Revista, v. 2, p. 165–198.

Sars, G. L., 1867, Histoire Naturelle des Crustacés d'eau Douce Norvège; C. Johnson, Kristiania, p. 1–145, 10 pl.

——, 1888, On *Cyclestheria hislopi* (Baird), a new generic type of bivalve Phyllopoda: Kristiania, Norske Vidensk Selsk. Forhandlinger aar 1887, v. 1, p. 1–65, pl. 1–8.

——, 1889, On some Indian Phyllopods: Archiv for mathematik og naturvidenskap, v. 22, no. 9, p. 3–27.

Shaeffer, B., 1972, A Jurassic fish from Antarctica: American Museum Novitates, no. 2495, p. 1–16.

Shen Yan-Bin, 1978, Leaid conchostracans from the Middle Devonian of south China with notes on their origin, classification, and evolution, *in* Papers for the International Symposium on the Devonian System: Nanking, China, Nanking Institute of Geology and Palaeontology, p. 1–9, 3 pl.

——, 1981a, Cretaceous conchostracan fossils from Eastern Shandong: Acta Palaeontologica Sinica, v. 20, no. 6, p. 524–526, 3 pl.

——, 1981b, Fossil conchostracans from the Chijinpu Formation (Upper Jurassic) and the Xinminpu Group (Lower Cretaceous) in Hexi Corridor, Gansu: Acta Palaeontologica Sinica, v. 20, no. 3, p. 20.

——, 1984, Occurrence of Permian leaiid conchostracans in China and its palaeogeographical significance: Acta Palaeontologica Sinica, v. 23, no. 4, p. 505–512, pl. 1, text fig. 1, 2.

Spence, J., 1957, The geology of part of the Eastern Province of Tanganyika: Geological Survey of Tanganyika Bulletin, v. 28, p. 18–32.

Spencer, B., and Hall, T. S., 1896, Crustacea, *in* Spencer, B., ed., Report, Horn

Scientific Expedition, central Australia, pt. 2, (Zoology): London, Dulan & Co., p. 238–244.

Sproll, W., and Dietz, R., 1969, Morphological continental drift fit of Australia and Antarctica: Nature, v. 222, p. 345–348, fig. 3.

Srivastava, R. N., 1973, Estheriids from the Lower Tals Mussoori Synform, Uttar Pradesh, Himalayas: Records of the Geological Survey of India, v. 105, no. 2, p. 193.

Stebbing, T.R.R., 1910, General catalogue of South African Crustacea: South African Museum, Annals, v. 6, p. 401–494.

Stockley, G. M., 1947, Report on the Geology of Basutoland, Masero, Basutoland (=Lesotho): London, Basutoland Government, p. 30–54.

Straśkraba, M., 1965a, Taxonomic studies on Czechoslovak Conchostraca, I. Family Limnadiidae: Crustaceana, v. 9, no. 3, p. 263–273.

—— , 1965b, Taxonomical studies on Czechoslovak Conchostraca, II. Families Lynceidae and Cyzicidae: Vest. cgl. spul. zool. (Acta Soc. Zool. Bohemoslovenicae), v. 29, no. 3, p. 205–214.

—— , 1966, Taxonomical studies on Czechoslovak Conchostraca, III. Family Leptestheridae: Hydrobiologia, v. 29, no. 3–4, p. 571–589.

Talent, J. A., 1965, A new species of conchostracan from the Lower Cretaceous of Victoria: Royal Society of Victoria (Australia) Proceedings, n.s., v. 79, no. 1, p. 197–203, pl. 26.

Talent, J. A., Duncan, P. M., and Handby, P. L., 1966, Early Cretaceous feathers from Victoria: Emu, v. 66, no. 2, p. 81–86.

Tasch, P., 1956, Three general principles for a system of classification of fossil conchostracans: Journal of Paleontology, v. 30, p. 1248–1257.

—— , 1958, Novojilov's classification of fossil conchostracans: A critical evaluation: Journal of Paleontology, v. 32, no. 6, p. 1094–1106, 33 text figs.

—— , 1961, Data on some new Leonardian conchostracans with observations on the taxonomy of the Family Vertexiidae: Journal of Paleontology, v. 35, no. 6, p. 1121–1129, pl. 133, 134.

—— , 1962, Taxonomic and evolutionary significance of two new conchostracan genera from the midcontinent Wellington Formation: Journal of Paleontology, v. 36, no. 4, p. 817–821, pl. 120.

—— , 1963a, Evolution of the Branchiopoda, *in* Whittington, H. B., and Rolfe, W.D.I., eds., Phylogeny and evolution of Crustacea: Cambridge, Harvard University, Museum of Comparative Zoology, Special Publication, p. 145–157.

—— 1963b, Paleolimnology, Part 3—Marion and Dickinson counties, Kansas, with additional sections in Harvey and Sedgwick counties; Stratigraphy and biota: Journal of Paleontology, v. 37, no. 6, p. 1233–1251, pl. 172–174, 5 text figs.

—— , 1965, The significance of "Serrate Dorsal Margin" in living and fossil conchostracans: Wichita State University Bulletin, University Studies No. 64, v. 4, no. 3, 7 p.

—— , 1968, A Permian trace fossil from the Antarctic Ohio Range: Kansas Academy of Science, Transactions, v. 71, p. 33–37.

—— , 1969, Branchiopoda, *in* Moore, R. C., ed., Treatise on invertebrate paleontology, Part R - Arthropoda 4, Crustacea (except Ostracoda): Boulder, Colorado, Geological Society of America and the University of Kansas, p. R128–R191.

—— , 1970a, Antarctic Leaiid Zone: Seasonal events: Gondwana correlations, *in* Amos, A. J., ed., Gondwana stratigraphy, International Union of Geological Sciences Symposium (Buenos Aires): Paris, United Nations Educational, Scientific, and Cultural Organization, p. 185–194.

—— , 1970b, Antarctic and other Gondwana Conchostracans and insects; New data; Significance for drift theory, *in* Haughton, S. H., ed., 2nd Gondwana Symposium Proceedings and Papers: Pretoria, South Africa, Council for Scientific Industrial Research, p. 589–592.

—— , 1971, Invertebrate fossil record and continental drift, *in* Quam, L. O., ed., Research in the Antarctic: American Association for the Advancement of Science Publication 93, p. 703–716.

—— , 1972, Paleobiogeography of Leaiid Conchostracans and modern drift theory, *in* Marek, L., ed., International Palaeontological Union, International Geological Congress, 23rd Session, Czechoslovakia, Proceedings, 1968: Warsaw, Instytut Geologiczny, p. 351–365.

—— , 1973, Jurassic beetle from southern Victoria Land, Antarctica: Journal of Paleontology, v. 47, no. 3, p. 590–591, 1 text fig.

—— , 1975, Non-marine Arthropoda of the Tasmanian Triassic: Royal Society of Tasmania, Papers and Proceedings, v. 109, p. 97–106, 1 text fig., 1 pl.

—— , 1976, Jurassic non-marine trace fossils (Transantarctic Mountains) and the food web: Journal of Paleontology, v. 50, no. 4, p. 754–758, 1 text fig., 1 pl.

—— , 1977a, Ancient Antarctic freshwater ecosystems, *in* Llano, G. A., ed., Adaptations within Antarctic Ecosystems, Third Scientific Committee for Antarctic Research Symposium on Antarctic Biology Proceedings, Washington, D.C.: Smithsonian Institution, p. 1077–1089, figs. 1–2, tables 1 and 2.

—— , 1977b, Data retrieval from growth lines of fossil crustaceans (Branchiopoda: Conchostraca). North American Paleontological Convention 2. Abstracts of Papers: Journal of Paleontology, v. 51 (Supplement to no. 2, pt. 3 of 3), p. 27–28.

—— , 1979c, Permian and Triassic Conchostraca from the Bowen Basin (with a note on a Carboniferous leaiid from the Drummond Basin), Queensland, *in* Tasch, P., and Jones, P. J., Carboniferous, Permian, and Triassic Conchostracans of Australia; Three new studies: Canberra, Australia Bureau of Mineral Resources, Geological and Geophysical Bulletin, Australian Government Publishing Service, v. 185, p. 33–44.

—— , 1980, New non-marine fossil links in Gondwana correlations and their significance: Washington, D.C., Division of Polar Programs, National Science Foundation, Antarctic Journal of the United States (1980 Review), p. 5–6.

—— , 1981a, Non-marine evidence for Paleozoic/Mesozoic Gondwana correlations: Update, *in* Cresswell, M. M., and Vella, P., eds., Gondwana five, 5th International Gondwana Symposium Selected Papers and Abstracts, Wellington, New Zealand: Rotterdam, A. A. Balkema, p. 11–14.

—— , 1981b, Significant new Gondwana data on Australian, South African, and South American conchostracans: Antarctic Journal of the United States, v. 16, no. 5, p. 4–5.

—— , 1982a, Conchostracan dispersal (Paleozoic/Mesozoic) between South America, Africa, and Antarctica: Antarctic Journal of the United States (Special Issue), p. 45–46.

—— , 1982b, Experimental valve geothermometry applied to fossil conchostracan valves, Blizzard Heights, Antarctica, *in* Craddock, C., ed., Antarctic geoscience, Symposium on Antarctic geology and geophysics, August 22-27, 1977, International Union of Geological Sciences: Madison, University of Wisconsin Press, p. 661–668.

—— , 1984, Biostratigraphy and paleontology of some conchostracan-bearing beds in southern Africa: Paleontologia Africania, Bernard Price Institute for Paleontological Research, Annals: Johannesburg, University of Witwatersrand, v. 15, p. 61–85.

Tasch, P., and Gafford, E. L., 1968, Paleosalinity of Permian non-marine deposits of Antarctica: Science, v. 160, p. 1221–1222.

Tasch, P., and Jones, P. J., 1979a, Carboniferous and Triassic conchostraca from the Canning Basin, western Australia, *in* Oldham, W. H., ed., Carboniferous, Permian, and Triassic Conchostracans of Australia; Three new studies: Canberra, Australia Bureau of Mineral Resources, Geological and Geophysical Bulletin, Australian Government Publishing Service, v. 185, p. 3–19.

—— , 1979b, Lower Triassic conchostraca from the Bonaparte Gulf Basin, northwestern Australia (with a note on *Cyzicus (Euestheria) minuta*(?) from the Carnarvon Basin), *in* Oldham, W. H., ed., Carboniferous, Permian, and Triassic Conchostracans of Australia; Three new studies: Canberra, Australia Bureau of Mineral Resources, Geological and Geophysical Bulletin, Australian Government Publishing Service, v. 185, p. 23–30.

Tasch, P., and Lammons, J. M., 1978, Palynology of some lacustrine interbeds of the Antarctic Jurassic: Palinologia, no. 1, p. 455–460, text chart and pl. 1.

Tasch, P., and Oesterlen, P. M., 1977, New data on the Phyllopod beds (Karroo System) northern Angola: South Central Geological Society of America (El Paso) Abstracts with Program, v. 9, no. 1, p. 77.

Tasch, P., and Riek, E. F., 1969, Permian insect wing from the Antarctic Sentinel Mountains: Science, v. 164, p. 1529–1530.

Tasch, P., and Shaffer, B. L., 1964, Conchostracans: Living and fossil from Chihuahua and Sonora, Mexico: Science, v. 143, no. 3608, p. 806–807.

Tasch, P., and Volkheimer, W., 1970, Jurassic conchostracans from Patagonia: University of Kansas Paleontological Contribution, Paper 50, 23 p.

Tasch, P., and Zimmerman, J. R., 1961, Comparative ecology of living and fossil conchostracans in seven county area of Kansas and Oklahoma: University of Wichita Bulletin, University Studies 47, p. 1–14.

Tasch, P., Sastry, M.V.A., Shah, S. C., Rao, B.R.J., Rao, C. N., and Ghosh, S. C., 1975, Estheriids of the Indian Gondwanas: Significance for continental fit, *in* Campbell, K.S.W., ed., Gondwana geology: Papers from the 3rd Gondwana Symposium, Canberra, Australia: Australian National University Press, p. 443–452.

Teichert, C., and Talent, J. A., 1958, Geology of the Buchan area, East Gippsland: Geological Survey of Victoria (Australia) Memoir 21, p. 1–56.

Teixeira, C., 1943, Sur l'*Estheria borgesi* nouvelle espèce du Karroo du Mozambique Portugais: Société Géologique de France Bulletin, Notes et Mémoirs, v. 13, p. 71–72, figs. 6, 7, pl. III.

——, 1947, Acerca des filôpodes fôsseis do Karroo da Escarpa do Quela (Angola): Lisboa, Estudos de Geologia e Paleontologia Anais, 1947, v. 2, p. 29–43, pl. 1–6.

——, 1960, Sur quelques fossiles du Karroo de la Lunda, Angola: Angola, Museo do Dundo, 97 p., 16 pl.

——, 1961, Paleontological notes on the Karroo of the Lunda (Angola): Revista da Junta Investigacôes do Ultramar: Lisboa, Garcia De Orta, v. 9, no. 2, p. 307–311.

Temperlye, B. N., 1952, A review of the Gondwana rocks of Kenya Colony: Alger, 19th Congrés Géologique International, Symposium sur Les Séries de Gondwana, p. 195–208.

Termier, G., and Termier, H., 1950, Invertébrés de l'Ere Primaire, Fascicule 4: Paris, Actualités Scientifique et Industrielles, 1095, Palaéontologie Marocaine, pt. 2, p. 9, 152–153, fig. 39–41, 42–44.

Times Atlas of the World, 1967: The Times of London in collaboration with John Bartholomew and Sons, Ltd., Boston, Houghton Mifflin Co., 272 p.

Tinoco, I., and Katoo, I., 1975, Conchostráceos da Formãco Sousa, Bacia do Rio do Peixe, Estado da Paraiba, *in* Atlas do Simposio de Geologia VII: Fortaleza, Brasil, p. 135–147.

Townrow, J. A., 1967, Fossil plants from Allan and Carapace nunataks and from Upper Mill and Shackleton glaciers, Antarctica: New Zealand Journal of Geology and Geophysics, v. 10, no. 2, p. 456–473.

Trumpy, D., 1943, Pre–Cretaceous of Colombia: Geological Society of America Bulletin, v. 54, p. 1281–1304.

Veevers, J. J., Jones, J. G., and Talent, J. A., 1971, Indo–Australian stratigraphy and the configuration and dispersal of Gondwanaland: Nature, v. 229, p. 383–388, text figs. 1–7.

von Alberti, F. A., 1834, Beitrag zu einer Monographie des Bunten Sandsteins, Muschelkalks und Keupers und die Verbindung dieser Gebilde zu einer Formation: Stuttgart und Tubingen, Verlag der J. G. Cotta'schen, p. 114; 119–120.

von Hillebrandt, A., 1973, Neue ergebnisse uber den Jura in Chile und Argentinien: Münster, Fortschritte der Geologie und Paläontologie, v. 31/32, p. 167–199.

von Zieten, G. H., 1833, Die Versteinerungen Württembergs: Stuttgart (Zusammengestellt von G. H. Zieten), p. 72, pl. 54, fig. 5.

Warth, M., 1969, Conchostraken (Crustacea, Phyllopoda) aus dem Keuper (Ob. Trias) Zentral–Württembergs: Stuttgart, Gesellschaft für Naturkunde in Württemberg, v. 124, p. 123–145.

Webb, J. A., 1978, A new *Palaeolimnadiopsis* (Crustacea: Conchostraca) from the Sydney Basin, New South Wales: Alcheringa, v. 2, p. 261–267, figs. 1 and 2A, B.

Webb, J. A., and Bell, G. D., 1979, A new species of *Limnadia* (Crustacea: Conchostraca) from the Granite Belt: Australia, Linnean Society of New South Wales Proceedings, v. 103, no. 4, p. 237–245.

Weiss, C. E., 1875, Notes on *Estheria (Estheriella) costata* and *Estheria (Estheriella) lineata* Weiss: Deutsche Geologische Gesellschaft Zeitschrift, v. 29, p. 710–712.

White, M. E., 1981, Revision of the Talbragar fish bed flora (Jurassic) of New South Wales: Australian Museum Records, v. 33, no. 15, p. 695–721.

Woodward, A. S., 1895, The fossil fishes of the Talbragar Beds (Jurassic?): Sidney, Geological Survey of New South Wales Memoirs, Paleontology 9., C. Potter, Government Printer, fold map, 27 p.

Zaspelová, V. S., 1965, New Lower Triassic Phyllopods of Genus *Cornia*: Paleontological Journal (American Geological Institute translation of Paleontologicheskiy Zhurnal), no. 4, p. 41–49.

——, 1966, Konkhostraki kontinental'nykh otlozhenii Karbona prinurinskikh mestorozhdenii Uglya tsentral; nogo Kazakhstana, *in* Nalivkin, D. V., and Martinson, G. G., eds., Kontinental'nyi Verkhnii Paleozoi I Mezozoi Sibiri I Tsentral'ngo Kazakhstana (Biostratigrafiya I Paleontologiya): Akademiya Nauk. SSSR. Izdat. "Nauka," Moskya–Leningrad, p. 98–149, pl. 1–11.

Zeuner, P. E., 1959, Jurassic beetles from Grahamland, Antarctica: Palaeontology, v. 5, p. 407–409.

Index

1. General

Index

2. Scientific Names
(Numbers in italics denote photographs)

Typeset by WESType Publishing Services, Inc., Boulder, Colorado
Printed in U.S.A. by Malloy Lithographing, Inc., Ann Arbor, Michigan